Longman Guide to World Science and Technology

LONGMAN GUIDE TO WORLD SCIENCE AND TECHNOLOGY

Series editor: Ann Pernet

Science and Technology in the Middle East*
by Ziauddin Sardar

Science and Technology in Latin America
by Latin American Newsletters Limited*

Science and Technology in the USSR
by Vera Rich

Science and Technology in Eastern Europe
by Vera Rich

Science and Technology in Israel
by Vera Rich

Science and Technology in the UK
by Anthony P. Harvey and Ann Pernet

Science and Technology in China
by Tong B. Tang

Science and Technology in Japan
by Alun M. Anderson

Science and Technology in South-East Asia
by Ziauddin Sardar

Science and Technology in the Indian Subcontinent
by Ziauddin Sardar

*Already published.

Science and Technology in Latin America

Latin American Newsletters Limited

Editors: Christopher Roper and Jorge Silva

Longman
London and New York

SCIENCE AND TECHNOLOGY IN LATIN AMERICA

Longman Group Limited,
6th Floor, Westgate House, Harlow, Essex CM20 1NE, UK

© F.H. Books Limited 1983

Distributed exclusively in the USA and Canada by Gale Research Company,
Book Tower, Detroit, Michigan 48226, USA.

First published 1983

British Library Cataloguing in Publication Data

Latin American Newsletters Limited
  Science and technology in Latin America.
  - (Longman guide to world science and technology)
  1. Research – Latin America – Directories
  2. Research, Industrial – Latin America – Directories
  I. Title
  607'.208 T175

ISBN 0 582 90057 3

Latin American Newsletters Limited
Boundary House,
91-93 Charterhouse Street,
London EC1M 6HR

Printed in Great Britain by
Butler and Tanner Ltd, Frome, Somerset

Contents

Introduction   vii
Map   x

Argentina   1
Belize   23
Bolivia   24
Brazil   32
Chile   58
Colombia   75
Costa Rica   92
Cuba   100
Dominican Republic   110
Ecuador   118
El Salvador   131
French Guiana   137
Guatemala   139
Guyana   147
Haiti   152
Honduras   155
Jamaica   159
Mexico   162
Nicaragua   186
Panama   194
Paraguay   203
Peru   207
Puerto Rico   222
Surinam   226

Uruguay 228
Venezuela 237

Appendices

Appendix 1 Regional statistics 249
Appendix 2 Directory of selected establishments 260

Indexes

314

Introduction

Images of Latin America in the rest of the world are generally false and misleading. It is generally lumped with Africa and Asia in the 'Third World'. In terms of per capita income this may be correct, but the great difference is that Latin America's dominant culture (like North America's) is European in origin. The first university to be founded in the New World was not in Boston, but in Santo Domingo.

Furthermore, Latin America's scientific culture did not begin with the arrival of the Spanish conquistadores. Mayan astronomical observations, Aztec and Inca plant breeders, Amazonian and Andean herbalists, and Inca engineers had reached levels of technical sophistication which at that time surpassed anything to be found in Europe, only comparable perhaps with the achievements of Chinese civilization.

Latin America remains a possible contender for future scientific leadership in the world. Argentine nuclear physicists and biologists, and Brazilian scientists are in the front rank of the world's scientific workers, despite political conditions at home which frequently drive them to find work in exile in North America or Europe.

In fact, two of the problems facing Latin America will be familiar to scientists from other parts of the world. The first is a lack of career opportunities for first rate scientific talents, leading to the famous 'braindrain' to the United States, and the second is the lack of mechanisms to link advanced scientific research to the necessities of the impoverished populations of Latin America.

The two problems are, needless to say, linked, and Latin American scientists have demonstrated a high degree of concern, which may yet produce innovative solutions to the problem of successfully locating the scientist in the decision making processes of society.

Much of the funding for scientific work in Latin America comes directly or indirectly from the United States and large numbers of Latin American students do their graduate work in North American universities. This inevitably shapes and colours the kind of work being done in the region. But to an increasing extent, as will be evident from this volume, the Latin American scientific community is setting its own agenda, considering the training of scientists in a context of national needs, and beginning to recover a sense of continuity with the great scientists, who must have lived in Peru and Mexico before the arrival of the Spanish.

All the countries of Latin America are wrestling with a series of problems which require a high degree of integration among science policy making, capital investment, and socio-economic goals. These problems include energy options, food production and distribution, pressures on local eco-systems, the rapid development of computers. The acuteness of the situation is directly linked to the current state of affairs in Latin America.

The improvement in communications, and the opening of the era of mass travel by plane, has meant that Latin America is an integral part of the known world in a way that was scarcely imaginable twenty years ago. Latin America, admittedly underpopulated in most countries, faces all the problems of a rapidly expanding population, and the growth figures mask an even more dramatic growth rate in the urban population. The world's two largest cities are now in Latin America, São Paulo and Mexico City.

The problems faced by Latin America cannot be solved by science. Strictly speaking they are not technical problems. They are social, political and economic, and need to be solved in that order. But as they are solved, it is inevitable that scientists will be called on to play an ever greater part in feeding the hungry, and providing energy for development.

In the race to provide jobs, Latin American policy makers have often allowed industrialists from North America and Europe to ignore consequent pollution. Problems of excessive pollution have not been long in appearing, as anyone who recently visited Mexico City will testify. After some spectacular environmental disasters in Brazil and Mexico, concern is growing.

Nuclear power looked like a quick solution for energy-short Brazil, but as it is turning out, the problems may outweight the benefits. Likewise, few Latin American countries took any serious interest in computing. These were large and convenient machines, to be bought abroad. As computers became smaller and cheaper through the 1960s and 1970s, Brazil, at least, realized that questions of national policy were involved, and Brazil became the first country in the third world to make a serious attempt to build its own computer manufacturing industry.

Venezuela has another response. It is making a sustained effort to improve the intelligence of its children by enriching the school curriculum, and through radio and television programmes. This sounded bizarre when first proposed, but is winning increasing critical attention. Demand for education in Latin America is so great, and the resources so slender, that countries have to think in terms which break with the old classroom model, and look increasingly to satellite television and small computers as future vectors of scientific culture in societies in which more than half the population is under fifteen years of age.

This book is not policy oriented. It is more a resource for researchers seeking points of reference and contact in Latin America. It brings together an enormous quantity of information from a wide range of sources, up-dating and collating data. There are no comprehensive sources of data on science and technology in Latin America. Inevitably there will be errors and omissions, and these will be corrected and filled by readers of this first edition.

The great bulk of the research was undertaken by Jorge Silva, formerly of the Chilean Air Force, who took enormous pains to ensure that the most up-to-date facts and figures were included in the text. This is extremely difficult in a continent where official documents and directories are not regularly produced, and may occasionally be invalidated overnight by a political earthquake. Scientific development does not take place in a vacuum, and Latin America currently faces major political transitions as the military governments which dominated the region in the 1970s give way to civilian governments.

As this introduction indicates, scientific policy making will remain high on the agenda of all Latin American governments, and the region remains rich in promise and potential.

Map showing the general area of Latin America.

Argentina

Demographic, Political and Economic Features

Argentina is a federal republic of South America situated on the Atlantic side of the Southern Cone. With an area of 2 776 889 square kilometres (1 072 157 square miles), it is the second largest country of the continent. The Cordillera de los Andes forms its western border, separating it from Chile; the northern border is shared with Bolivia, Paraguay, Brazil and Uruguay. The east coast from the River Plate to the Río Negro in the south forms a belt of sand banks, shallow beaches and sandy stretches. From the Río Negro to the south the coast is almost completely made up of sheer precipices interrupted only by the estuaries of rivers.

The most important waterway system of the country is the River Plate, formed by its direct tributaries – Paraná, Uruguay and the Salado Sur – and by rivers flowing into these tributaries – Iguazú, Paraguay, Pilcomayo, Bermejo and Salado Norte.

Argentina is predominantly a vast prairie land with a temperate climate. The north-west is a region of high ground traversed by mountain chains and valleys that run in a north–south direction; this region is known as La Puna. The arid central Andes, whose peaks are perpetually covered with snow, change towards the south to a mountainous region lower in altitude than that of the north, and of luscious vegetation. The country has a population of 27 862 771, 80 per cent of which is urban population. Buenos Aires, the capital of the Republic, situated on the southern side of the River Plate, has a population of 3 000 000 and this figure reaches almost 10 000 000 if the suburbs which form 'Greater Buenos Aires' are included. Other important cities include: Tucumán, Córdoba, Santa Fé, Rosario, Mendoza and Bahía Blanca.

During most of the twentieth century, Argentina has been under military governments with only a few sporadic democratic experiences. Since March 1976 a military junta has exercised political power; this junta is made up of the commanders-in-chief of the armed forces and the president of the republic is nominated by the junta. The legislative power is in the hands of the *Comisión de Asesoramiento Legislativo* (Commission of Legislative Counselling); which comprises nine high officials designated by the junta. The various areas of national or international matters are covered by the ministries of planning; culture and education; defence; social services; interior; foreign affairs; labour and justice.

The Argentinian economy relies mainly on the central prairie land. The breeding of 61 000 000 head of cattle and 35 000 000 head of sheep make Argentina one of the main meat producers. Agriculture, with an area of 38 911 000 hectares of sown ground, is one of the main economic activities. The principal crops are cereals and flax, fodder, industrial crops, fruit and vegetables. Argentina exports a considerable amount of wine. The country has a vast amount of forest land (63 million hectares) and large mineral resources: coal (Río Turbio), iron (Jujuy, La Rioja, Mendoza, San Luís and Chubut) and important deposits of lead, zinc, tin, copper, uranium, antimony, sulphur and gypsum in the Andean region. In Patagonia (Tierra del Fuego) there are oil wells and deposits which produce annually (1976) 23 129 300 cubic metres of crude oil.

Argentina, like other Latin American countries, started its industrial development at the beginning of the 1920s and reached a high level of development at the end of the 1950s. The automobile and machinery industry has allowed Argentina to become one of the main suppliers for the countries in the area. Argentina also exports manufactured items of leather and textiles. The remaining industrial products, such as chemicals, processed food, electronics and by-products of oil, show a high level of development and make Argentina one of the most industrialized countries in South America.

Organization of Science and Technology

The majority of the organizations dedicated to science and technology are located within the university sector. More than a third of researchers work in this sector. The decentralized institutions of the state account for practically all the remaining researchers, so that 90 per cent of research is concentrated in these groups (see Figure 1). The figure shows only administrative relationships between the various organizations; the functions and activities can be found in the relative sections.

Policies and Financing in the Field of Science and Technology

Although the ministries have a certain autonomy in carrying out research programmes in subjects of interest to them, these must be consistent with the general objectives laid down by the Secretariat of Science and Technology. These general objectives and the five national programmes form the body of government policy for science and technology.

General Objectives

The establishment of a national system of science and technology, considering the national interests and the objectives of national reorganization of the system in all fields.

To continue the work initiated in the national programmes and to transfer the results to the sectors of production, health and others.

ARGENTINA

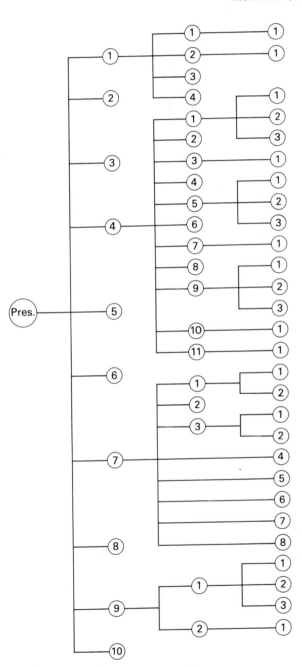

Fig. 1 Organization of science and technology

Fig. 1 - *continued*

1	Ministerio de Cultura y Educacion (Ministry of Culture and Education)
1.1	Secretaría de Ciencia y Technologia (Secretariat of Science and Technology)
1.1.1	Consejo Nacional de Investigación Científica y Técnica (CONICET) (National Council of Scientific and Technical Research)
1.2	Secretaría de Educación (Secretariat of Education)
1.2.1	Universidades Nacionales (National Universities)
1.3	Secretaría de Cultura (Secretary of Culture)
1.4	Comisión Nacional Argentina para la UNESCO (Argentinian National Commission for UNESCO)
2	Ministerio de Planeamiento (Ministry of Planning)
3	Ministerio del Interior (Ministry of the Interior)
4	Ministerio de Economía (Ministry of the Economy)
4.1	Secretaría de Programación y Coordinacion Económica (Secretariat of Economic Programming and Coordination)
4.1.1	Instituto Nacional de Estadística y Censo (National Institute of Statistics and Census)
4.1.2	Instituto Nacional de Programación Económica (National Institute of Economic Programming)
4.1.3	Corporación de Empresas Nacionales (Corporation of National Companies)
4.2	Secretaría de Hacienda (The Exchequer)
4.3	Secretaría de Desarrollo Industrial (Secretariat of Industrial Development)
4.3.1	Instituto Nacional de Tecnología Industrial (INTI) (National Institute of Industrial Technology)
4.4	Secretaría de Comunicaciones (Secretariat of Communications)
4.5	Secretaría de Agricultura y Ganadería (Secretariat of Agriculture and Livestock)
4.5.1	Instituto Nacional de Technología Agropecuaria (INTA) (National Institute of Arable and Livestock Technology)
4.5.2	Instituto Nacional Forestal (National Forestry Institute)
4.5.3	Servicio Nacional de Parques (National Park Service)
4.6	Secretaría de Energía (Secretariat of Energy)
4.7	Secretaría de Minería (Secretariat of Mining)
4.7.1	Servicio Nacional Minero Geológico (National Service of Geological Sources)
4.8	Secretaría de Comercio Exterior y Negociaciones Económicas Internacionales (Secretariat of Foreign Trade and International Economic Negotiations)
4.9	Secretaría de Transportes y Obras Públicas (Secretariat of Transport and Public Works)
4.9.1	Instituto Nacional de Prevencion Sísmica (National Institute of Seismic Prevention)
4.9.2	Dirección Nacional de Vialidad (National Governing Board of Highway Engineering)
4.9.3	Instituto Nacional de Ciencia y Técnica Hidraúlicas (National Institute of Hydraulic Science and Technology)
4.10	Secretaría de Comercio (Secretariat of Commerce)
4.10.1	Instituto Nacional de Vitivinicultura (National Institute of Viticulture and Viniculture)
4.11	Secretaría de Intereses Marítimos (Secretariat of Maritime Affairs)
4.11.1	Servicio Nacional de Pesca (National Service of Fisheries)
5	Ministerio de Relaciones Exteriores y Culto (Ministry of Foreign Affairs and Religion)
6	Ministerio del Trabajo (Ministry of Labour)
7	Ministerio de Defensa (Ministry of Defence)
7.1	Comando en Jefe de la Armada (Commander-in-Chief of the Navy)
7.1.1	Servicio de Hidrografía Naval (SHN) (Naval Hydrographic Service)
7.1.2	Servicio Naval de Investigación y Desarrollo (SENID) (Naval Service of Research and Development)
7.2	Comando en Jefe del Ejército (Commander-in-Chief of the Army)
7.3	Comando en Jefe de la Fuerza Aérea (Commander-in-Chief of the Air Force)
7.3.1	Instituto de Investigación Aeronáutica y Espacial (IIAE) (Institute of Aeronautical and Space Research)
7.3.2	Comisión Nacional de Investigación Espacial (CNIE) (National Commission of Space Research)
7.4	Dirección General de Investigación y Desarrollo (DIGID) (National Governing Board of Research and Development)

7.5   Instituto de Investigaciones Científicas y Técnicas de las Fuerzas Armadas (Institute of Scientific and Technical Research of the Armed Forces)
7.6   Dirección Nacional de Fabricaciones Militares (National Governing Board of Military Construction)
7.7   Instituto Geográfico Militar (Military Geographical Institute)
7.8   Dirección Nacional del Antártico (National Governing Board of the Antarctic)
8   Ministerio de Justicia (Ministry of Justice)
9   Ministerio de Bienestar Social (Ministry of Social Affairs)
9.1   Secretaría de Salud Pública (Secretariat of Public Health)
9.1.1   Instituto Nacional de Microbiología (National Institute of Microbiology)
9.1.2   Servicio Nacional Integrado de Salud (National Integrated Health Service)
9.1.3   Instituto Nacional de Farmacología y Bromatología (National Institute of Pharmacology and Food Science)
9.2   Secretaría de Coordinación y Promoción Social (Secretariat of Social Coordination and Advancement)
9.2.1   Comisión Nacional de Toxicomanía y Narcóticos (National Commission of Toxic Substances and Narcotics)
10   Comisión Nacional de Energía Atómica (National Commission of Atomic Energy)

To detect priority areas and to generate new programmes deemed necessary within these priority areas.

To stimulate direct or indirect participation of private industry in research and development in important areas.

To maintain a constant evaluation of the system.

To establish bilateral and multilateral agreements of cooperation for scientific and technical research in subjects of national interest.

To maintain an active participation in the programmes of international organizations in science and technology.

To stimulate the activities of national academies and scientific associations.

To encourage scientific publications.

To maintain the existing scientific and technological work-force at its present level and to promote the training of new researchers and scientists.

To negotiate a salary structure for researchers which reflects the inflationary spiral, in order to maintain a level which allows their work to continue unhindered.

To provide the resources necessary to obtain an adequate infrastructure for the development of science and technology.

To lay down a policy which embraces a programme of re-equipping the scientific and technical sector.

To formulate a system of information and documentation which supplies the needs of research and development.

To facilitate the import of scientific instruments and materials.

To continue the present system of regionalizing science and technology.

To increase the number of regional centres and to maintain in each centre a staff of scientists and researchers.

To coordinate the use of national technology and try, where possible, to substitute it for imported technology.

To study and standardize the technology in use.

To ensure legal means to facilitate the transference of technology.

National Programmes

In order to benefit from the general objectives, an inter-sector commission has been set up. The staff of the various scientific institutions work in co-ordination with the programmes to draw up a strategy in each area.

The national programmes are centred on the following areas:

Electronics
Food technology
Endemic diseases
Housing technology
Non-conventional energy

The objective of the national programmes is the coordination of all the efforts (scientific, administrative, management, equipment, etc.) made by the different institutions working at all levels of technology.

Like most Latin American countries, Argentina is confronted with large social and economic problems which generally limit the resources available for science and technology.

Of the total funds available for scientific and technical activities (1976), 87.03 per cent was provided by the state, 7.14 per cent by companies from the private industrial sector and 3 per cent by international institutions. These funds were distributed as follows: 61.12 per cent went to the general service sector; 31.3 per cent to higher education; 4.48 per cent to the state or mixed productive sector; and 3.1 per cent to the private productive sector. (The remainder of the last sector was totally financed by private firms.)

In 1978 of the total budget allocated to science and technology, 34.84 per cent went to the Ministry of Culture and Education; 9 per cent to the National Commission for Atomic Energy; 17.94 per cent to the Ministry of Defence; 0.20 per cent to the Navy; 1.72 per cent to the Air Force; 0.34 per cent to the Secretariat of Industrial Development; 34.64 per cent to the Secretariat of Agriculture; 0.04 per cent to Public Health and 0.76 per cent to Housing and Urban Development.

The main areas that these funds finance can be seen in Table 1.1, showing percentages for 1976.

In 1981 the general budget of the National Administration for Science and Technology (in millions of pesos) was as shown in Table 1.2.

Human resources in research and development have shown a positive up-turn in the last few years, not only in the number of researchers but also in those entering into research in specialized areas and experimental research. The total of scientists and technicians in 1980 was estimated to be in the region of 530 000, of which about 22 000 are involved in research and development.

Table 1.3 shows the human resources in the various sectors of science and technology from 1970. 1985 and 1990 are projections based on previous years.

In spite of the growth in human resources in science and technology as a whole, the number of researchers and technicians entering the field of research is declining steadily. With the objective of overcoming this

Table 1.1

	Percentages	
Advanced technology (civil sector)	7.07:	
Nuclear		4.07
Space		2.00
Computer sciences		1.00
Improving the economy	15.04:	
Productivity and industrial technology		10.46
Infrastructure and services		4.58
Productivity and agricultural technology	27.86:	
Services to the community	29.98:	
Health		21.40
Contamination		3.59
Others		4.99
Scientific progress	13.15:	
Natural sciences and engineering		5.50
Social sciences and humanities		4.67
Others		2.98
Other civil activities of research and development	—	
Defence	6.90	
	100.00	

Table 1.2

	Totals	Running expenses	Capital goods
National universities	153 584	150 134	3 450
Public bodies	1 695 451	1 106 390	589 061
Total	1 849 035	1 256 524	592 511

Table 1.3 Scientists and technicians

Year	Number of scientists	Natural sciences	Engineering	Agriculture	Medicine	Social sciences	Technicians in R & D
1970	6 500	2490	790	920	1450	850	9 800
1975	7 500	2775	1050	1050	1650	975	10 800
1980	9 500	3500	1370	1330	2100	1200	13 300
1985	12 000	4400	1800	1700	2500	1600	16 000
1990	15 000	5400	2350	2250	2900	2100	20 000

situation, the *Consejo Nacional de Investigación Científica y Técnica (CONICET)* (National Council for Scientific and Technical Research), through agreements with the universities, is developing specialist research institutes with the aim of recruiting more researchers and technical staff.

Science and Technology at Government Level

As can be seen in Figure 1, research and development activities are distributed throughout practically all government organizations. However, there are specialized institutions whose main functions are research and development.

Secretaría de Estado de Ciencia y Tecnología (State Secretariat of Science and Technology)
This is the highest governmental executive body that carries out the co-ordination and planning of policies for research and development as well as formulating operative programmes in technical and administrative areas.

Consejo Nacional de Investigación Científica y Técnica (CONICET) (National Council for Scientific and Technical Research)
This body is concerned with promoting, coordinating and orientating research carried out in the field of pure and applied sciences; it advises the executive power and other bodies on their needs as regards science and technology. The council also presents the executive power with the methods necessary to promote scientific and technical research. These functions are carried out mainly through the distribution of grants and subsidies for the development of scientific research, scientific information services, scientific coordination, teaching of sciences, the forming of international scientific relations, the establishment of international scientific exchange and the creation of conditions that facilitate the expansion and higher levels of achievement in the field of science.

The council has been responsible for the establishing of many research institutes and bilateral agreements with universities and other higher education bodies. In 1978 the council maintained relations with almost seventy-eight institutes and research laboratories. Among the institutes created by the council and which have direct dependency are:

Centro de Documentación Científica (Centre of Scientific Documentation)
Instituto Nacional de Sismología (National Institute of Seismology)
Instituto de Radioastronomía (Institute of Radio-astronomy)
Centro Nacional de Radiación Cósmica (National Centre of Cosmic Radiation)
Instituto Nacional de Oceanografía (National Oceanographic Institute)
Laboratorio de Análisis de Rocas y Minerales (Laboratory for the analysis of rocks and minerals)
Laboratorio de Microanálisis de Elementos Orgánicos (Laboratory for the micro-analysis of organic elements)
Laboratorio de Datación de Radiocarbono (Laboratory for radio carbon dating)

The council is governed and administered by a directive which comprises the following advisory commissions: internal commissions (grants and subsidies); regional commissions corresponding to the different geographical zones.

Instituto Nacional de Geología y Minería (National Institute of Geology and Mining)
Founded in 1963, its main activity is the drawing up of geo-economic maps, the evaluation of mineral resources by exploration and experimental exploitation and the giving of technical assistance to public and private institutions. The institute also provides training for technicians and workers. Various specialized services and laboratories attached to the institute carry out basic research for the development of projects belonging to other institutions.

Instituto Nacional de Tecnología Agropecuaria (INTA) (National Institute of Arable and Livestock Technology)
Administratively this institute depends on the Ministry of Agriculture; a more detailed description can be found in the section below entitled 'Agricultural and Marine Science and Technology'.

Instituto Nacional de Tecnología Industrial (INTI) (National Institute of Industrial Technology)
The institute depends on the Secretariat of Industrial Development. A more detailed description can be found in the section below entitled 'Industrial Science and Technology'.

Comisión Nacional de Energía Atómica (National Commission for Atomic Energy)
This body is described under 'Nuclear Science and Technology'.

The rest of the institutions which are dependent on the ministries are described in the sections corresponding to their main field of activity.

Science and Technology at an Academic Level

The universities are the first and most important centres of research. The university institutes normally carry out work in conjunction with governmental organizations and these specialized institutes are usually situated within the universities themselves.
In Argentina there are three categories of university: national (or federal), financed by a federal budget; provincial, financed by a provincial budget; private, created and financed privately but with state authorization to function.

Industrial Science and Technology

Research activities in this field are mainly carried out in institutes dependent on the state in collaboration with some of the specialized centres of the universities.

Instituto de Mecánica Aplicada y Estructuras (Institute of Applied and Structural Mechanics)
Founded in 1962, it is dedicated to the study of resistance of materials; analysis; standards and structures. The institute maintains a library of 1000 volumes and publishes various publications irregularly.

Instituto Nacional de Tecnología Industrial (INTI) (National Institute of Industrial Technology)
Established in 1957, it has a library of 20 000 volumes, 3000 reviews and 15 000 standards. A technical bulletin is published irregularly.

Instituto Argentino de Racionalizacíon de Materiales (IARM) (Argentine Institute of Standards)
Founded in 1935, it has produced 1650 dissertations and maintains a library of 440 874 standards and publishes *Dinámica IARM* (Dynamics) monthly. *Normas IARM* (Standards) and *Catálogo General de Normas IARM* (General Catalogue of Standards) are published every two years. *IARM Tecnología y Gestión* (Technology and Management) is published three times a year.

Agricultural and Marine Science and Technology

Instituto Nacional de Tecnología Agropecuaria (INTA) (National Institute of Arable and Livestock Technology)
Created in 1956 with the objectives of improving and expanding agricultural and livestock technology and to raise the standard of living among the rural population. The institute publishes *Revista IDIA* (Review), *Revista de Investigaciones Agropecuarias* (Review of Arable and Livestock Farming), *Colección Científica* (Scientific Collection) and *Colección Agropecuaria* (Arable and Livestock Farming Collection).

Centro de Investigaciónes de Recursos Naturales (Natural Resources Research Centre)
Founded in 1944, it was reorganized in 1970 and forms part of *INTA* (see above). The centre carries out research in soil conservation, fertility, botany, afforestation, meteorology, and agronomy. There is a staff of ninety-one researchers and a library of 10 000 volumes. It publishes a series of technical publications: *Flora de la República Argentina IDIA* (Flora of the Republic of Argentina), *Supplemento Forestal* (Forestry Supplement).

Estación Experimental Agropecuaria de Salta - INTA (The Salta Experimental Station of Arable and Livestock Farming)
Created in 1960 as a dependency of the Secretariat of State for Agriculture, it possesses a library of 1500 volumes and 174 periodicals.

Instituto de Microbiología e Industrias Agropecuarias - INTA (Institute of Microbiology and Arable and Livestock Farming)
This institute publishes various reviews.

Estación Experimental Agropecuaria Mendoza (Mendoza Experimental Station of Arable and Livestock Farming)
Created in 1958 as a section adjoining *INTA* (see above).

Estación Experimental Regional Agropecuaria (Regional Experimental Station of Arable and Livestock Farming)
Founded in 1912, it subsequently joined *INTA*. The station maintains a library of 40 000 volumes. Research is carried into agriculture – cultivation of wheat, maize, flax, marigolds and fodder crops, and livestock breeding; there are facilities for extension courses in agriculture, and the station also has a statistical department and another of economics and sociology. About forty annual publications in series; a *Informe Técnicó* (Technical Report) and a *Boletín de Divulgación Técnica* (Bulletin of Technical Information) are published.

Instituto Agrario Argentino de Cultura Rural (Argentine Agrarian Institute of Rural Culture)
Established in 1937, it maintains a library of 2000 volumes and publishes *Reseñas Argentinas* (Argentine Review), *Reseñas* (Review) and *Comunicados* (Communications).

Chacra Experimental de Barrow (Barrow Experimental Farm)
A provincial research station of the Ministry of Agriculture which was founded in 1923 and has a library of 15 000 volumes. This experimental farm is dedicated to genetic research on cereals and the breeding of thoroughbred animals.

Estación Experimental Agro-Industrial (Agro-Industrial Experimental Station)
Founded in 1909, it possesses the 'Alfredo Guzman' library and publishes *Revista Industrial* (Industrial Review) and *Agrícola de Tucumán* (Agriculture of Tucumán), three times a year, a bulletin and circulars at irregular intervals.

Instituto Nacional de Vitivinicultura (National Institute of Viticulture and Viniculture)
Founded in 1959, it controls production standards of wine and promotes research into viticulture and viniculture. There is a library of 6718 volumes.

Instituto de Suelos y Agrotécnica (Institute of Soil and Agrotechnology)
Established in 1944, it carries out research into soil fertility and studies on erosion. The institute publishes *Técnicas* (Technical Notes), *Apartados de Artículos* (Extracts) and *Tiradas Internas* (Internal Circulars).

Centro de Investigación de Biología Marina (Research Centre of Marine Biology)
Founded in 1960, it has a staff of thirty-six members and a library of 3500 volumes and publishes *Contribuciones Técnicas* (Technical Contributions) and *Contribuciones Científicas* (Scientific Contributions).

Instituto Nacional de Desarollo Pesquero (National Institute of the Development of Fisheries)

Established in 1977 with a staff of 200 members the institute carries out research in marine biology and ecology; fishery technology; micro-plankton; and marine parasites. It publishes *Contribuciones* (Contributions), *Revista* (Review), and *Memorial Anual* (Annual Memorandum).

Other institutes conducting agricultural research include:

Instituto Nacional de Limnología (National Institute of Limnology)

Estacion Hidrobiológica (Hydro-biological Station)

Centro de Investigaciones Bella Vista (Bella Vista Research Centre)

Medical Science and Technology

Studies and research activities in this field are carried out in specialist institutes of the faculties of medicine and in hospital centres.

Instituto Nacional de Microbiología (National Institute of Microbiology)
Founded in 1957, it develops activities to improve biological products of sanitary importance, to control biological products, to examine the problems encountered in this field and to teach staff.

Instituto de Hematología, Instituto Nacional de la Salud (Haematology Institute, National Health Institute)

Instituto de Investigaciones Médicas (Institute of Medical Research)

Instituto de Biología y Medicina Experimental (Institute of Biology and Experimental Medicine)
Founded in 1944, it has a library of 15 000 specialized volumes.

Servicio de Endocrinología y Metabolismo (Endocrinology and Metabolism Centre)

Fundación Cosio (The Cosio Foundation)
Established in 1957, it develops research activities, studies, and seminars related mainly to cardiology and immunology.

Nuclear Science and Technology

Comisión Nacional de Energía Atómica (CNEA) (National Commission for Atomic Energy)
Founded in 1950, it centralizes all research activities in this field. Its functions permit the control of production, stock, commercialization and use of the materials essential to atomic energy. The body is dependent on the *Comisión de los Centros Atómicos de Constituyentes Eseiza* (Commission for Atomic Centres of the Eseiza) and the *Instituto Balseiro en Bariloche* (Balseiro Institute in Bariloche). *CNEA* carries out feasibility studies in the *Central Nuclear de Atucha* (Atucha Nuclear Centre) and publishes various bulletins.

Aerospace Science and Technology

Comisión Nacional de Investigaciones Espaciales (CNIE) (National Commission of Space Research)
Established in 1960, it carries out research on the peaceful use of space; this research is in conjunction with the various institutions specializing in space technology, solar energy, atmospheric physics, etc. There is a staff of 400 and a library of 15 000 specialized volumes and the commission publishes *Informe Argentino de Actividades COSPAR* (Argentinian Report on COSPAR Activities) annually.

Centro Espacial Vicente López (Vicente López Space Centre)

Centro Espacial San Miguel – Observatorio Nacional de Física Cósmica (San Miguel Space Centre – National Observatory of Cosmic Physics)
Founded in 1935, it carries out research into aerospace, solar physics, non-conventional energy, atmospheric electricity, etc. It has a library of 65 000 specialized volumes.

Military Science and Technology

Junta de Investigaciones y Experimentaciones de las FFAA (Council of the Armed Forces for Research and Experimentation)
This is a body concerned with the coordination and direction of subjects related to the policy of research in military matters. The council depends on the National Ministry of Defence.

Instituto de Investigaciones Científicas y Técnicas de las FFAA (Institute of the Armed Forces for Scientific and Technical Research)
Established in 1954, it carries out research into standards, materials, equipment and basic research. Its main areas of work concern space projects and pure technological research such as the development of techniques, materials and industrial processes related to security and defence. It is dependent on the National Ministry of Defence.

Instituto de Investigación Aeronáutica y Espacial (IIAE) (Institute of Aeronautic and Space Research)
Founded in 1957. Apart from its function of general research into subjects relating to space, the institute also carries out studies with a military emphasis such as the construction of electronic rockets for military use.

Other institutes which undertake research and studies for military and civil application are spread among the different branches of the armed forces. Of these, two merit a special mention:

Instituto Geográfico Militar (Military Geographical Institute) created in 1879, and *Departamento de Estudios Históricos Navales* (Department of Historical Naval Studies).

Scientific and Technological Information Services

The information services are mainly established in libraries and specialized centres, public or private. The *Instituto Nacional de Estadística y Censos* (National Institute of Statistics and Census) publishes a major part of the statistics which form the basis for demographical studies.

In Argentina there are thirty important libraries, apart from the university libraries mentioned in the section on Science and Technology at an Academic Level. Of these the main ones include the *Biblioteca Nacional* (National Library) in Buenos Aires, which has 1 600 000 volumes, and the *Biblioteca del Servicio Geológico Nacional* (National Geological Service Library) in Buenos Aires, which is noted for its collection of 15 000 maps and charts, 45 000 specialized pamphlets and 150 000 volumes.

Various centres of information form part of the information services section of the United Nations.

Centro de Investigación Documentario del Instituto Nacional de Tecnología Industrial – UNIDO (Centre for Documentary Research of the National Institute of Industrial Technology)

Departamento Efectos del Medio Contaminado, Dirección Nacional de Estudios y Proyectos – UNEP (Department for Measuring the Effects of Pollution, National Direction of Studies and Projects)

Subsecretaria de Planeamiento Ambiental – UNEP (Sub-secretariat of Environmental Planning)

Centro Argentino de Información Científica y Tecnológica – UNESCO (Argentinian Centre of Scientific and Technical Information)

Centro Nacional de Información Educativa – UNESCO (National Centre of Educational Information)

Centro de Información de las Naciones Unidas (Information Centre of the United Nations)

Meteorology, Environmental Pollution and Astrophysics

Observatorio Astronómico (Astronomical Observatory)
Founded in 1871 in the city of Córdoba, it is dependent on the University of Córdoba. It publishes *Resultados* (Results).

Servicio Meteorológico Nacional (National Meteorological Service)
Founded in 1872 in Buenos Aires, it publishes weather maps, hydrological annual reports, and climatological annual reports.

International Cooperation in Science and Technology

Argentina plays an active role in international cooperation in science and technology, principally in the interchange of scientists and experts, grants and teaching, and training of human resources, as well as the exchange of information. Cooperation is carried out on an inter-governmental basis.

A major part of multilateral cooperation is in coordination with the specialized system of the United Nations: UNESCO, UNDP, UNIDO, FAO, WHO, etc.

Argentina's participation in the United Nations Development Programme (UNDP) has led to the development of national projects such as: that aimed at increasing the national capacity for scientific research and technological development, and also regional projects including those on replenishable energy resources in Latin America; multinational technological enterprises and scientific and technological research in Latin America, etc.

Technical cooperation between the developing countries (CTPD) has been carried out at different levels. March 1980 saw the beginnings of a closer relationship between Argentina and the countries of the Caribbean: Barbados, Guyana, Haiti, Jamaica, Dominican Republic, Santa Lucía and Trinidad and Tobago, on a scientific, technological and cultural basis. Argentina has also established links with African countries: Cameroon, Congo, Gabon, Ghana, Equatorial Guinea, Senegal, Togo and Zaïre.

Various projects are being executed with UNESCO, among them are an international study to collate information on the organization and yield of the scientific and technological research units (ICSOPRU) and studies to determine priority areas in the different regions of the country.

In regional cooperation, the Regional Programme of Scientific and Technological Development (*PROCYT*) of the Organization of American States (OAS) is undertaking ordinary and specialized projects. These are financed by the Mar del Plata Fund and are incorporated into the *Consejo Interamericano para la Educación, Ciencia y Cultura (CIECC)* (Inter-American Council for Education, Science and Culture). During 1980/81 Argentina had twenty-nine national and multinational projects under way in the field of basic and applied sciences, political and technological development and the exchange of technology.

At inter-regional level, Argentina participates actively in the *Comisiones Económicas para Africa (CEPA)* (African Economic Commissions) and *Comisiones Económicas para America Latina (CEPAL)* (Latin American Economic Commissions) with the object of attaining a high level of cooperation between these regions.

In the field of bilateral cooperation, Argentina has established agreements with developed and developing countries, among them France (telecommunications, physics, computing, cotton cultivation, atomic energy); Saudi Arabia (red meat production); Honduras (exchange of information in grain production); Japan (mineral resources, fish processing plants, Chagas Mazza nursing school); Italy (industrial technology) and Colombia (gamma-ray experiments).

Societies and Professional Associations in the General Field of Science and Technology

Apart from the institutes already mentioned, the following are not classified, and are dedicated fundamentally to study and research in the social sciences.

Instituto Torcuato di Tella (Torcuato di Tella Institute)

Instituto de Desarrollo Económico y Social (Institute of Economic and Social Development)

Centro de Estudios Urbanos y Regionales (CEUR) (Centre of Urban and Regional Studies)

Instituto de Planeamiento Regional y Urbano (IPRU) (Institute of Regional and Urban Planning)

Instituto Nacional de Estudios del Teatro (National Institute of Theatre Studies)

Numerous societies and associations promote research and development activities in science and technology. Among them are:

Academia Nacional de Ciencias de Buenos Aires (National Academy of Sciences of Buenos Aires)

Academia Nacional de Medicina (National Academy of Medicine)

Academia Nacional de Agronomía y Veterinaria (National Academy of Agronomy and Veterinary Sciences)

These bodies act as patrons for national programmes for the development of scientific knowledge.

National and Provincial Universities

Universidad de Buenos Aires (University of Buenos Aires)
Founded in 1821 and situated in the city of Buenos Aires. It has a teaching and research staff of 3440 and a student population of 187 000. Its faculties comprise: law and social sciences, economic sciencies, natural and exact sciences, architecture and urban planning, philosophy and arts, engineering, medicine, agriculture, dentistry, pharmacy and biochemistry, and veterinary sciences.

The following institutes are affiliated to the university: *Escuela Superior de Comercio 'Carlos Pellegrini'* (Carlos Pellegrini Higher School of Commerce), *Colegio Nacional de Buenos Aires* (National College of Buenos Aires); *Centro de Investigaciones Médicas* (Centre of Medical Research), and the *Instituto Bibliotecnológico* (Institute of Librarianship). For the development of its activities there is a library dispersed among the faculties with a total of 2 080 000 volumes. The university publishes *Boletín* (Bulletin), *Revista* (Review) and *Serie de Publicaciones* (Series of Publications). The faculties publish various other works.

Universidad Nacional de Catamarca (National University of Catamarca)
Established in 1972 in the town of Catamarca, it has a staff of 304 teach-

ers and researchers and student population of 1666. Its departments include: agricultural sciences; economics; education; health sciences and technology. The university publishes the review *Aportes* (Contributions).

Universidad Nacional del Centro de la Provincia de Buenos Aires (Central National University of the Province of Buenos Aires)
Founded in 1975 with a staff of 675 teachers and researchers and 2345 students. Its faculties are: agronomy, economics, sciences, humanities, engineering, and veterinary science. *Temas* and *Revista* (Themes and Review) are published.

Universidad Nacional del Comahué (National University of Comahué)
Established in 1971 and situated in the town of Neuquén. With a staff of 698 teachers and researchers and a population of 2950 students it has regional centres in Bariloche and Viedma. Its faculties comprise: social sciences, education, agriculture, economics and business studies, humanities, engineering, and tourism. Also affiliated to the university are the *Escuela Superior de Biología* and *Escuela Superior de Idiomas* (Higher School of Biology and the Higher School of Languages).

Universidad Nacional de Córdoba (National University of Córdoba)
Founded in 1613, situated in the city of Córdoba, it has a staff of 4967 teachers and researchers and 43 463 students. Its faculties are in law and social sciences, medicine, exact sciences, physics and natural sciences, economics, philosophy and humanities, architecture and urban planning, dentistry and chemical sciences. Also affiliated to the university are the schools of science and information, social services, and higher languages. The university also possesses institutes of agronomy and mathematics. *Revista* (Review) is published.

Universidad Nacional de Cuyo (National University of Cuyo)
Established in 1939 in the city of Mendoza, it has a staff of 844 teachers and researchers and a student population of 7579. Faculties comprise agriculture, economics, medicine, political and social sciences, engineering, and philosophy and the arts.

Universidad Nacional de Jujuy (National University of Jujuy)
Founded in 1972 and situated in the town of Jujuy. There is a staff of 253 teachers and researchers and 996 students. Its faculties are: economics, engineering, and agriculture. There are also institutes of physics and mathematics, chemistry and geology.

Universidad de la Pampa (University of La Pampa)
Established in 1958 with a staff of 522 teachers and researchers, it has a population of 1614 students. There are faculties in economics, human sciences, exact and natural sciences, agronomy and veterinary sciences.

Universidad Nacional de La Plata (National University of La Plata)
Founded in 1884 in the city of La Plata, its staff comprises 1209 teachers and researchers and 22 000 students. Its faculties are: architecture, agriculture, economic sciences, engineering, exact sciences, humanities and

education, juridical and social sciences, medical sciences, natural sciences, veterinary sciences, dentistry and fine arts. The following institutes and schools are affiliated to the university: *Colegio Nacional* (National College), *Escuela Graduado 'Joaquin V. González'* (Joaquin V. González Graduate School), *Escuela de Periodismo* (School of Journalism), *Escuela Practica de Agricultura y Ganaderia 'María Cruz y Manuel L. Inchausti'* (María Cruz and Manuel L. Inchausti School of Practical Agriculture and Animal Husbandry), *Instituto Superior del Observatorio Astronómico* (Higher Institute of the Astronomical Observatory) and the *Departamento de Cinematografía* (Department of Cinematography). The university possesses a library of 450 000 volumes and publishes *Revista de la Universidad* (University Review).

Universidad Nacional del Litoral (National University of the Littoral)
Established in 1919, it is situated in the city of Santa Fé. The teaching and research staff numbers 700 and the student population 8591. Its faculties are: law and social sciences, chemical engineering, economics, biochemistry and biology, agronomy, and veterinary sciences. A university review is published three times a year.

Universidad Nacional de Mar del Plata (National University of Mar del Plata)
Founded in 1961 in Mar del Plata in the province of Buenos Aires, it has a staff of 1600 teachers and researchers, and a population of 5500 students. Its faculties are: architecture and planning, economics and social sciences, economics, engineering, humanities, exact sciences, natural sciences and biology, agriculture and law, and there are schools and institutes in occupational therapy, health sciences, marine research, physical education and sport, methodology and the philosophy of science. The university publishes three times a year *Revista de Letras* (Arts Review).

Universidad Nacional del Nordeste (National University of the North-East)
Established in 1957, it is situated in the town of Corrientes. Its staff numbers 2332 teachers and researchers and 25 000 students. Faculties comprise: engineering, architecture and urban planning, economic sciences, humanities, dentistry, medicine, agricultural sciences, veterinary sciences, law, social and political sciences, science and land surveying, agricultural engineering and natural resources. The university publishes various works in collaboration with the faculties.

Universidad Nacional de la Patagonia San Juán Bosco (San Juán Bosco National University of Patagonia)
Founded in 1980 in the town of Comodoro Rivadavia, it has a staff of 240 teachers and researchers and 1049 students. Its faculties are: sciences, economic and social sciences (Trelew), art and education (Río Gallegos), with an oceanographic department (Trelew). There are also schools of social work, art and design, nursing and forestry (Esquel).

Universidad Nacional de Río Cuarto (National University of Río Cuarto)
Established in 1962, it is situated in the town of Río Cuarto, and has

faculties in agriculture, engineering, economics, education, humanities, nursing and veterinary medicine.

Universidad Nacional de Rosario (National University of Rosario)
Founded in 1968 in the town of Rosario, it has a staff of 3542 teachers and researchers and 28 891 students. Its faculties are: economic sciences, medical sciences, humanities and art, law, agricultural sciences, odontology, biochemistry, architecture, pure sciences and engineering, political sciences and international relations, and veterinary medicine.

Universidad Nacional de Salta (National University of Salta)
Established in 1972, it is situated in the town of Salta. Its staff numbers 716 teachers and researchers and there is a student population of 3830. There are departments in humanities, health sciences, technological sciences, economics, natural science, and exact sciences. The *Consejo de Investigación* (Council of Research) is affiliated to the university.

Universidad Nacional de Santiago del Estero (National University of Santiago del Estero)
Founded in 1973 in the town of Santiago del Estero, it has a staff of 400 teachers and researchers and 1700 students. It possesses departments in basic sciences, social sciences, civil engineering and natural resources. The *Centro Educativo Rural* (Rural Centre of Education) is a dependent institute of the university.

Universidad Nacional del Sur (National University of the South)
Established in 1956 in the town of Bahía Blanca, it has a staff of 460 teachers and researchers and 4515 students. Its departments comprise: agriculture, economics, exact sciences, natural sciences, social sciences and engineering. It has institutes of oceanography, biochemistry research, mathematics, and a pioneering project of chemical engineering. The university possesses a library of 85 000 volumes.

Universidad Nacional de Tucumán (National University of Tucumán)
Founded in 1914, it is situated in the town of San Miguel de Tucumán and has a staff of 2367 teachers and researchers and 17 723 students. Its faculties are: exact sciences and technology, agriculture and animal breeding, philosophy and the arts, architecture and urban planning, biochemistry, chemistry and pharmacy, economics, medicine, dentistry, law and social sciences, and natural sciences. The university possesses a central library of 100 000 volumes and publishes *Memoria Anual* (Annual Review).

Universidad Tecnológica Nacional (National University of Technology)
Established in 1959 in the city of Buenos Aires, it has regional campuses in ten different towns. The university runs evening courses in construction, electrical engineering and electronics, chemistry, metallurgy, mechanics, nautical engineering and textiles.

Escuela de Ingeniería Aeronáutica (School of Aeronautical Engineering)
Founded in 1947, it is situated in the city of Córdoba and is dependent on the Argentinian Air Force. It runs courses of three years' duration for civil and military personnel.

Private Universities

Universidad del Aconcagua (University of Aconcagua)
Established in 1968 in the city of Mendoza, it has a staff of 251 teachers and researchers and 884 students. There are faculties in social sciences and business studies, economics and commerce and psychology, and there is also a school of phono-audiometry.

Universidad Argentina de la Empresa (The Argentine University of La Empresa)
Founded in 1962 in the city of Buenos Aires, its staff numbers 800 teachers and researchers and 6500 students. Its faculties are: economic sciences, business studies, law and social sciences, and engineering. The university runs pre-university and extension courses. The *Instituto de Investigación Financiera y Económica* (Institute of Financial and Economic Research) and the *Centro de Estudio de Energía* (Centre of Energy Studies), are affiliated to the university. Its library has 13 213 volumes and *Informe* (Report), *Bancar* (Banking), and *Energía* (Energy) are published.

Universidad Argentina 'John F. Kennedy' (Argentine University of John F. Kennedy)
Founded in 1961, it is situated in Buenos Aires and has a staff of 200 teachers and researchers and 2000 students. It has schools of art and science, business studies, demography and tourism, dramatic art, education, journalism, political science, public relations, psychology, and sociology.

Universidad de Belgrano (University of Belgrano)
Established in 1964 in the city of Buenos Aires, it has a staff of 1300 teachers and researchers and a student population of 8800. Its faculties are: humanities, economics, law and social sciences, architecture, technology, and graduate studies. It possesses the following institutes: sociology and history, psychology, architectural planning, public rights, civil rights, international studies, strategic studies, technological research, economic and financial research, business studies and accountancy. The university has a library of 22 000 volumes and publishes monthly *Vigencia* and *La Situación Internacional* (The International Situation).

Universidad Católica Argentina 'Santa María de los Buenos Aires' (Santa María Argentinian Catholic University of Buenos Aires)
Founded in 1958 in the city of Buenos Aires, it has a staff of 2900 teachers and researchers and 11 000 students. Its faculties are: philosophy and the arts, law and political sciences, economics and social sciences, theology, physics, mathematics and engineering, humanities and education (Mendoza), social sciences (Rosario) and chemistry (Rosario). It possesses the following institutes: cultural studies, health sciences, university extension courses and pre-university studies. The university publishes *Anuario* (Annual Report), *El Derecho* (Law), 'Universitas' and 'Sapientia'.

Universidad del Museo Social Argentino (University of the Social Museum of Argentina)

Founded in the city of Buenos Aires in 1961, it possesses faculties in sociology, journalism, humanities, politics, law, economics, and media studies.

Universidad del Salvador (University of Salvador)
Established in 1959 in the city of Buenos Aires, it has a staff of 1002 teachers and researchers and 4350 students. Its faculties are: legal sciences, philosophy, history and arts, medicine, psychology, social sciences, theology, psychology in teaching, education sciences and social communication, specialist engineering, and human relations. There is a library of 50 000 volumes and the university publishes *Anales* (Annals), 'Signos Universitarios' and 'Atenea'.

Universidad Católica de Córdoba (Catholic University of Córdoba)
Founded in 1956 in the city of Córdoba, it has a staff of 700 teachers and researchers and 4100 students, and faculties in: architecture, agriculture, economics and business studies, law and social sciences, philosophy and humanities, engineering, medicine, chemical sciences, and political science and international relations. The university has a library of 80 500 volumes.

Universidad Católica de Cuyo (Catholic University of Cuyo)
Established in 1953, it is situated in the town of San Juán and has a staff of 310 teachers and researchers and 1100 students. Its faculties are: law and social sciences, economic sciences, philosophy and humanities and nutrition. There is also a university nursing school. The university publishes *Cuadernos* (Textbooks) and *Boletín Anual* (Annual Bulletin).

Universidad Católica de La Plata (Catholic University of La Plata)
Founded in 1968 in the city of La Plata, it has faculties in law, sociology, architecture, statistics, and economics.

Universidad Notarial Argentina (Notarial University of Argentina)
Established in 1965, it is situated in the city of La Plata and has a staff of fifty-five teachers and researchers and a student population of 3326. It publishes *Cuardernos Notariales* (Notarial Textbooks) and *Ediciones UNA*.

Universidad de Mendoza (University of Mendoza)
Established in 1960 in the town of Mendoza, it has a staff of 229 teachers and researchers and 1932 students. Its faculties are: law and social sciences, architecture and urban planning, electrical engineering and electronics. The university publishes *Idearium.*

Universidad 'Juan Agustín Maza' (Juan Agustín Maza University)
Founded in 1960, it is situated in Mendoza and has a staff of 260 teachers and researchers and 875 students. Its faculties are: engineering, pharmacy and biochemistry, physics and mathematics, journalism, and a faculty devoted to oenology, fruit cultivation and horticulture. The School of Nutrition is also affiliated to this university. *Anuario* (Annual Report) is published by the university.

Universidad de Morón (University of Morón)
Established in 1960, it is situated in Buenos Aires. It has a staff of 1350

teachers and researchers and 11 800 students. Its faculties are: law and social sciences, engineering, exact sciences, chemistry and natural sciences, philosophy and the arts, agronomy, economics and architecture. Also affiliated to the university are the Higher Institute of Technology, Higher Institute of Tourism and Diocesan School of Social Services. The university publishes *Periódico* (Periodical) monthly and *Revista de la Universidad* (University Review) every six months.

Universidad del Norte Santo Tomás de Aquino (Santo Tomás de Aquino University of the North)
Founded in 1965, it is situated in the town of San Miguel de Tucumán and has a staff of 235 teachers and researchers and 1115 students. Its faculties are: humanities, law and social sciences, economics and business studies, and industrial engineering. There is also a centre located in Concepción.

Universidad del Patagonia 'San Juán Bosco' (San Juán Bosco University of Patagonia)
Established in 1961, it is situated in Comodoro Rivadavia. It has a staff of 114 teachers and researchers and 710 students. It possesses schools in sciences and humanities and publishes *Anales* (Annals).

Universidad Católica de Salta (Catholic University of Salta)
Founded in 1967 in the town of Salta, its staff numbers ninety teachers and researchers. Its faculties comprise: art and sciences, economics and business studies and engineering, and there are schools of social services and law.

Universidad Católica de Santa Fé (Catholic University of Santa Fé)
Established in 1959, it is situated in Santa Fé with a staff of 300 teachers and researchers and 1782 students. Its faculties are: law, economic sciences, education, philosophy, history, arts, architecture, and soil sciences.

Universidad Católica de Santiago del Estero (Catholic University of Santiago del Estero)
Founded in 1960 in the town of Santiago del Estero, it has a staff of 219 teachers and researchers and a student population of 1515. Its faculties are: economics, education, and legal, social and political sciences, and its departments include philosophy and theology, and applied mathematics.

Argentina also has eleven establishments of higher education and six schools of art and music.

Belize

Belize lies in Central America on the Caribbean coast, with Mexico to the north-west and Guatemala to the south-west. The total area is about 8867 square miles for an estimated population of 152 000 (mid 1978).

Most of the country is covered by forest, of which 50 per cent is high rain forest, 15.5 pine forest and dry savannah, 5.5 wet savannah and mangrove forest, and the remaining 20 per cent existing or recently abandoned cultivation.

The economy is based on agriculture, which accounts for 60 per cent of foreign exchange earnings. The main crops are sugar, citrus fruit and rice. Banana cultivation has also been revived in the south, mainly for export to Europe. In 1980 1 050 000 metric tons of sugar cane were produced, 46 000 metric tons of oranges, 16 000 tons of grapefruit and pomelos and 6000 tons of paddy rice. There were 21 000 metric tons of bananas produced in that year. Lobster tails are exported to the USA and forestry is again emerging.

Primary education is principally carried on through subsidized denominational schools under government control. In 1980 there were also twenty secondary schools, four technical colleges, four vocational schools and a teacher-training college. The Belize College of Arts, Science and Technology, founded in 1980, is the first stage in the planned University of Belize. At present it has forty students and eighteen teachers. There is an extra-mural branch of the University of the West Indies in Belize. Other colleges are: Fletcher College, which has eleven teachers and 150 students, and Wesley College, which has twenty-six teachers and 476 students.

Education is compulsory between the ages of six and fourteen years.

There is a National Library Service which is conducted by the Government through the Ministry of Education and a statutory Library Board. The Central Library and Headquarters are in the Bliss Institute, Belize City. The Central Library includes a National Collection.

Bolivia

Geographical, Demographical, Political and Economic Features

Bolivia is one of the two land-locked countries of Latin America, with an area of 1 098 581 square kilometres. The legal capital and judicial seat is Sucre, though the seat of government and the most important commercial centre is La Paz. The borders to the north and east are with Brazil, with a frontier of 3 125 kilometres; to the south-east with Paraguay, to the south with Argentina, to the south-west with Chile and to the west with Peru.

Estimates in 1980 showed the population to be 5 599 592 with a projection for 1985 of 6 656 000. The density per square kilometre, according to the same estimates, is 5.1. Of the population 44.4 per cent live in urban areas while 51.4 per cent live on the Altiplano – 27.1 per cent in the valleys and 21.5 per cent in the tropical region. The composition of the population is 54 per cent indigenous, 31.2 per cent mestiza (mixed blood) and 14.8 per cent white. About 1 million Indians speak Quechua; 660 000 speak Aymara and 150 000 other languages.

Government

Bolivia was proclaimed a republic on 6 August 1825, and the first constitution was adopted on 19 November 1826. Political history reveals that Bolivia has been constantly affected by insurrections, coups and military counter-coups; the republic has had more than 185 revolutions since its independence in 1825. The last constitutional reform was adopted in 1967 and paved the way for the election of a president for a period of four years and a bicameral congress. Universal suffrage does exist, although since 1966 the country has not been able to hold elections. The executive is represented by the President of the Republic and a cabinet of fifteen ministers.

Remote Sensing Project in Bolivia

Bolivia's Geological Service (GEOBOL) is concluding a successful project designated to give the Department of Oruno the capability to evaluate its natural resources using data generated by three United States satellites. The Oruro project saw creation of a permanently integrated Geographical Information System (GIS) which will be used to prepare base maps of soil characteristics, land use, geomorphology, geology, water resources and hydrology. The system will be used in land planning, particularly as regards agriculture. Oruro's most important economic activity is tin and tungsten mining, with agriculture limited to small areas. Further development of agriculture has been held back by problems associated with overgrazing, soil erosion, lack of water, salinity and poor land management. Plans are now under way to extend a second phase of the project to Bolivia's Department of Potosi. The Oruro project is the first of its kind in Latin America. It is being financed in part by the IDB.

Natural Resources

AGRICULTURE

The region to the east of the Andes provides about three-quarters of the country's agricultural production. Since the agrarian reform in 1959 this area has been developing programmes for the production of rice, sugar-cane and cotton, the objective being to satisfy national demand. Also produced are oats, bananas, potatoes, cacao, coffee, barley, onions, and citrus fruits, among others.

FORESTRY

The forests, which occupy about 46 344 000 hectares, produce oak, walnut, cedar, mahogany, laurel, ochoo, tarara, jitchuturugui, copaiba, quebracho, campana, urundel, lapacho, balsa, etc. The quantity of cut wood for 1977 reached 154 000 cubic metres.

LIVESTOCK

Figures for 1976 show 2.9 million head of cattle. There is an important market for vicuña, chinchilla and red fox skins, although hunting vicuña is prohibited.

MINERALS

Minerals are Bolivia's most important exports, constituting 55 per cent of foreign earnings. Almost a half of the country's mineral reserves are tin, although exploitation is costly as a result of climate and the height of the mines (12 000 to 18 000 feet above sea level), which makes transportation difficult. The mineral is extracted from very deep shafts and out of the total mined only 35 per cent is profitable.

There are also deposits of gold in Alto Buin, which produce about 100 kilos annually. Foreign consortiums are prospecting for uranium deposits and the Bolivian government hopes that more companies will show interest.

Other reserves include oil, natural gas, zinc, lead, antimony, cadmium, tungsten, copper, manganese, asbestos, sulphur and gypsum.

INDUSTRIAL PRODUCTION

Main products include beer, cigarettes, wine and by-products of oil. Also sugar, cement, flour, coffee, edible oils, alcohol, spinning yarn, soap, detergents, powdered milk, cut wood, and woollen manufacture.

Education

Primary education is free and obligatory between the ages of six and fourteen, and illiterate people between the ages of fifteen and fifty must attend reading classes. Estimates in 1976 show that 63.3 per cent of the population is literate.

International Relations

Bolivia is a member of the United Nations and the Organization of American States.

Organization of Science and Technology

In 1977 by Decreto Supremo No 15.111 basic political guidelines were established for the building up of a *Sistema Nacional para el Desarrollo Científico y Tecnológico (SINDECYT)* (National System for Scientific and Technological Development), and the *Dirección de Ciencia y Tecnología* (Board of Science and Technology), dependent on the Ministry of Planning and Coordination, in conjunction with other organizations, was given the task of formulating operative plans in this field.

Figure 2 shows the organization of institutions directly involved in science and technology.

Financing and Policies in Science and Technology

Although there is no explicit policy structure in science and technology, research activities are carried out in two fundamental directions:

(a) basic research activities in which Bolivia, because of its favourable conditions, can make significant headway, eg biology at high altitudes; astronomy, etc.

(b) applied research in priority areas for the socio-economic development of the country, eg agricultural sector; agro-industry; energy; metal-mechanics industry; iron and steel manufacture.

One of the first tasks accomplished by the Board of Science and Technology was the compilation of a census of human resources. Out of a list of 673 institutions, 302 were selected that were directly involved in scientific and technological research. In these institutions 610 people were directly involved in research; this figure differs significantly from a similar study made in 1974, in which the number of scientists was stated to be

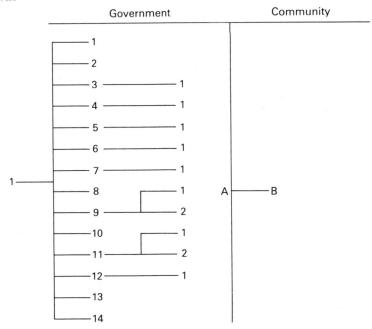

Fig. 2 Organization of science and technology

1	Presidencia (Presidency)
1.1	Ministerio de Vivienda y Urbanismo (Ministry of Housing and Town Planning)
1.2	Ministerio de Transporte y Comunicaciones (Ministry of Transport and Communications)
1.3	Ministerio de Defensa (Ministry of Defence)
1.3.1	Instituto Geográfico Militar (Military Geographical Institute)
1.4	Ministerio de Prevision Social y Salud (Ministry of Social Planning and Health)
1.4.1	Institutos de Nutrición, Biología de Altura (INSO) (Institutes of Nutrition and the Study of Biology at Altitude)
1.5	Ministerio de Industria y Comercio (Ministry of Industry and Commerce)
1.5.1	Dirección de Normas y Tecnología (Board for Standards and Technology)
1.6	Ministerio de Planificación y Coordinación (Ministry of Planning and Coordination)
1.6.1	Dirección de Ciencia y Tecnología (Board of Science and Technology)
1.7	Ministerio de Agropecuarios (Ministry of Agriculture and Livestock)
1.7.1	Instituto de Biología Animal – Estaciones Experimentales (Institute of Animal Biology – Experimental Stations)
1.8	Ministerio de Prevision Social y Trabajo (Ministry of Social Planning and Labour)
1.9	Ministerio de Educación (Ministry of Education)
1.9.1	Academia Nacional de Ciencias (National Academy of Science)
1.9.2	Consejo Nacional de Ciencia y Tecnología (National Council of Science and Technology)
1.10	Ministerio de Interior (Ministry of the Interior)
1.11	Ministerio de Minería y Metalurgia (Ministry of Mining and Metallurgy)
1.11.1	Instituto de Investigaciones Minero-Metalúrgicas (Institute of Mining and Metallurgic Research)
1.11.2	Comisión Nacional de Energía Nuclear (National Commission for Nuclear Energy)
1.12	Ministerio de Energía e Hidrocarburos (Ministry of Energy and Hydrocarbons)
1.12.1	Centro de Tecnología Petrolera (Centre for Petroleum Technology)
1.13	Ministerio de Finanzas (Ministry of Finance)
1.14	Ministerio de Relaciones Exteriores (Ministry of Foreign Relations)
A	Consejo Nacional de Alta Educación (National Council of Higher Education)
B	Universidades (Universities)

1790, but without any indication whether they were involved in research. The work on the census was started in 1977 and to date there are no more accurate figures available.

A similar study was carried out for financial resources. In 1974, in an inquiry made into 296 institutions, 247 had a total budget of $150 million, but the amount destined for research is not stated. More recent data for eighteen institutions indicate a budget of $15.7 million and in the San Andrés University $1.8 million was allocated to scientific and technological research in 1978; this represents 12 per cent of its total budget.

Science and Technology at Government Level

Of the governmental organizations involved in scientific research, the following are among the most important:

Dirección de Ciencia y Tecnología (Board of Science and Technology)
Previously mentioned.

Dirección de Normas y Tecnología (Board of Standards and Technology)
A dependent organization of the Ministry of Industry and Commerce.

Consejo Nacional de Ciencia y Tecnología (National Council of Science and Technology)
A dependant of the Ministry of Education, involved in policy making and advising at government level.

The following institutes hold important positions for disseminating basic information for many research projects:

Instituto Geográfico Militar y de Catastro Nacional (Military Institute of Geography and National Property Register)
The institute draws up geographical maps and charts and land surveys. *Boletín Informativo* (Information Bulletin) is published.

Instituto Nacional de Estadística (National Statistics Institute)
Founded in 1936, the institute carries out demographic surveys and is responsible for the national census. *Boletín Estadística* (Statistical Bulletin) and *Indice de Precios al Consumidor* (Consumer Price Index) are published regularly.

The other organizations are mentioned in the relevant chapters.

Science and Technology at an Academic Level

In accordance with a law passed in June 1972 the new Bolivian universities are administered by the *Consejo Nacional de Alta Educación de Bolivia* (National Council for Higher Education in Bolivia). Bolivia is divided into three geographical zones. In each zone there are three universities and courses are run complementary to the particular geographical zone.

Universidad Boliviana Mayor Real y Pontificia de San Francisco Javier (San Francisco Javier Royal and Pontifical University of Bolivia)
Established in 1624 by the Papal Seal of Gregory XV and by Royal Charter of Felipe III in 1622. Students number 4210 and there are 203 teachers. The university is made up of the Institutes of Biology, Cancer Research, Nuclear Medicine, and various institutes appertaining to the humanities.

Universidad Boliviana Mayor de 'San Andrés' (San Andrés University)
Founded in 1830, it has a student population of 17 000 with 900 teachers for its faculties of pure and natural sciences, health sciences, technology and the Polytechnic Institute, as well as the humanities.

Universidad Mayor de 'San Simón' (San Simón University)
A state-run university founded in 1833, it has 10 000 students and 523 lecturers in its faculties of health sciences, agriculture and animal husbandry, sciences and the humanities.

Universidad Técnica de Oruro (Oruro Technical University)
Established in 1862, it has 260 teachers for its 6100 students with faculties of engineering, economics, and law, with a Polytechnic Institute.

Universidad Boliviana 'Tomás Frías' (Tomás Frías University)
Founded in 1892, students number 2500 with 207 teachers and *Revista Científica* (Scientific Review) is published. Faculties are: pure and natural sciences, and technology, with an affiliated Polytechnic Institute. There are also faculties in the arts and humanities.

Universidad Boliviana Mayor 'Gabriel René Moreno' (Gabriel René Moreno University)
Founded in 1880, it has a library of 35 000 volumes to serve its 1900 students and 270 teachers. *Universidad* (University) and *Revista Universitaria* (University Review) (four-monthly) are published by the university. Faculties are: health sciences, pure and natural sciences, technology and tropical agriculture. Also affiliated to the university is the Polytechnic Institute, as well as the faculties of arts and humanities.

Universidad Boliviana 'Juan Misael Saracho' (Juan Misael Saracho University)
Established in 1946, it has 750 students and 104 teachers for its faculties of pure and natural sciences, technology and economics. Affiliated to the university are the Department of Dentistry and the Polytechnic Institute.

Universidad Boliviana 'Mariscal Jose Ballivián' (Mariscal Jose Ballivián University)
A state university founded in 1967 which presently has a student population of 500 and forty-four teachers for its faculties of agriculture and animal husbandry and the Institute of Livestock Research.

Universidad Católica Boliviana (Bolivian Catholic University)
Founded in 1960, it has departments of communications, economics and psychology.

Industrial Research and Technology

The major part of activities in this field is concentrated in the governmental institutes as well as in the university faculties – for example the Faculty of Technology and the Polytechnic Institute of the San Andrés University play significant roles in the development of industrial research.

Centro de Tecnología Petrolera (Centre of Petroleum Technology)
A dependant of the Ministry of Energy and Hydrocarbons.

Instituto Boliviano del Petróleo (Bolivian Oil Institute)
An autonomous institute founded in 1959 which develops and coordinates scientific and technical studies in the field of oil and petroleum. An informative bulletin is published and there is a specialized library housing 1000 volumes.

Agricultural Science and Technology

The main research centre is the *Instituto de Biología Animal* (Institute of Animal Biology) with the experimental stations affiliated to the institute.

Instituto Boliviano de Tecnología Agropecuaria (IBTA) (Bolivian Institute of Agriculture and Livestock Technology)
The universities also have important research centres in this field; one predominant in this area is the *Universidad Boliviana Mariscal Jose Ballivián* in the region of Beni, with its faculty of agriculture and animal husbandry and Institute of Livestock Research.

Medical Science and Technology

The main centres are in the universities:

Instituto de Investigación del Cancer (Institute of Cancer Research) and the *Instituto de Medicina Nuclear* (Institute of Nuclear Medicine) of the San Francisco Javier Royal and Pontifical University in Sucre.

Instituto de Nutrición (Institute of Nutrition)
Work in this field is also carried out in the specialized departments of hospitals.

Nuclear Science and Technology

Comisión Boliviana de Energía Nuclear (Bolivian Commission for Nuclear Energy)
Founded in 1960, advises the government in specialized areas and carries out research for the peaceful use of nuclear energy. Various publications are issued.

Comisión Nacional de Energía Atómica (National Commission for Atomic Energy)

Astrophysical and Meteorological Research

Observatorio 'San Calixto' (San Calixto Observatory)
Established in 1892, publishes a seismological bulletin.

Professional Associations and Societies

There are thirty-three such bodies in the country, among which are:

Academia Nacional de Ciencias de Bolivia (Bolivian National Academy of Sciences)

Instituto Médico Sucre (Sucre Medical Institute)
Carries out research and produces vaccine and serum.

Instituto de Cancerología 'Cupertino Arteaga' (Cupertino Arteaga Cancer Institute)

Servicio Geológico de Bolivia (Bolivian Geological Service)
Reorganized in 1966, has ten specialized laboratories. Geological maps and informative bulletins are published.

Information Services in Science and Technology

There are thirteen institutes involved in documentation and seven museums and specialized historical archives:

Centro Nacional de Documentación Científica y Tecnológica (National Centre for Scientific and Technological Documentation)
Founded in 1967 in collaboration with FAO, WHO and ILO, it houses a wide range of documentation in various fields. *Actualidades* (News) is published.

Centro Nacional de Documentación e Información Educativa (National Centre for Educational Documentation and Information)
Affiliated to the Ministry of Education.

Biblioteca Central de la Universidad Mayor de 'San Simón' (Central Library of San Simón University)
Founded in 1925, it houses 41 544 volumes.

Biblioteca Central de la Universidad Mayor de 'San Andrés' (Central Library of San Andrés University)
Founded in 1930 has 121 000 volumes.

Biblioteca del Congreso Nacional (Library of the National Congress)
Established in 1912, at present has 15 000 volumes.

Brazil

Geographical, Demographic and Economic Features

Brazil was discovered on 22 April 1500 by the Portuguese Admiral Pedro Alvarez Cabral, and subsequently became a Portuguese colony. On 7 September 1822 it became independent.

The Atlantic Ocean forms a natural border to the east of the country and its frontiers to the north-west and south-west are shared with all the South American countries with the exception of Chile and Ecuador. From north to south Brazil has a lineal distance of 4320 kilometres and running from east to west, 4328 kilometres. On the map Brazil appears as a large inverted triangle.

According to the census of 1 September 1970 the population is estimated to be 116 393 100, with a density of fourteen inhabitants per square kilometre. From the same census the population comprises 46 331 343 men and 46 807 694 women. In 1950 the urban and suburban areas accounted for 36.2 per cent of the population while in 1960 the percentage rose to 45.1 per cent and to 55.9 per cent in 1970.

The capital city was transferred to Brasília in 1960. The principal cities in order of population are: São Paulo with 7 198 608 inhabitants; Rio de Janeiro (4 857 716); Belo Horizonte (1 557 464); Recife (1 249 821); Salvador (1 237 373); Fortaleza (1 109 837) and Porto Alegre with 1 043 964.

Government

On 24 January 1967 both branches of the congress approved the new constitution which came into effect on 15 March of the same year. In October 1969 the constitution was amended and has remained in this form to the present date. The constitution stipulates the indirect election of the President and the Vice-President of the Republic and this is carried out by the *colegio electoral* (electoral college) made up of members of the congress and delegates of the legislative branches. The constitution grants the president special powers to issue decrees and laws in matters of the economy and national security. The president also has the right to intervene in any of the twenty-one states without consulting the congress and to impose martial law or to govern by decree.

The election of the president and the vice-president must, according to the constitution, be carried out every six years, and a president may not be re-elected after resigning from office. The Senate must be elected every eight years and the Chamber of Deputies every four.

The name of the country has experienced several changes. From the United States of Brazil it changed to Brazil and then to the Federal Republic of Brazil. The official language is Portuguese.

Voting is obligatory for both sexes between the ages of eighteen and sixty-five. Members of the armed forces and the illiterate do not have the right to vote in any elections.

Administratively, Brazil is made up of twenty-one states, four federal territories and one federal district. Each state has its own legislative and judicial authorities, constitution and laws; these must, however, be in accordance with the national constitution.

Natural Resources

AGRICULTURE

The 44.7 per cent of the population who live in the rural areas provide Brazil's agricultural exports and 75 per cent of the foreign revenue.

The four states of São Paulo, Paraná, Minas Gerais and Espírito Santo are the main coffee-producing regions. Large plantations with more than 100 000 trees are the norm. Output for 1974 was 3.22 million tons from 2 269 738 hectares of plantation. Exports for 1977 were 512 391 tons and for 1980 exports totalled 1 070 000 tons.

The main centre for cocoa production is Bahia, where two crops are raised annually. After the United States of America, Brazil is the second world producer of citrus fruits, especially oranges. Rubber is another important product and the main centres are Acre, Amazonas and Para. Brazil is also the main world producer of 'carnauba' wax, which is used in electrical insulation.

LIVESTOCK

Figures for 1976 show 12.3 million head of cattle; 7.4 million pigs; 700 000 sheep; 400 000 goats and 335.2 million poultry.

FISH

The Brazilian fishing fleet numbers 200 000 and fish production for 1975 reached a total of 759 792 tons.

MINERALS

Brazil is very rich in mineral resources and those mentioned below form only part of the immense wealth. Brazil is the only country in the world which has quartz crystal in commercial quantities and exports for 1977 totalled 1610 tons. Diamonds are another important commodity, and for the same year exports reached 215 grams. Brazil is the fifth world producer of mica and the second for chromium, of which 886 514 tons were exported in 1976 and there are 4 million tons in reserve. Brazil also ranks first as world producer of beryllium, graphite, titanium and magnesium. In the vicinity of Rio de Janeiro, deposits of thorium have been found and it is estimated that reserves may total 100 000 tons. In Rio Grande do Sul, Santa Catarina, São Paulo and Paraná there are deposits of coal and their reserves are estimated to be in the region of 5000 million tons. Production for 1976 reached 7.88 million tons. The manganese mines located in the Amapa region are estimated to contain only 10 million tons. Lead production for 1976 totalled 282 688 tons and asbestos production in the same year was 1 442 223 tons.

Education

Primary education is obligatory and figures from the 1970 census indicate that 47 864 531 inhabitants from the age of five years are literate. This represents 60.33 per cent of this age group. Of this figure 50.9 per cent are male.

International Relations

Brazil is a member of the United Nations Organization, the Organization of American States and the Latin American Free Trade Association.

Organization of Science and Technology

The *Secretaría de Planificación (SEPLAN)* (Secretariat of Planning) which is dependent on the Presidency of the Republic, is the body responsible for the general planning and development in Brazil and under its jurisdiction it has established the *Consejo Nacional de Pesquisas (CNPq)* (National Research Council). In order to assist SEPLAN in the effective coordination of scientific and technological development, the *Banco Nacional para el Desarrollo (BNDE)* (National Development Bank); *la Agencia para el Financiamiento de Estudios y Proyectos (FINEP)* (Agency for the Financing of Studies and Projects); and the *Instituto Brasileiro de Geografía y Estadística (IBGE)* (Brazilian Geographical and Statistics Institute) have been placed under the administration of SEPLAN together with CNPq.

In 1974, the year of institutional reorganization, the functions of the CNPq were extended and the latter became the *Conselho Nacional de Desenvolvimento Científico e Tecnológico* (National Council of Scientific and Technological Development) having legal status as a private enterprise but coming under the jurisdiction of SEPLAN. In the same law, 6.036, the following functions were assigned to CNPq: '. . . analysis of projects and programmes in the various sectors of science and technology, such as the formulation and promotion of policies for the development of science and technology as established by the Federal Government. . . .'

In 1975 a system for the development of science and technology was created and the task of supervising its progress was assigned to CNPq. The system is endeavouring to centralize the bodies which carry out research and development activities in the field of science and technology, whether these be in the private sector or governmental. Since 1975 intermediary bodies have been created whose objective it is to maintain links between organizations which execute work in similar fields. One of these bodies is the *Conselho Científico Tecnológico (CCT)* (Scientific and Technological Council), which is made up of a body of scientists and representatives of institutions such as the *Academia de Ciências* (Academy of Sciences), representatives of the ministries and other governmental organizations. The main function of the CCT is to advise the CNPq on the coordination of the *Sistema Nacional de Desarrollo Científico e Tecnológico* (National System of Scientific and Technological Development).

Figure 3 is a simplified version of the network of institutions which form the National System.

Financing and Policies in Science and Technology

The promotion of research and development activities in science and technology is guided by the *Planes Básicos* (Basic Plans) for the development of science and technology. These plans were initiated in 1972, the year in which several policies were drafted to be implemented by the ministries. The execution of these plans was the responsibility of the *Secretaría de Planificación* (Secretariat of Planning) together with the CNPq. The second Basic Plan was drafted for the period 1975–79 and its text comprised ten chapters. The first chapter, 'Policies for Science and Technology', dealt with the definition and progression of the role played by CNPq within the National System of Scientific and Technological Development. The second chapter reported on the subsidies and grants made available for science and technology. The third chapter, entitled 'New Technologies' indicated the activities which should be promoted, such as: nuclear energy programmes, space investigation, study into marine resources, research into sources and forms of non-conventional energy. The fourth chapter, 'Infrastructure', dealt with the present development of various sectors such as electrical energy, oil, transport and communications. The fifth chapter, entitled 'Industrial Technology', provided a strategy with the following objectives: the establishment of an industrial standardization programme, the modernization and consolidation of the System of Industrial Patents, the establishment of a technological information system, the improvement of consultative capacity and the organization of national firms. The sixth chapter covered the means of research and development in the agricultural and livestock sector with a view to increasing productivity and development in this field. The seventh chapter analysed the alternatives which would provide a more effective distribution of funds and reduce the inequality between the different sectors. The eighth chapter, 'The Preparation of Research Staff', stated the methods which could be implemented in this field, such as the contracting of foreign lecturers, programme of grants, and the participation of national researchers in international events. The institutional organization of science and technology was analysed in

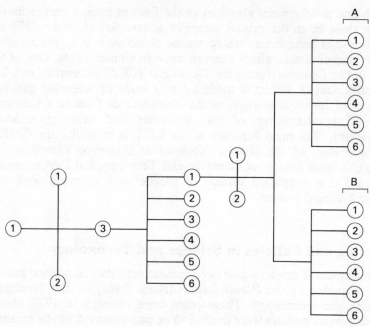

Fig. 3 Organization of science and technology

1	Presidencia de la República (Presidency of the Republic)
1.1	Ministerior (Ministers)
1.2	Otros organismos asesores de la Presidencia (Other advisory bodies of the Presidency)
1.3	SEPLAN
1.3.1	CNPq
1.3.1.1	CCT
1.3.1.2	Otros organismos asesores (Other advisory bodies)
A	Unidades Técnico Administrativas (Technical Administrative Units)
A.1	Superintendencia para Administración de Programas (Superintendency for Administration of Programmes)
A.2	Superintendencia para el Desarrollo de los Recursos Humanos (Superintendency for Development of Human Resources)
A.3	Superintendencia de Planificación (Superintendency for Planning)
A.4	Superintendencia para la Administración y Financiamiento (Superintendency for Administration and Financing)
A.5	Superintendencia para la Cooperación Internacional (Superintendency for International Cooperation)
A.6	Superintendencia para el Desarrollo Científico (Superintendency for Scientific Development)
B	Unidades subordinadas (Subordinate units)
B.1	Instituto Nacional de Investigaciones Amazónicas (INPA) (National Institute for Amazon Research)
B.2	Instituto Brasileño para la Información en Ciencia y Tecnología (IBICT) (Brazilian Institute for Information in Science and Technology)
B.3	Instituto de Investigaciones Espaciales (INPE) (Institute for Space Research)
B.4	Instituto de Matemáticas Puras y Aplicadas (IMPA) (Institute of Applied and Pure Mathematics)
B.5	Centro Brasileño de Investigaciones Físicas (Brazilian Centre for Physical Research)
B.6	Observatorio Nacional (ON) (National Observatory)

the ninth chapter. The last chapter studied the activities connected with research and development such as scientific and technological information, computerized systems and international cooperation.

Human resources in the field of science and technology are extensive in Brazil when compared to other Latin American countries. An estimate made in 1976 by CNPq on the resources in this field shows the following:

Table 2.1

Science area	Number of centres	Per cent	Number of researchers	Per cent
Biomedical	219	45	3251	40
Agricultural	52	11	1323	16
Social	85	17	499	6
Exact sciences	65	13	1649	20
Technology	67	14	1472	18
Total	488	100	8194	100

Preliminary data in 1977 estimated that for the execution of the second basic plan the following financial resources were required during the three-year period 1975–77:

Table 2.2 Development of new technology

	1975/77 (thousands of Cruzeiros)
Development of new technology	3 482 883
Infrastructure	5 284 771
Industrial technology	3 045 248
Agricultural development	4 190 777
Applied technology for regional and social development	1 246 274
Scientific development and implementation	4 310 900
Support and promotion (information, etc.)	2 691 498
Technical development in national firms	814 557
	25 866 907

As a result of the inadequacy of specific organizations to carry out the accountancy for each project, which would give a more realistic figure of expenditure in research and development, efforts have been increased since 1978 to make more rigorous checks in this area.

International Cooperation

International cooperation is developed through the combined work of the *Superintendencia de Cooperación Internacional (SCI)* (Superintendency of International Cooperation); the *Ministerio de Asuntos Exteriores* (Ministry of Foreign Affairs) through its *Departamento de Cooperación Cultural, Científica y Tecnológica (DCT)* (Department of Cultural, Scientific and

Technological Cooperation); the *Secretaría de Planificación* (Secretariat of Planning) and the CNPq.

The university system for the development of science and technology, especially through the Fondo Nacional (National funds), has begun to give more importance to international research activities at an academic level. It is worth mentioning here the universities of São Paulo, Campinas, and Paraíba, the Federal University of Rio de Janeiro and the Catholic University of Rio de Janeiro.

The CNPq, for its part, has established bilateral agreements with various organizations such as the *Consejo Nacional de Ciencia y Tecnología de Mexico* (National Council for Science and Technology of Mexico); the *Consejo Nacional para la Investigación Científica y Tecnológica de Costa Rica* (National Council for Scientific and Technological Research of Costa Rica). Agreements for aid and cooperation in the field of science and technology have been established with Germany, Canada and France, and bilateral agreements are being negotiated with Japan, Australia, Kenya, Italy, Chile, Colombia and India.

The CNPq has established, directly or through other organizations, connections with specialist associations within the United Nations, for example, the *Centro Latinoamericano de Física* (Latin American Centre of Physics) which was founded by UNESCO, has its base in Brazil and receives aid from the institutions of the *Ministerio de Relaciones Exteriores* (Ministry of Foreign Relations), as well as grants through the CNPq. The *Superintendencia para la Cooperación Internacional* (Superintendency of International Cooperation), dependent on the CNPq, has drawn up a manual for the use of the Brazilian institutions which wish to be included in programmes of scientific and technological cooperation under bilateral agreements with foreign organizations.

Science and Technology at Government Level

The complex of organizations that make up the *Conselho Nacional de Desenvolvimento Científico e Tecnológico (CNPq)* (National Council for Scientific and Technological Development), is without doubt the principal network in science and technology. The government, through state companies like PETROBRAS, are developing large research programmes in the field of technology; and at the level of scientific research, institutes like *Comissão Nacional de Energia Nuclear* (National Commission for Atomic Energy) and others maintain state interest in these activities.

Most of the state institutions have been classified according to their area of specialization and these are mentioned later in this chapter, but some are not subject to special classification, such as:

Departamento Nacional de Produção Mineral (National Department of Mineral Production), which is a dependant of the *Ministerio de Minas y Energía* (Ministry of Mines and Energy).

Research in Science and Technology at an Academic Level

Conselho de Reitores das Universidades Brasileiras (Council of Rectors of the Brazilian Universities)
Founded in 1966. It studies the problems of higher education.

Associação Brasileira de Escolas Superiores Católicas (Association of Catholic Higher Education in Brazil)
This association supervises the work of eleven universities and various faculties.

Associação de Educação Católica de Brasil (Association of Catholic Education in Brazil)
Coordinates the work of 1750 schools and seminaries throughout the country.

Instituto Brasileiro de Educação, Ciência e Cultura (Brazilian Institute for Education, Science and Culture)
This body is a national commission of UNESCO.

Fundação Centro de Pesquisas e Estudos (Foundation Centre for Research and Studies)
This institute deals with education and 'refresher' courses for technicians employed in state enterprise.

Centro Brasileiro de Pesquisas Educacionais (Brazilian Centre for Educational Research)
Affiliated to the *Instituto Nacional de Estudos Pedagógicos* (National Institute for Pedagogic Studies). There are regional centres in São Paulo, Bahia, Minas Gerais, Rio Grande do Sul and Pernambuco.

Centro de Pesquisas de Geografia do Brasil (Centre for Geographical Research of Brazil)

Centro Brasileiro de Pesquisas Físicas (Brazilian Centre for Physics Research)

Science and Technology at an Academic Level

Universidade Federal do Acre (Federal University of Acre)
Founded in 1971. Has 1447 students and 243 teachers. Its faculties are law, economics, education, languages and literature, geography and history, mathematics and statistics, social sciences and philosophy, health sciences, physical education and sports, natural sciences and technology and agrarian sciences. Its library houses 21 858 volumes.

Universidade Federal de Alagoas (Federal University of Alagoas)
The university was established in 1961 and falls under state control. Its academic years, comprising two semesters, runs between March and December. Lectures are carried out in Portuguese. The teaching staff numbers 788 with 5244 students. The university publishes a periodical bulletin

entitled *UFAL (Universidad Federal de Alagoas)*. Its various departments are known as Centres and comprise the following: Centre of Exact Sciences and Natural Sciences; Centre of Human Sciences; Humanities and Art; Centre of Biological Sciences; Centre of Technology; Centre of Agrarian Sciences; Centre of Health Sciences; Centre of Applied Social Sciences.

Universidade do Amazonas (Amazonas University)
Founded in 1965. Has a student population of 6613 and 794 teachers. Its faculties comprise health sciences, social sciences, education, technology, humanities, exact sciences, biological sciences.

Universidade Federal da Bahia (Federal University of Bahia)
The university has a total of 1642 teachers and 15311 students. Its central library has thirty-one departments with a total of 231952 volumes. The university publishes *Boletim de Pessoal, Afro, Asia Universitas* and *Estudos Baianos* (Bahian Studies). Its faculties comprise architecture, economics, law, medicine, dentistry, education, pharmacy and philosophy. Besides these faculties the university also houses the *Escuela Politécnica* (Polytechnic), which has departments in fine art, librarianship, nursing, business studies, agriculture, veterinary medicine, dietetics, music and landscape art, as well as institutes of mathematics, physics, chemistry, biology, geology, humanities and health sciences; and three centres of education: of Afro-Oriental studies, of 'Bahian' studies and of inter-disciplinary studies for public service.

Universidade Regional de Blumenau (Regional University of Blumenau)
Established in 1968, it has 242 lecturers and 4000 students developing their studies under the control of the *Fundação Universidade Regional de Blumenau* (Regional University Foundation of Blumenau). Its library houses 46052 volumes and 28675 periodicals. Publications include *Boletim Universitario* (University Bulletin); *Boletim Bibliografico* (Bibliographical Bulletin), and *FURB – Revista de Divulgacão Cultural* (Regional University Foundation of Blumenau Cultural Review). Its faculties include philosophy, physical education and sport, economic sciences, law and engineering.

Universidade de Brasília (University of Brasília)
This university, founded in 1961 and inaugurated in 1962, functions under the control of the *Fundaçao Universidade Brasília* (University of Brasília Foundation). Teaching is carried out in Portuguese and the academic year is divided into two semesters. There are 700 teachers and 10000 students. There are five institutes offering courses of study as follows: Institute of Exact Sciences (physics, geo-sciences, mathematics, chemistry and statistics); Institute of Biological Sciences (cellular biology, plant biology, animal biology and psychology); Institute of Human Sciences (economics, geography, history and social sciences); Institute of Expression and Communication (communication, literature and linguistics); Institute of Architecture and Urban Planning (architecture, urban planning and design).

The university faculties of technology (for agricultural, civil, electrical and mechanical engineering); health sciences (for complementary medicine, general and community medicine, physical education, and specialist medicine); applied social studies (business studies, librarianship, political sciences and international relations); and education (basic education theory, methods and techniques, business studies and planning).

Pontifícia Universidade Católica de Campinas (Pontifical Catholic University of Campinas)
This university was established in 1941 and functions under private control, having an academic year that runs between March and December. The lecturers number 928 and there are 20 000 students. The following publications are produced by the university: *Revista de Universidade* (University Review); *Revista Geomorfológica* (Geomorphological Review); *Noticia Bibliográfica e Historica* (Bibliographical and Historical Notice) and *Revista Reflexao* (Meditative Review). The university possesses eleven faculties (law, economics, dentistry, engineering, nursing, social sciences, education, librarianship, physical education, architecture and urban planning, and medicine) and seven institutes offering courses in the art of communication, biological sciences, exact sciences, philosophy and theology, human sciences, literature and psychology.

Universidade Estadual de Campinas (State University of Campinas)
This university functions under state control and its academic year runs from March to June, and from August to December. It has 1400 lecturers and 8638 students. Institutes include art, biology, philosophy and human sciences, physics, language studies, mathematics, statistics and computing sciences, and chemistry. Its faculties are as follows: medicine, education, agriculture and food cultivation in Campinas and Limeira; dentistry, logic, epistemology and history of science. The *Colegio Técnico Industrial* (Industrial Technical College) in Campinas is affiliated to the university as is also the *Colegio Técnico Industrial* of Limeira.

Universidade de Caxias do Sul (University of Caxias do Sul)
Founded in 1967; functions under private control, having 411 teachers and 9027 students. The university publishes: *Cuadernos* (Notebooks); *Revista Chronos* (Chronicle); *Cuadernos de Pesquisas* (Research Notes); *Jornal* (Journal); *Guia Académico* (Academic Guide). There are four centres: Centre of Humanities and Arts, Centre of Applied Social Studies, Centre of Science and Technology, which has departments of exact sciences, and engineering; Centre of Biological Sciences and Health, with departments of nursing, clinical surgery, clinical medicine, biological sciences, and physical education.

Universidade Federal do Ceará (Federal University of Ceará)
The university was founded in 1955 and has 1428 teachers and 12 627 students. The following publications are produced: *Boletim Mensal* (Monthly Bulletin); *Catálogo Anual* (Annual Catalogue); and others from the various departments. There is a Centre of Humanities and a Centre of Sciences which has departments in mathematics, statistics, physics, industrial chemistry, geography, geology, biology and data processing. The Centre of Technology includes departments in architecture and mechanical

and electrical engineering. The Centre for Agricultural and Food Sciences possesses departments in fish, domestic sciences, and nutrition sciences. The Centre of Health Sciences has departments in medicine, dentistry and pharmacy. Finally there is a Centre of Applied Social Studies.

Universidade Federal do Espírito Santo (Federal University of Espírito Santo)
Established in 1961. It has 1390 teachers and 8555 students and publishes *Boletim Informativo* (Informative Bulletin) and *Revista de Cultura da UFES* (Cultural Review of the University of Espirito Santo). There are seven centres: Arts Centre, Biomedical Centre, Juridical and Economic Sciences Centre, General Studies Centre, Physical Education and Sports Centre, Pedagogic Centre, Agricultural and Animal Husbandry Centre.

Universidade Federal Fluminense (Federal University of Fluminense)
The university was established in 1960 and has an academic year divided into two periods – March to July and August to December. There are 1987 teachers and a student population of 16 571. The university publishes *Catálogo Geral* (General Catalogue); *Catalogo de Teses e Disertações* (Catalogue of Theses and Dissertations); *Cursos de Pós-Graduação* (Postgraduate Courses); *Pesquisas en Andamento* (Research in Progress) (annual); *Linguagem* (Linguistics); *Cadernos de Geociências* (Geo-scientific Notes). There is a Centre of General Studies which possesses institutes in physics, geo-sciences, mathematics and chemistry. There is also a Centre of Applied Social Sciences and a Centre of Medicine with institutes in bio-medicine, nursing, pharmacy, medicine, veterinary medicine and dentistry.

Universidade de Fortaleza (University of Fortaleza)
This university has functioned under private control since 1973, the date it was established, and is affiliated to the *Fundação Educacional Edson Queiroz* (Edson Queiroz Educational Foundation). There are 464 teachers and 9536 students.

Universidade Gama Filho (Gama Filho University)
Founded in 1972, it has 1500 teachers and 23 300 students. Its library houses 60 000 volumes and there are centres of science and technology, biological and health sciences, social sciences, and human sciences.

Universidade Federal de Goiás (Federal University of Goiás)
Established in 1960, it has 985 teachers and 7645 students. Its publications include: *Boletim do Pessoal; Informacao Goiana* (monthly) (University Information); *Anais da Escola de Agronomia e Veterinária* (annual) (Annals of the Agronomical and Veterinary School); *Boletim Estatistico* (twice yearly) (Statistical Bulletin); *Revista Goiana de Medicina* (University Medical Review), which is published every four months. In the area of science the university has the following faculties: medicine, pharmacy, odontology, engineering, agriculture and veterinary medicine. Institutes include mathematics and physics, tropical pathology, biological sciences, chemistry and geo-sciences.

Universidade de Itaúna (Itauna University)
This university has functioned under private control since its inauguration in 1965 and has an academic year divided into two periods, March to July and August to December. There are 159 teachers and 2261 students. Publications include *Odonto-Itaúna* and *Cuadernos de Extensão* (Extension Notes). Its faculties are dentistry, engineering, law, economics, education, languages and social sciences.

Universidade Federal de Juiz de Fora (Juiz de Fora Federal University)
Founded in 1960 and comes under federal control. Its academic year is in two sessions, March to July and August to December. Its student population numbers 6345 with 916 lecturers. Publications include: *Boletim da Reitoria* (Rector's Bulletin); *Boletim Alemao de Pesquisas* (monthly) (Bulletin of Research); *Boletim do Instituto de Ciências Biológicas e de Geociências* (bi-monthly) (Bulletin of the Institute of Biological and Geo-sciences); *Revista do Hospital Escola* (quarterly) (Hospital School Review); *Lumina Spargere*; and *Tabulae*. Its faculties comprise medicine, dentistry, pharmacy and biochemistry, engineering and law, and there is an Institute of Exact Sciences.

Universidade Estadual de Londrina (State University of Londrina)
The university is administered by the *Fundação Universidade Estadual de Londrina* (State University Foundation of Londrina) since the date of its inauguration in 1971. Its library houses 43 000 volumes for a student population of 8047 and 712 teachers. Publications include *Catálogo Geral* (annual) (General Catalogue); *Manual do Candidato* (Candidate's Manual); *Boletim Oficial* (monthly) (Official Bulletin); *Semina* (bi-monthly). University departments are: biological sciences, exact sciences, applied social studies, health sciences, rural sciences and technology.

Universidade Mackenzie (Mackenzie University)
Founded in 1952, it functions under private control. The academic year is divided into two periods – March to July and August to December. Students number 11 000 and there are 700 lecturers. Publications include *O Picareta; Revista de Engenharia* (Engineering Review); *Folha Mackenzista* (Review of Mackenzie University); *Jornal das Universidades* (University Journal). The faculties are as follows: architecture, education, economics, law, technology, and post-graduate studies.

Universidade Federal do Maranhão (Federal University of Maranhão)
Established in 1966, it has faculties in nursing, pharmacy, biochemistry, medicine, odontology, industrial design, design, geography, mathematics, chemistry, communications, industrial chemistry, electrical engineering, law, and the arts.

Universidade Estadual de Maringá (State University of Maringá)
The university was founded in 1970 and is run as a private concern. The academic year runs between March and December. The staff comprises 517 lecturers for its 4760 students. *UNIMAR* (University of Maringa) is published by the university. Apart from the humanities departments there

is a department of geography, Centres of Biological Sciences and Health Studies with departments of biology, agrarian sciences, pharmacy and biochemistry, a Centre of Exact Sciences with departments of statistics, and physics, and a Centre of Technology where courses are offered in civil engineering and computing.

Universidade Estadual de Mato Grosso do Sul (State University of the Southern Mato Grosso)
Founded in 1970. This is a state university with 300 teachers for its 2500 students. *Boletim Geral* (General Bulletin) is published. Its faculties are pharmacy, dentistry, physiology, pathology, medicine, and veterinary sciences.

Universidade Federal de Mato Grosso (Federal University of the Mato Grosso)
Founded in 1970, it has 760 teachers and 5485 students. Its faculties include agrarian sciences, health and biological sciences, exact sciences, technology and technologists' training, as well as faculties in the humanities.

Universidade Católica de Minas Gerais (Catholic University of Minas Gerais)
Founded in 1958, it has a student population of 11 516 and 713 teachers. Apart from the various humanities departments, there are departments in engineering, dentistry and biological sciences. This is a private university.

Universidade Federal de Minas Gerais (Federal University of Minas Gerais)
This is one of the oldest universities in Brazil, dating back to 1927. Its library contains 357 356 books and 10 347 periodicals. Among its publications are *Arquivos da Escola de Veterinária* (Archives of the Veterinary School); *Arquivos do Centros Estudos da Faculdade de Odontologia da UFMG* (Archives of the Faculty of Odontology of the University of Minas Gerais). There is a staff of 2194 for its 15 029 students. Apart from the humanities departments, there are institutes of biological sciences, pure sciences and geo-sciences. Also affiliated to the university are the schools of architecture, economic sciences, nursing, engineering, medicine, dentistry, veterinary sciences, and chemistry and pharmacy.

Universidade de Mogi das Cruzes (Mogi das Cruzes University)
Founded in 1973 and run under private control. Its students number 13 318 with 994 teachers, and there is a library of 30 000 volumes. *Boletim* (Bulletin) is published monthly and *Catálogo Geral* (General Catalogue) is published annually. Apart from the faculties of humanities there is an Institute of Sciences, Faculty of Engineering, Faculty of Operational Engineering, an Institute of Biology, a Faculty of Medicine, Faculty of Dentistry, and a School of Nursing.

Universidade Regional do Nordeste (Regional University of the North-East)
Established in 1966, it has an academic year which runs between March and December. Publications are *Catálogo Geral* (General Catalogue), published annually; *Roteiro; Signun;* and *INFURNE.* Its library contains 65 000 volumes for its student population of 7200 and 450 teachers. Its faculties include the humanities and a centre for technical and scientific teaching with departments of statistics, physics, geo-sciences and chemistry.

There is also a centre of biology and health with departments of biology, dentistry, nursing and physiotherapy.

Fundação Norte Mineira de Ensino Superior (North Mineira Foundation for Higher Education)
The foundation is privately run and is officially recognized. The academic year is divided into two periods, March to June and August to December. There are 183 lecturers and 1962 students. Its library houses 17 366 volumes and faculties include philosophy, law, business studies and finance, and medicine.

Universidade Federal de Ouro Prêto (Federal University of Ouro Prêto)
This is a state-run university and was founded in 1969. Its staff numbers 178 teachers for its 1106 students. University publications include *Revista da Escola de Minas* (monthly) (Mining School Review); *Revista da Escola de Farmácia* (Pharmacy School Review); *Espeleologia* (Speleology); and a *Boletim de Geologia* (Geology Bulletin). The university has two constituent institutes. There is also a School of Mining which offers courses in civil engineering, geology and mining and metallurgy; the school employs 158 teachers and 1050 students. The second institute is the School of Pharmacy.

Universidade Federal do Pará (Federal University of Pará)
Established in 1957 and has a staff of 1485 teachers and 12 329 students. Publications include: *Revista* (Review); *Boletim do Serviços* (Bulletin of Services) and *Boletim Informativo* (Informative Bulletin) which is published monthly. There are departments in exact and natural sciences, and centres of biology, bio-medicine, technology, geological and geophysical sciences. There is also a Centre of Advanced Amazonian Studies as well as faculties in humanities and arts.

Universidade Federal da Paraíba (Federal University of Paraíba)
The academic year extends between March and December. The university was founded in 1955 and has a total of 2635 teachers and 20 639 students. Publications are *Boletim de Noticias* (News Bulletin); *Boletim de Servicos* (monthly) (Bulletin of Services); *Revista Nordestina de Biologia* (Biology Review); *Revista CCS* (CCS Review); *Revista Horizonte* (Horizon Review) among others. Its faculties are divided among seven different campuses as follows:

Campus I 'João Pessoa'
Which holds the Centre of Health Sciences and Health, with departments of surgery, health sciences, maternity and infant health, internal medicine, nursing, clinical dentists, restorative dentistry, pharmaceuticals, morphology, physiology and pathology. There is also a Centre of Exact and Natural Sciences where courses are offered in mathematics, chemistry, physics, biology and geo-sciences. The Centre of Technology has departments of civil engineering, mechanical engineering, chemistry, food technology, and architecture.

Campus II 'Campina Grande'
This campus possesses the Centre of Science and Technology which is

made up of the departments of electrical engineering, civil and mechanical engineering, computing, geology and mining, agricultural engineering, atmospheric sciences, chemical engineering, physics, mathematics and statistics. The Centre of Biological Sciences and Health has departments in basic sciences and health, preventive medicine, social medicine, maternity and infant health. Finally there is a Centre of Human Sciences, a Centre of Applied Social Sciences, a Centre of Education, and a Centre of Literature and Arts.

Campus III 'Areia'
Its Centre of Agricultural Sciences includes rural and soil engineering, animal husbandry, crop cultivation technology, and social and basic sciences.

Campus IV 'Bananeiras'
With its Centre of Technological Training, this campus offers courses in rural technology and basic and social sciences.

Campus V, VI, VII
These cover education, humanities and business studies.

Universidade Católica do Paraná (Catholic University of Paraná)
Founded in 1959, it has an academic year divided into two periods between the months of March and November. The staff totals 489 teachers with 7069 students. There is a Centre of Humanities and Theology, a Centre of Legal and Social Studies, a Centre of Pure and Applied Sciences, and a Centre of Bio-medical Sciences.

Universidade Federal do Paraná (Federal University of Paraná)
Founded in 1912, it is one of the oldest in Brazil. There are 1744 teachers for its 14 393 students. Its publications include *Acta Biológica Paranaense* (Biology Review of the University); and *Anais de Medicina* (Medical Annals), among others. Its faculties include agriculture, biological sciences, exact sciences, applied social sciences, health sciences, humanities, literature and art.

Universidade de Passo Fundo (Passo Fundo University)
Established in 1968, it is run under private control with an academic year between March and November (two semesters). Its teaching staff totals 400 with a total of 6300 students. Two publications are issued: *Guia Académica* (Academic Guide); and *Pesquisa sobre Aveia* (Research Review). The university has schools of engineering, agronomy, dentistry, medicine, biological sciences, and an Institute of Exact and Geo-sciences, as well as schools of arts and literature.

Universidade Católica de Pelotas (Catholic University of Pelotas)
Established in 1960, it has a library of 34 800 volumes, a student population of 5192 and 450 teachers. There are three main centres: Centre of Health and Biological Sciences; Centre of Humanities and Centre of Exact Sciences and Technology. Apart from the Department of Humanities, there are departments of biology, science, civil engineering, electrical

engineering, pharmacy and bio-chemistry, mathematics, medicine and chemistry.

Universidade Federal de Pelotas (Federal University of Pelotas)
Founded in 1883 as the *Universidade Federal Rural de Rio Grande do Sul* (Rural Federal University of Rio Grande do Sul); changed its name in 1969. Its staff totals 849 teachers with 5023 students. Only one publication is issued, *Boletim Administrativo da Reitoria* (Administrative Bulletin). Apart from the Faculty of Literature and Arts, there are faculties of agronomy, veterinary studies, dentistry, medicine, and domestic sciences with an Institute of Chemistry and Geo-sciences, and a team who carry out research into agricultural technology.

Universidade Católica de Pernambuco (Catholic University of Pernambuco)
Founded in 1951, it has an academic year that runs between the months of March and December. There are basically three centres: Social Sciences and Theology; Human Sciences; Science and Technology. There is a staff of 539 teachers and 11 176 students.

Universidade Federal de Pernambuco (Federal University of Pernambuco)
Established in 1946, it has an academic year that runs between March and December, and publishes *Boletim Oficial* (Official Bulletin). Students number 18 500 with a staff of 2277 teachers. The university is divided into centres. Apart from arts and literature these centres include: the Centre of Natural and Exact Sciences, with departments in chemistry, physics, statistics and mathematics; the Centre of Biological Sciences, with departments of anatomy, histology and embryology, biochemistry, biophysics and radiobiology, physiology and pharmacology, general biology, mycology, and antibiotics; the Centre of Health Sciences, with departments of clinical medicine, surgery, tropical medicine, neuro-psychiatry, pathology, social medicine, odontology, preventive dentistry, orofacial and prosthetic surgery, nutrition, nursing, and rehabilitation; and the Centre of Technology with departments of cartography, civil engineering, electrical and mechanical engineering, mining, geology, electronics and systems, nuclear energy, oceanography, chemical engineering and industrial chemistry.

Universidade Federal Rural de Pernambuco (Rural Federal University of Pernambuco)
Founded in 1954, it has a library of 23 300 volumes for its 360 teachers and 4000 students. Its departments include agriculture, veterinary medicine, animal husbandry, fish culture, physics and mathematics, biology, chemistry, physiology and morphology, as well as the usual humanities and arts departments.

Universidade Católica de Petrópolis (Catholic University of Petropolis)
Established in 1961, it has a staff of 298 for its 4200 students. Apart from the humanities departments there is also the Institute of Exact and Natural Sciences.

Universidade Federal do Piauí (Federal University of Piauí)
Founded in 1968, it is controlled by the *Fundação Universidade do Piauí*

(University Foundation of Piauí). Its library contains 64 812 volumes for its 5698 students and 805 teachers. Faculties include medicine, dentistry, law philosophy and business studies.

Universidade Estadual de Ponta Grossa (State University of Ponta Grossa)
The academic year runs from March to June and from August to November. Founded in 1970, the student population numbers 3517 with 324 lecturers. Departments are as follows: mathematics and physics, chemistry, geosciences, engineering, biology, odontology, pharmaceutical and clinical analysis, as well as the faculties in humanities.

Fundação Universidade do Rio Grande (University Foundation of Rio Grande)
Established in 1969, it functions under private control with 3476 students and 490 lecturers. Its library houses 56 000 volumes and its departments include chemistry, physics, mathematics, geo-sciences, material and construction studies, oceanography, surgery, internal medicine, morphology (biology), physiology, pathology and the humanities.

Universidade Federal do Rio Grande do Norte (Federal University of Rio Grande do Norte)
Under federal control, it was founded in 1958. The academic year runs from March to June and August to December. Apart from the departments of arts and literature there are departments in the following areas: Exact sciences (chemistry, physics, mathematics and geology); technology (agriculture, architecture, mechanical, chemical, electrical and civil engineering); biological sciences (biology, oceanography and limnology); health sciences (nutrition, orthopaedics, nursing, surgery, general dentistry, paediatrics, clinical dentistry, restorative dentistry, pharmacy, food technology, clinical analysis, toxicology, gynaecology, pathology, public health, and clinical medicine.

Pontifícia Universidade do Rio Grande do Norte (Pontifical University of Rio Grande do Norte)
This university is run privately and was founded in 1934. Its academic year falls between March and December. The teaching staff numbers 1540 with 21 000 students, and the following publications are issued: *Veritas* (Truth); *Letras de Hoje*; *Boletim Informativo de IESPE* (Informative Bulletin). Apart from the faculties of humanities, faculties include medicine, dentistry, and zootechnology, and there are institutes of psychology, geosciences, physics, mathematics, chemistry, bio-sciences, and computing.

Universidade Federal do Rio Grande do Sul (Federal University of Rio Grande do Sul)
Founded in 1934, it is run under federal control, with an academic year from March to November and a student population of 17 000 with 2247 lecturers. *Informativa da UFRGS* (Informative Bulletin of the University) is published, and there are the following science faculties: medicine, dentistry, pharmacy, agriculture, veterinary medicine, in addition to the following: School of Engineering, a Nursing School, an Institute of

Biological Sciences, Institute of Chemistry, Institute of Food Technology, Institute of Geological Sciences, Institute of Hydraulic Research, Institute of Mathematics, and Institute of Physics.

Pontifícia Universidade do Rio de Janeiro (Pontifical University of Rio de Janeiro)
Established in 1940, it is a private university with an academic year divided between March to July and August to December. Apart from the humanities departments, there are the following departments: mathematics, physics, chemistry, computer sciences, civil engineering, electrical and mechanical engineering, metallurgy, and industrial engineering. Affiliated institutes are: applied psychology, telecommunications research, computing centre, dentistry, and the Post-graduate Medical School.

Universidade do Estado do Rio de Janeiro (State University of Rio de Janeiro)
Founded in 1950, it has a staff of 1126 teachers and 12 059 students, and is run under state control. The academic year runs from March to December. *Jornal* (Journal) and *Boletim* (Bulletin) are published. The university is made up of the following centres: bio-medicine, science and technology, education and humanities, and social sciences. The faculties of the first two centres are: nursing, medicine, engineering, dentistry, mathematics and statistics, physics, social medicine, biology, chemistry and geo-sciences.

Universidade Federal do Rio de Janeiro (Federal University of Rio de Janeiro)
Founded in 1920 it has a library and twenty-two faculties with a total of 3241 lecturers and 37 678 students. Two publications are issued: *Anais* (Annals) and *Boletim* (Bulletin). Among the science faculties are: the Centre of Mathematics and Natural Sciences, with institutes of biology, physics, geo-sciences, mathematics, and chemistry; Centre of Technology, with engineering, chemistry, electronics, and an experimental meteorological unit; and the Centre of Medical Sciences, with bio-medicine, nursing, pharmacy, medicine, microbiology, nutrition, dentistry, biophysics, gynaecology, neurology and psychiatry. There is also an Institute of Phthisiology and Pneumology.

Universidade Federal Rural do Rio de Janeiro (Rural Federal University of Rio de Janeiro)
Created in 1944 as the *Universidad Rural del Brasil* (Rural University of Brazil), it is run under state control. Its library contains 46 343 books for its 611 teachers and 4500 students. *Agronomia y Archivos* (Agronomy and Archives) is published every two months. Institutes include: veterinary sciences, animal husbandry, biology, forestry, technology, exact sciences, and agronomy, as well as the arts and humanities.

Universidade Católica do Salvador (Catholic University of Salvador)
A private university inaugurated in 1961. Teaching is carried out in Portuguese, French and English. The staff totals 550 lecturers with 11 000 students. There is an Institute of Sciences, and an Engineering School as well as the various arts and humanities departments.

Universidade Federal de Santa Catarina (Federal University of Santa Catarina)
Run under federal control, it has a total of 9113 students and 1249 lecturers. The following publications are issued: Graduate Catalogue; Post-Graduate Catalogue; Rector's Annual Report and Personnel Bulletin. Apart from the Arts and Humanities there are centres of technology and engineering, agrarian sciences, biomedical sciences, physics and mathematics, and biological sciences.

Universidade para a Desenvolvimento do Estado de Santa Catarina (Development University for the State of Santa Catarina)
Founded in 1966, it has a staff of 300 teachers and 2523 students. Its faculties are as follows: education, business studies, engineering (Joinville), and the School of Advanced Veterinary Medicine.

Universidade Federal de Santa Maria (Federal University of Santa Maria)
Functions under the control of the Federal Government. It was founded in 1960 and has an academic year between March to December (two semesters). Its library houses 61 129 volumes and the staff numbers 1328 teachers, with 9374 students. There are centres of technology, rural sciences, and natural and exact sciences, as well as the humanities.

Universidade Federal de São Carlos (Federal University of São Carlos)
Run under federal control since its inauguration in 1970. The academic year runs from March to December. The student population is 1798 with 403 lecturers. Departments are as follows: computing, stress engineering, chemical engineering, production engineering, physics, mathematics, chemistry, education technology, biology and health.

Universidade de São Paulo (São Paulo University)
A state-run university founded in 1934 whose academic year runs between March and November. The staff numbers 4461 lecturers with 44 159 students. Many publications are issued, among which are: *Revista de Farmácia y Bioquimica da USP* (Pharmacy and Biochemistry Review of the University of São Paulo); *Revista de Saúde Pública* (Public Health Review); *Folia Clínica e Biológica* (Clinical and Biological Notebook); *Revista da Faculdade de Medicina Veterinária e Zootecnica* (Review of the Faculty of Veterinary Medicine and Zootechnology); *Revista da Faculdade de Odontologiade USP* (Review of the University Faculty of Odontology); *Revista da Escola de Enfermagen da USP* (Nursing School Review); *Anais da Escola Superior de Agricultura Luis de Queiroz* (Annals of the Luis de Queiroz High School of Agriculture); *Anuario Astronómico* (Astronomical Annual). Apart from the faculties of arts and humanities, there are faculties of medicine, pharmacy, dentistry, veterinary medicine and zootechnology, public health, a School of Engineering and institutes of biosciences, biomedical sciences, physics, geophysics and astronomy, mathematics and statistics, chemistry, psychology, geo-sciences, mathematical sciences, physics and chemistry, and oceanography; and faculties of pharmacy and dentistry. Dependent institutes comprise those of technological research, atomic energy and electro-technics.

Universidade Estadual Paulista 'Julio de Mesquita Filho' (Julio de Mesquita Filho State University)
Founded in 1976, incorporating the existing faculties in the state of São Paulo. Faculties are: pharmacy, dentistry, chemistry, agricultural sciences, medicine, veterinary medicine, and biosciences; with institutes of geosciences and exact sciences, agrarian and veterinary sciences and engineering, and a Centre of Technology.

Pontifícia Universidade Católica de São Paulo (Pontifical University of São Paulo)
Established in 1946, it develops its activities between March and December with a staff of 1480 and 14 817 students. *Revista* (Review) and *Revista de Psicologia Normal e Patológica* (Review of Normal and Pathological Psychology) are published by the university. Apart from the faculties of arts and humanities there are centres of technological sciences, physics and mathematics, medicine and biology, and faculties of biological sciences and medicine.

Universidade Federal de Sergipe (Federal University of Sergipe)
Inaugurated in 1967 with a staff of 277 teachers and 2700 students, and the following centres: biological and health sciences, applied social sciences, education and humanities; exact sciences and technology.

Universidade Federal de Uberlândia (Federal University of Uberlândia)
Founded in 1969, its academic year runs from February to June and August to December. Its library contains 28 000 volumes for its total of 6416 students and 647 teachers. Apart from the faculties of arts and humanities there are centres of exact sciences and technology, including basic engineering, mathematics, civil, electrical, chemical, and mechanical engineering, and a Centre of Biomedical Studies, composed of the departments of biology, medicine, odontology and veterinary medicine.

Universidade do Vale do Rio dos Sinos (University of the Rio dos Sinos Valley)
A private university founded in 1969 with a staff of 650 teachers for its 24 000 students. Among its publications are *Pesquisas y Boletim Informativo* (Research and Informative Bulletin). Faculties are as follows: health sciences, positive sciences, education and, technological sciences.

Universidade Federal de Viçosa (Federal University of Viçosa)
Founded in 1948 as the *Universidade Rural do Estado de Minas Gerais* (Rural University of the State of Minas Gerais). Its staff numbers 614 teachers with 5567 students. The following publications are issued by the university among others: *Revista Ceres* (Review); *Experientia* (Experience); *Boletim Informativo* (Informative Bulletin); *Revista Brasileira de Armamento* (Brazilian Equipment Review); *Revista da Sociedade Brasileira de Zootecnica* (Brazilian Zootechnological Society Review). The university is divided into schools of advanced agricultural studies, forestry, domestic sciences, exact sciences, biological sciences, and there is a Council of Research.

Industrial Technology and Research

Centro de Pesquisas e Desenvolvimento (CEPED) (Research and Development Centre)
Created in 1969, carries out research into food technology; petrochemicals; metallurgy; metal smelting; environmental engineering; construction materials; quality studies and the analysis of materials. The centre has a library of 13 000 volumes.

Fundação Instituto Tecnológico de Estado de Pernambuco (Technological Institute of the State of Pernambuco)

Instituto Brasileiro de Petróleo (Brazilian Petroleum Institute)
The institute carries out studies into the petrochemical industry and is responsible for the improvement of standards for machinery and petroleum products.

Instituto de Pesquisas Tecnológicas (Institute of Technological Research)
Research is carried out into material resistance and stress; standardization and the efficiency of human resources. There is a library of 40 000 volumes.

Instituto Nacional de Tecnologia (National Institute of Technology)

Centro de Pesquisas e Desenvolvimento 'Leopoldo A. Miguez de Mello' (PETROBRAS) (Leopoldo A. Miguez de Mello Research and Development Centre)
The centre carries out research into the exploiting and refining of oil. There is a staff of 300 researchers and a library of 15 500 volumes.

Agricultural and Marine Science and Technology

Thirty-five organizations are dedicated to research and development activities in the field of agriculture, veterinary and marine science and livestock farming. Of these the following are worth a mention:

Centro Nacional de Pesquisa de Mandioca e Fruticultura – EMBRAPA (National Centre for Cassava and Fruit Research)
Affiliated to the Ministry of Agriculture, the centre carries research into the cultivation of citrus fruits, bananas, pineapples and mangoes. There are forty-eight researchers.

Centro de Pesquisa Agropecuária de Tropico Umido – EMBRAPA (Centre for Research in Agriculture and Livestock Farming in Tropical Zones)
The centre carries out research on agriculture and livestock resources in the tropical zones with special emphasis on soils; climate; natural vegetation and socio-economic factors. Other subjects of research are the production and cultivation of guarana, palm oil, black pepper, Brazil nuts, tropical fruit, rice, beans, maize, fibre-producing plants and livestock. There is a staff of eighty-one researchers and technicians.

Serviço Nacional de Levantamento e Conservação de Solos – EMBRAPA
(National Service for the Surveying and Conservation of Soil)
Affiliated to the Ministry of Agriculture, the service has a staff of forty-seven and a specialized library of 29 000 volumes.

The Ministry of Agriculture supervises research enterprises in the various states, for example:

Emprêsa Pernambucana de Pesquisa Agropecuária (Agricultural and Livestock Research Enterprise of the State of Pernambuco)
Founded in 1935, it has a staff of 120 and a library of 5700 volumes. There are experimental stations in Campos; Itagui; Rio Grande; Fitotecnica de Taquari and the Hacienda Regional de Criacao.

Emprêsa de Pesquisa Agropecuária da Bahia (Enterprise for Agricultural and Livestock Research for the State of Bahia)
Founded in 1976, it has a staff of eighty-eight and a library of 2000 volumes.

Emprêsa de Pesquisa Agropecuária de Minas Gerais (EPAMIG) (Enterprise for Agricultural and Livestock Research for the State of Minas Gerais)
Founded in 1947, it carries out research into animal production and grains including pisciculture and forestry. Its library houses 15 000 volumes.

Instituto de Pesquisas do Experimentação Agropecuário do Nordeste (IPEANE) (Experimental Agricultural and Livestock Research of the Northeast)
The institute includes the experimental stations of Itapirema and Cana de Azucar del Curado.

Instituto Agronómico (Institute of Agronomy)
Founded in 1887, it researches into agricultural engineering, industrial plants and food plantations. The institute has nineteen experimental stations throughout the country, a staff of 210 researchers and a library of 135 000 volumes.

Instituto Brasileiro do Café (Brazilian Coffee Institute)
This is the official organization of the Brazilian coffee industry. There is a library of 19 000 volumes.

Instituto Florestal – Estado de São Paulo (São Paulo State Forestry Institute)
Founded in 1896, it has a staff of seventy-five and a library of 5000 volumes.

Instituto Brasileiro de Desenvolvimento Florestal (Brazilian Institute of Forestry Development)

Serviço de Pesquisa e Experimentação de Cancer (Cancer Research Centre)

Nuclear Science and Technology

Comissão Nacional de Energia Nuclear (CNEN) (National Commission for Nuclear Energy)

Founded in 1956, the commission studies and proposes standards for Brazilian nuclear policy and promotes the peaceful use of nuclear energy and the development and progress of scientists and experiments in this field. There are three reactors in use and two in construction.

Instituto de Energia Atómica (Atomic Energy Institute)
Established in 1956, the institute carries out applied research in the peaceful use of atomic energy.

Instituto de Engenharia Nuclear (Nuclear Engineering Institute)
Founded in 1962, it carries out research into the use of atomic energy.

Instituto de Pesquisas Radioativas (Radioactive Research Institute)
Established in 1963, it researches into the application and development of atomic energy in industry.

Centro de Energia Nuclear na Agricultura (CENA) (Centre of Nuclear Energy in Agriculture)
Founded in 1966, it carries out research into radiogenetics; soil study; radio-chemistry; immunology; animal nutrition; plant nutrition, etc. Its staff numbers thirty-four.

Aerospace Research

Instituto de Pesquisas Espaciais (INPE) (Institute of Space Research)
Founded in 1971, the institute carries out research into satellite communication; systems engineering; space and aerial environment. The staff numbers 1000 with a library of 15 000 volumes.

Meteorological and Astrophysical Research

Asociação Internacional de Lunologia (International Association of Lunology)
Established in 1969: publishes a review covering lunar research carried out in various countries.

Departamento Nacional de Meteorología (National Department of Meteorology)
Founded in 1921: publishes *Boletín Diáriodo Tempo* (Daily Weather Reports); *Boletim Agrometeorológico* (Agro-meteorological Bulletin), etc.

Instituto Regional de Meteorología 'Coussirat Araújo' (Coussirat Araújo Regional Institute of Meteorology)
Founded in 1909; publishes *Boletim Meteoro-Agrícola do Rio Grande do Sul* (Meteorological/Agricultural Bulletin of Rio Grande do Sul).

Observatorio Nacional do Brasil (National Observatory of Brazil)
Founded in 1827, carries out astronomical and astrophysical research. There are two magnetic and astrophysical observatories.

Professional Associations and Societies in the General Field of Science and Technology

There are seventy-four organizations that carry out activities related to science and technology. Among them are:

Academia Brasileiro de Ciências (Brazilian Academy of Sciences)

Academia Nacional de Medicina (National Academy of Medicine)

Instituto 'Nami Jafet' para o Progresso da Ciência e Cultura (Nami Jafet Institute for the Advancement of Science and Culture)
Founded in 1961; it awards diplomas and grants and was established in memory of the industrial professor Nami Jafet.

Instituto Geológico (Institute of Geology)
Founded in 1886; has a library of 100 000 volumes and 30 000 geological maps.

Sociedade Brasileira para o Progresso da Ciência (Brazilian Society for Scientific Progress)

Sociedade Científica de São Paulo (Scientific Society of São Paulo)
Established in 1939; has departments in law, chemical technology, geology and mineralogy, astronomy, architecture, genealogy, philology, medicine, history and pre-history, topography.

Associação Brasileira de Química (Brazilian Association of Chemistry)
Founded in 1951; has regional branches in: Amazonas; Campinas; Minas Gerais; Para; Rio Grande do Sul; São Paulo; Ceara; Maranhao; Rio de Janeiro and Pernambuco.

Instituto de Engenharia de São Paulo (Engineering Institute of São Paulo)

Information Services in Science and Technology

In Brazil there are over 100 organizations connected with information services and the storing of data; most of these organizations are included in the sixty-one libraries and forty-six museums of the country. Among them are:

Biblioteca Nacional (National Library)
Founded in 1810, with 1 800 000 volumes. Among this total are 60 000 volumes of the *Biblioteca Real de Ajuda* (Royal Library of Ajuda), which was brought to Brazil in 1808 by the Portuguese royal family.

Biblioteca Central, Universidade de Brasília (Central Library of the University of Brasília)
Established in 1962, it has a total of 170 000 volumes.

Biblioteca do Ministério das Relações Exteriores (Library of the Ministry of Foreign Relations)
With 270 000 volumes.

Centro de Documentação e Informação da Cãmara dos Deputados (Documentation and Information Centre of the Chamber of Deputies)
With 235 000 volumes.

Serviço de Documentação Geral da Marinha (Documentation Centre of the Navy)
With 80 000 volumes.

Biblioteca Municipal Mario de Andrade (Mario de Andrade Municipal Library)
Located in São Paulo with a central library of 896 000 volumes and twelve branches with a total of 206 000 volumes.

Centro de Tecnologia Agrícola e Alimentar da EMBRAPA (Agricultural and Food Technology Centre)
Research is carried out into grains, cereals, essences and oils. There is a staff of thirty.

Medical Science and Technology

There are thirteen institutes for research and eleven medical societies in Brazil. Most of the institutes are affiliated to the Ministry of Public Health and to the universities through the faculties of medicine.

Instituto 'Adolfo Lutz' (Adolfo Lutz Institute)
Founded in 1892, it is the central laboratory for public health for the state of São Paulo. There is a library of 40 000 volumes.

Instituto Brasileiro de Estudos e Pesquisas de Gastroenterologia (Brazilian Institute for Gastroenterological Research)
This institute carries out research into gastroenterology, nutrition and psychosomatic medicine. Courses are for post-graduate study.

Instituto Butantan (Butantan Institute)
This is an institute of public health and deals with the production of vaccine and serum and also carries out research into ophiology and biomedicine. The institute is renowned for its snake farm and its studies into accidents with scorpions, snakes and spiders. The library houses 73 200 volumes.

Instituto Evandro Chagas (Evandro Chagas Institute)
Research is conducted into bacteriology, parasitology, pathology and mycology.

Instituto 'Oscar Freire' (Oscar Freire Institute)
Teaching and research into forensic medicine is carried out. The institute is affiliated to the University of São Paulo. Its library contains 4200 volumes.

Instituto Oswaldo Cruz (Oswaldo Cruz Institute)
Founded in 1900, the institute has a staff of 130 researchers and a library
180 000 volumes.

Laboratorio Central Gonçalo Monis (Gonçalo Monis Central Laboratory)
This laboratory researches into parasitology, public health and infectious
diseases. Its library has 40 000 volumes.

The following medical societies are worth mentioning:

Associação Brasileira de Psiquiatria (Brazilian Psychiatry Association)

Associação Brasileira de Odontologia (Brazilian Odontology Association)
Founded in 1907, it has 1500 members and a library of 4000 volumes.

Asociação Médica Brasileira (Brazilian Medical Association)
Founded in 1951, the association has 35 000 members and publishes
various technical works.

Chile

Geographical, Demographical, Political and Economic Features

The Republic of Chile is situated on the extreme south-western side of the continent of South America, and is a long and narrow belt of land of 4329 kilometres in length, compared with 445 kilometres wide at the widest point. Borders to the north are shared with Peru; to the south with the Southern Ocean; to the east with Bolivia and Argentina and to the west with the Pacific Ocean with a coastline of 10 000 kilometres.

According to official estimates made in 1980 the population is 11 104 293, with a projection for 1985 of 12 386 000; the population density is 5.5 per square kilometre. Eighty-one per cent of the population is urban with 35 per cent living in Santiago and the suburbs.

Government

Until 11 September 1973, the country was governed by the constitution of 1925, when Augusto Pinochet, the current President of the Republic, took power through a coup d'état. Parliament was dissolved and in June 1976 the Consejo de Estado (Council of State) was inaugurated, a consultative body made up of eighteen members without legislative powers. In March 1977, the government announced that the Consejo had been reorganized into a *Camara Legislativa* (Legislative Chamber). A new constitution was adopted on 11 September 1980.

Natural Resources

AGRICULTURE

Agriculture and forestry make up approximately one-tenth of the national production, although one-third of the population is involved in this sector. Main agricultural products are wheat, barley, oats, maize and rice. In 1955 there were more than 300 large estates, each one of more than 12 250 acres, while 500 000 peasants worked parcels of land of less than 4 acres per family. The military government endeavoured to increase the number of private estates and during 1974 about 5000 deeds of property were granted covering 138 500 hectares, most of which were exploited as cooperatives.

LIVESTOCK

Animal production in 1976 reached (in 1000s of tons) 162 cattle; 32 sheep; 57 pigs; 60 poultry; 1 849 000 eggs and 35 000 kilos wool. According to the same estimates, in 1976, figures showed 3.34 million head of cattle; 450 000 horses; 5.61 million sheep; 800 000 goats; 892 000 pigs and about 20 million poultry.

FISH

Fish is one of the most important national products with 1 285 000 tons being exploited in 1980.

MINERALS

The economy of Chile is based mainly on mineral exploitation with copper, the main resource, totalling almost 51 per cent of total exports. Copper production in Chile represents 44 per cent of world production. The copper industry, expropriated by the government in July 1971, has five large consortiums coming under its control in the provinces of Tarapaca, Antofagasta and O'Higgins. In 1977 their production reached 1.06 million tons valued at US$994.4 million. and the smaller companies reached a total of US$193 million, which together totalled US$1187.4 million of exports.

Chile also has deposits of high grade iron, with an estimated capacity of 1000 million tons in the provinces of Atacama and Coquimbo. Production in 1980 reached 574 000 metric tons. Coal reserves exceed 2000 million tons, although production for 1977 was only 1 138 800 tons.

Other minerals include molybdenum, zinc, manganese and lead. Chile also has oil, discovered in 1945, in the south of the province of Magallanes in the southern region of the country. In 1978 production was 1 020 000 metric tons of light oils and 1 690 000 metric tons of heavy oils.

INDUSTRY

Steel is one of the principal industrial products. In Huachipato, in the region of Concepción in the south of Chile, there is a large state-owned steel plant where production in 1977 was 509 300 000 tons of ingots. Cellulose and wood pulp are two sectors of industry experiencing remarkable growth with the first exports of cellulose totalling $90.9 million in 1977.

Education

Primary education is obligatory and free up to fifteen years of age. Literacy in 1977 was estimated to be 90.8 per cent of the population.

International Relations

Chile is an active member of the United Nations Organization, the Organization of American States and the Latin American Free Trade Association.

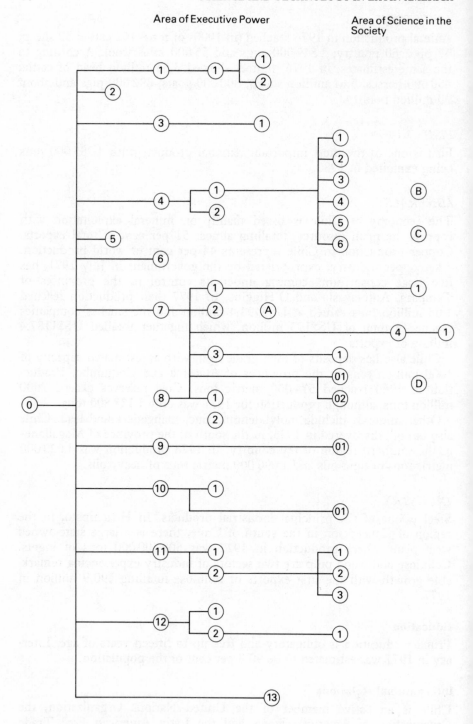

Fig. 4 Organization of science and technology

1	Ministerio del Interior (Ministry of Internal Affairs)
1.1	Intendencias Regionales (Regional Administration)
1.1.1	Consejo Regional de Desarrollo (Council for Regional Development)
1.1.2	Secretaría Regional de Planificación y Coordinación (Regional Secretariat of Planning and Coordination)
2	Oficina de Planificación Nacional (National Planning Office)
3	Ministerio de Relaciones Exteriores (Ministry of Foreign Affairs)
3.1	Instituto Antártico (Antarctic Institute)
4	Ministerio de Economía, Fomento y Reconstrucción (Ministry of the Economy, Public Works and Reconstruction)
4.1	Corporación de Fomento de la Producción (Corporation for the Development of Production)
4.1.1	Comisión de Energía Geotérmica (Commission for Geothermic Energy)
4.1.2	Instituto de Investigación de Recursos Naturales (Institute for the Research of Natural Resources)
4.1.3	Instituto de Investigaciones Tecnológicas (Institute of Technological Research)
4.1.4	Instituto de Fomento Pesquero (Institute for the Development of Fisheries)
4.1.5	Instituto de Investigaciones Geológicas (Institute of Geological Research)
4.1.6	Instituto Forestal (Forestry Institute]
4.2	Comisión Nacional de Aguas (National Commission of Water Supply)
5	Comisión Nacional de la Reforma Administrativa (National Commission for Administrative Reform)
6	Ministerio de Hacienda (The Exchequer)
7	Ministerio de Educacion Publica (Ministry of Public Education)
7.1	Comisión Nacional de Investigación Científica y Technológica (CONICYT) (National Commission of Scientific and Technological Research)
7.1.1	Fundación Chile (Chile Foundation)
7.2	Consejo de Rectores de las Universidades (Council of University Rectors)
7.3	Universidades del Estado (State Universities)
7.3.1	Centros Universitarios de Investigación (University Centres of Research)
7.4	Universidades Privadas (Private Universities)
7.4.1	Centros Universitarios de Investigación (University Centres of Research)
A	Instituto Chile (Chilean Institute)
8	Ministerio de Defensa Nacional (Ministry of National Defence)
8.1	Consejo de Investigaciones de las Fuerzas Armadas (Council of Research of the Armed Forces)
8.2	Comite Oceanográfico Nacional (National Oceanographic Committee)
8.01	Instituto Geográfico Militar (Military Geographical Institute)
8.02	Instituto Hidrográfico de la Armada (Hydrographical Institute of the Navy)
9	Ministerio de Obras Publicas (Ministry of Public Works)
9.01	Instituto Nacional de Hidraulica (National Institute of Hydraulics)
10	Ministerio de Agricultura (Ministry of Agriculture)
10.1	Corporación Nacional Forestal (National Forestry Corporation)
10.01	Instituto de Investigaciones Agropecuarias (Institute of Agricultural and Livestock Research)
11	Ministerio de Salud Pública (Ministry of Public Health)
11.1	Comisión Nacional de Alimentación y Nutrición (National Commission of Food and Nutrition)
11.2	Servicio Nacional de Salud (National Health Service)
11.2.1	Instituto Bacteriológico (Bacteriological Institute)
11.2.2	Instituto de Higiene del Trabajo y Contaminación Ambiental (Institute of Occupational Hygiene and Environmental Contamination)
12	Ministerio de Minería (Ministry of Mine Works)
12.1	Comisión Chilena del Cobre (Chilean Copper Commission)
12.2	Corporación del Cobre (Copper Corporation)
12.2.1	Centro de Investigaciones Minero y Metalúrgicas (Centre for Mining and Metallurgical Research)
13	Comisión Chilena de Energía Nuclear (Chilean Commission for Nuclear Energy)
B	Centro de Estudios de Medición y Certificados de Calidad (Centre for Measurement Studies and Certificates of Quality)
C	Sociedades Científicas y Profesionales (Scientific and Professional Societies)
D	Científicos Individuales (Scientific Individuals)

Organization of Science and Technology

The present institutional network of science and technology shows two groups of organizations that carry out activities in research and development. The first group is made up of planning organizations and the second of organizations executing activities in research and development. The first group consists of the ministries and their secretariats and the other planning bodies. The highest body is the *Comisión Nacional de Investigaciones Científicas y Tecnológicas (CONICYT)* (National Commission of Scientific and Technological Research). Among its main functions are:

(a) to advise the President of the Republic and the various ministries in matters of studies and planning in science and technology;
(b) to maintain specialized information services for the scientific community in general;
(c) to promote science and technology activity;
(d) to establish national development plans and to ensure these are fulfilled.

Four operative units are affiliated to *CONICYT*: *Dirección de Planificación* (Directorate of Planning); *Dirección de Asistencia Técnica Internacional* (Directorate of International Technical Assistance); *Dirección de Información y Documentación* (Directorate of Information and Documentation) and the *Departamento de Fomento* (Department of Development). *CONICYT* is at present in the process of redefining its tasks and objectives in accordance with the development policy in force.

The second group of organizations is dedicated to the execution of research and development. These consist of the university sector, state technological institutes and private technological institutes.

The university sector is made up of some 380 research units located in the national universities, where there is considerable research capacity in the areas of basic sciences, biology, medicine and general engineering. This sector also develops a high level of applied research into specialized areas. The main driving force behind the university research institutes is the *Consejo de Rectores de las Universidades Chilenas* (Council of Rectors of Chilean Universities).

The sector comprising the state research institutes is formed by thirteen research and development organizations in applied science and technology. The main areas of research are aimed at resolving technological problems in agriculture, marine resources, forestry, copper mining, mining in general, nuclear energy, geothermal energy, manufacturing industry, experimental industry, geology, hydraulics, atmospheric contamination and the Antarctic Territory. These institutes are generally dependent on their respective ministries, and six of them are affiliated to the *Corporación de Fomento (CORFO)* (Development Corporation).

The sector of the private research institutes has strengthened over the last five years with the growth of private companies and the transformation of state concerns into private entities. Out of the research units at company level, twenty correspond to public companies in the areas of copper mining, oil production, chemical production, sugar beet and others. In the private company sector, there are forty research units whose work is orientated towards agro-industry, forestry, mining, construction, the fishing industry, metallurgy, textiles, the chemical industry and electronics. The transformation of state organizations to the private sector, as part of the present government policy, has led to many important units gaining autonomy, including the *Instituto de Investigaciones Agropecuarias (INIA)* (Institute of Agricultural and Livestock Research), which formerly came under the Ministry of Agriculture, and the *Centro de Estudio, Medición y Certificación de Calidad (CESMEC)* (Centre of Measurement and Quality Certification) formerly affiliated to *CORFO*.

Policies and Financing in Science and Technology

As a result of economic policy, most institutes, especially those affiliated to *CORFO*, are obliged to finance themselves as well as paying part of their profits to the state. This policy embraces all areas of research and development. The objectives in matters of science and technology are:

(a) To improve the internal and external market in science and technology. One of the priority areas to meet these ends is the development of an adequate information system available to all branches of science and technology.
(b) To promote state support in projects whose results are of major social benefit. With this objective in mind the state must establish a more integrated system of basic science and technology; subsidize research projects where social benefits clearly override private gain; encourage applied research projects aimed at solving key problems in local or regional development; and promote and finance projects in areas where the private sector is unable to meet its requirements in agriculture, fishing and small-scale mining, etc.

Regarding human resources in science and technology the university centres employ the major part of the researchers (72.8 per cent) followed by state research institutes (20.4 per cent), private companies (3.5 per cent) and others (3.3 per cent). With respect to the problem of the 'exodus of the intelligentsia', during the period 1969 to 1976 the number has been steadily declining to insignificant levels.

In Table 4.1 the total number of scientists and engineers on a full-time basis for the year 1980 can be studied.

The governmental mechanisms for programming and financing in science and technology are:

(a) Institutional financing: each institute in the public sector (universities or state research centres) receives a grant, whether directly (in form of fiscal subsidy) or indirectly (transfer of funds) from the superior organizations on which it depends.
(b) Sectorial financing: financing in this area has been granted only to priority sectors such as agriculture, fishing, etc.

Table 3.1 Human resources in science and technology

Year	Population (millions)	Scientists and engineers working in research and development					
		Total number (units)	Breakdown by field of specialization (in units)				
			Natural sciences	Engineering and technology	Agricultural science	Medical science	Social science
1980	11.2	4823	1381	1113	712	777	840

Source: Informe Nacional de Chile (National Information of Chile), CONICYT. La Paz, Bolivia, October 1981.

(c) Regional financing: financing is restricted to self-investment.
(d) Apart from the mechanisms mentioned above, a study is being made into the possibility of creating a fund for applied research projects whose results do not justify private investment but could have social implications, and to granting financial aid to individuals.

Table 3.2 Expenditure in research and development. Breakdown by source and sector of activity, 1980 (unit, eg thousands, millions of pesos, cruzieros, etc.)

Sector of activity	Source National		Foreign	Total (millions of pesos)
	Government funds	Other funds*		
General governmental services	2305.1	1042	568.0	3915.1
Higher education	1802.4	529.3	—	2331.7
Companies of the productive sector	—	291.2	—	291.2
Total	4107.5	1862.5	568.0	6538.0

* Various sources of funds (eg private funds, funds of the productive sector, special funds, foundations, etc.).

Source: Informe Nacional de Chile (National Information of Chile). CONICYT, La Paz, October 1981.

Science and Technology at Government Level

Government research activities are mainly organized under a decentralized administrative system. The research carried out by these institutes represents the major part of the national technological research. Research activities are mainly centres in institutes belonging to the Ministries of Economy, Public Education, National Defence, Agriculture, Public Health and Mine Works. More detailed descriptions of these organizations can be found in the sections relating to their particular field of activity.

Classified in this section are institutes that play an important role in providing basic information, but because of their nature are not mentioned in the specialized sections.

Instituto de Investigaciones Geológicas (IIG) (Institute of Geological Research)
Founded in 1957, this body has as its main function the drawing up of the geological chart of Chile and research programmes which are commissioned by the state or other bodies. The institute has a library of 3000 volumes, 450 periodicals and an archive of maps and photographs. Periodically the institute publishes the *Carta Geológica de Chile* (Geological Chart of Chile) and *Revista Geologica de Chile* (The Geological Review of Chile). From the administrative point of view, the institute relies on the *Corporación de Fomento Nacional* (Corporation of National Public Works).

Instituto de Investigación de Recursos Naturales (Natural Resources Research Institute)
This body was established in 1964. Its principal aim is to produce an aerophotometric survey of Chile and to analyse and present the results of research to the different organizations that commissioned the work. Its functions are instigated through a government convention with international organizations with the objective of drawing up aerophotometric charts of the national territory. The institute has a staff of thirty-seven and possesses a library of 3000 volumes and 700 documents, besides various aerophotometric charts. As in the case of the Geological Institute, it depends on the Corporation of National Public Works.

Instituto Antártico Chileno (Chilean Antarctic Institute)
This centre was founded in 1963 to focus on scientific and technological research and studies which developed from, or are related to, the Antarctic territory of Chile. The centre possesses a library of 1350 volumes as well as 400 periodicals. *Serie Científica* (Scientific Series) and *Boletín de Difusión* (News Bulletin) are published annually. Administratively this centre depends on the Ministry of Foreign Affairs.

Other institutes which are not classified in the following sections are:

Centre for Mining and Metallurgical Research
Which is a dependent of the Copper Corporation.

Commission for Geothermic Energy
Which relies on the Corporation of National Public Works *(CORFO)* and is involved with carrying out studies into the exploitation of energy produced by volcanoes and geysers.

Science and Technology at Academic Level

There are thirteen universities and eleven professional institutions in the country with a total of 128 428 students.

Universidad Austral de Chile (Southern University of Chile)
Founded in 1954, the university has its campus in Valdivia in the south of the country. Its student population is 4760 and its teaching and research staff numbers 610. The university has faculties in medicine, veterinary

medicine, agriculture, forest technology, fine art, philosophy and social sciences, natural sciences, education and teaching, and physics and mathematics. Research is carried out in the various departments of the faculties incorporating the subjects listed above. The university publishes works in philological studies, environmental studies, health sciences, teaching studies, forestry and veterinary medicine.

Universidad Católica de Chile (Catholic University of Chile) – Private
Established in 1888; has its main campus in Santiago and regional centres in Talcahuano, Talca, Temuco and Villarrica. It has approximately 15 309 students and 2 393 teachers. Its central library has 300 000 volumes and an unspecified number of publications and periodicals. This university has faculties in theology, philosophy and learning (aesthetic and historical); exact sciences; biology; architecture; geography and urban planning; fine arts; social sciences; medical and health sciences; education; law; agronomy; engineering; economics and business studies.

The various institutes connected to the faculties, such as *CEPLAN* (Centre of Planning), *CIDU* and *IMPUR* (Inter-disciplinary Centre for Urban Development and the Institute for Urban Planning), carry out research in science and technology and many produce publications.

Universidad Católica de Valparaiso (Valparaiso Catholic University) – Private
Founded in 1928; its campus is in Valparaiso with a student population of 7145 and 444 full-time teaching staff. Its faculties comprise architecture and urban planning, engineering, social sciences, law, philosophy and education, and basic science and mathematics. It also possesses institutes for the social sciences and technology; and centres for scientific and technical research, and for research in computer sciences and systems analysis.
For the development of these activities there is a library of 155 000 volumes and various publications in the form of Informative Bulletins are published.

Universidad de Chile (University of Chile)
Established in 1738 as the *Universidad Real de San Felipe* (Royal University of San Felipe), it was inaugurated in 1843 as the University of Chile. It has 54 000 students and 10 000 teaching staff on all its campuses, and has faculties in social sciences and law, physics and mathematics, medicine and dentistry, economics, philosophy, music, fine arts, forestry, architecture, education, veterinary medicine, chemistry and pharmacy, agronomy, pure sciences, humanities, mathematics and natural sciences. There are campuses in Africa, Antofagasta, Iquique, La Serena, Valparaiso, Talca, Nuble, Temuco and Osorno. Scientific and technological research is developed within the various departments of the faculties. The university possesses a library of 1 000 000 volumes and among its most important publications is *Anales de la Universidad de Chile* (Annals of the University of Chile) plus another seventy-four publications.

Universidad de Concepción (University of Concepción) – Private
This university was founded in 1919 and its main campus is in Con-

cepción which is 500 kilometres south of Santiago. The student population is 12 500 and there are approximately 1800 full-time teachers. About a quarter of the students live on the university campus. There is a school of medicine and a nursing school, besides faculties in biology, chemistry, obstetrics, dentistry, civil engineering, physics and mathematics, economics, business studies and computer sciences, physical education, languages, anthropology, art, agronomy, forestry, veterinary medicine, and agricultural engineering. The university has campuses in Chillan and Los Angeles. Its library comprises 300 000 volumes and the following publications are produced: *Alenea*, a monthly on sciences, art and education, and law; and, four times a year, *Gayana* published by the Institute of Biology, *Paideia* by the institute of Education and an informative publication from the rectorship directed at public relations.

Universidad del Norte (University of the North) – Private
Established in 1965, this is one of the youngest universities in Chile. Its principal campus is in Antofagasta, 1000 kilometres north of the capital. It has a staff of 557 for its 6521 students, who come from various parts of the country. Its faculties include science, social sciences, humanities, and engineering and technology. It also has campuses in Arica and Iquique and research centres in Coquimbo. There is a library of 60 000 volumes and the university publishes chronicles and reviews periodically.

Universidad Técnica del Estado (State Technical University) – State
Founded in 1947, this was formerly known as the School of Art and Business Studies. It has a staff of 590 and 10 000 students, and faculties in sociology and philosophy, mathematics, physics, chemistry, electronics, mechanics and construction, mining studies, and metallurgy. Its principal campus is established in Santiago.

Universidad Técnica 'Federico Santa María (Federico Santa Maria Technical University) – Private
Established in 1926, with its campus in Valparaiso, 308 teachers and a student population of 3222. This university is involved with developing education in the basic technical disciplines through its faculties of civil, electrical, mechanical and chemical engineering. It possesses a library of 710 000 volumes, and publishes quarterly *Scientia* and an informative bulletin called *Boletín Informativo* monthly.

Other institutes of higher education include:

Escuela Agrícola 'El Vergel' (El Vergel School of Agriculture)

Escuela Militar 'General Bernardo O'Higgins' (Bernardo O'Higgins Military School)

Escuela Naval (Naval School)

Escuela de Aviación (School of Aviation).

Industrial Science and Technology

The research activities in this field are carried out mainly in state institutes in collaboration with certain specialized centres of the universities. Development of research in the scope of private manufacturing enterprises is usually carried out with the technical assistance of private or university organizations, these being national or international, and also from state bodies. Two important institutes which deal with the training of technical staff and skilled workers for the industrial sector are *Instituto Nacional de Capacitación Profesional (INACAP)* (The National Institute of Professional Training) and *El Servicio de Cooperacion Técnica (SERCOTEC)* (The Technical Cooperation Service).

Instituto Nacional de Investigaciones Tecnológicas y Normalización (INDITECNOR) (The National Institute of Technological Research and Standards)
This institute was founded in 1944 and was initiated by *CORFO*, the University of Chile and various professional and commercial associations. In the beginning the institute carried out various activities to promote technological research but it is now dedicated solely to the study of technical norms. There is a specialized library of 5000 volumes.

Instituto de Investigaciones y Ensayo de Materiales (IDIEM) (Institute of Research and Examination of Materials)
A dependant of the University of Chile, this institute was founded in 1898. The work carried out here is research commissioned by other organizations for publication or for private use. These research programmes include research into the resistance of materials to pressure, traction, fire, corrosion, organic agents, etc. Complex studies into the seismic resistance of buildings have been performed by this institute. The library houses 5500 volumes and a review, *Revista*, is published three times a year.

Instituto de Investigaciones Tecnológicas (Institute of Technological Research)
This organization is dependent on *CORFO* and has been created to solve the problems of regression in industry which are tending to develop. The institute carries out research and studies on new production processes and new products.

Centro de Estudios, Medición y Certificación de Calidad (CESMEC) (Centre of Measurement and Quality Certification)
A private concern founded in 1977, which sells technological services to the industrial sector, such as chemical analysis, mechanical tests, stress testing in construction materials, weights and measures, the certification of ships, quality control and others.

Agricultural and Marine Science and Technology

Instituto de Investigaciones Agropecuarias (Institute of Agricultural and Livestock Research)

Established in 1964, the institute is dependent on the Ministry of Agriculture. Its central objectives are to carry out studies to obtain better national agricultural production. Research has been completed on soil, fruit and other crop cultivation, animal husbandry, irrigation, etc. It has a staff of 170 researchers and maintains seven experimental stations across the country, at Carillanca, Cauquenes, Human, Kampenaike, La Platina, Quilamapu and Remehue. The institute publishes various works: *Agricultura Técnica* (Technical Agriculture) (every three months); *Memoria Anual* (Annual Report); *Investigación y Progreso Agrícola* (Research and Agricultural Progress) (yearly); and irregular publications of *Boletín Técnico* (Technical Bulletin) and *Boletín Divulgativo* (Informative Bulletin).

Instituto Forestal (Forestry Institute)
Founded in 1965, this is dependent on *CORFO* and carries out research in relation to the development, conservation and utilization of the forestry resources of the country. There is a staff of seventy and the institute publishes technical bulletins and maps as well as an annual *Boletín Estadístico* (Statistical Bulletin).

Instituto de Fomento Pesquero (Institute of Fisheries Development)
Established in 1964, it relies administratively on *CORFO*. Its objectives are the study of water and its living organisms. It has a staff of 200 and a library of 2250 specialized volumes. Publications include *Boletín Científico* (Scientific Bulletin); *Circular* (Circular); *Investigación Pesquera* (Fishery Research) and *Informes Pesqueras* (Fishery Information).

Instituto Agrario de Estudios Económicos (INTAGRO) (Agrarian Institute of Economic Studies)
This institute was created in 1960 and carries out research into agricultural problems and socio-economics. It provides technical assistance to agricultural societies and to individuals, and publishes dissertations and studies on agro-economic problems.

Estación Experimental 'Las Vegas' de la Sociedad Nacional de Agricultura (Las Vegas Experimental Station of the National Society of Agriculture)
Founded in 1924, it provides, through its research, technical advice to members of the society and to agriculturists in general. It publishes monthly the review *El Campesino* (The Farmer).

Instituto Científico de Lebu (Lebu Scientific Institute)
Dates back to 1945 and carries out research in marine biology in the Gulf of Arauco.

Science and Technology in the Medical Field

Research in this field is performed mainly in the various institutes belonging to the faculties of medicine and in the specialist institutes run by the principal hospital centres of the country.

Fundación Gildemeister (Gildemeister Foundation)
Founded in 1947 to cooperate with public and private institutes in medical research and in particular to assist in the circulation of techniques in thoracic surgery, cardiology, neurology and neuro-surgery. The foundation publishes *Revista Memoria* (Informative Review).

Instituto Bacteriológico de Chile (Bacteriological Institute of Chile)
Established in 1929, the institute is the headquarters of the network of laboratories which come under the National Health Service. The institute carries out research programmes and teaching. There is a staff of 720 with a specialized library of 3360 volumes plus a non-specified number of national and international publications. *Boletín* (Bulletin) is published regularly.

Instituto de Medicina Experimental del SNS (Institute of Experimental Medicine of the National Health Service)
Founded in 1937, the institute carries out research in physiology, neurology, endocrinology and cancer. It has a library of 6200 volumes and a staff of fifteen.

Nuclear Science and Technology

Comisión Chilena de Energía Nuclear (Chilean Commission for Nuclear Energy)
The commission was founded in 1964 with the objective of advising the government in matters of nuclear energy and promoting the development of research in this field. There is a specialized library of 5000 volumes and 140 000 reports.

Centro Nacional de Estudios Nucleares (National Centre of Nuclear Studies)
This centre was created to centralize research activities, education and technical assistance in the field of nuclear science and technology.

Military Science and Technology

The larger part of studies and research is carried out in the academies and schools in the branches of the armed forces, but there are at least two organizations which are performing very advanced research. These are *La Academia de Guerra* (The War Academy) and the *Instituto de Estudios de la Seguridad Nacional* (Institute of Studies for National Security). Included in this section are various institutes of the armed forces which give technical services to the general community. These are:

Instituto Hidrográfico de la Armada (Hydrographical Institute of the Navy)
Founded in 1874, it performs various activities concerning hydrographical and oceanographical problems in general. The institute draws up nautical charts and produces studies of the continental shelf. It houses a centre of oceanographical data and a library of 10 000 specialized volumes. Tide

tables are published annually as well as a yearly hydrographic report. Shipping notices are published twice weekly.

Instituto Geográfico Militar (Military Geographical Institute)
Established in 1922, this is the official Governmental organization for the publishing of maps and charts, and it carries out various geographical studies for military use as well as for the public in general. These studies are on subjects related to the cartography of Chile. It has a staff of 507 and a library of 3500 volumes of specialized literature and 3000 maps, some of which are on sale to the public. An informative bulletin is published every four months.

Information Services in Science and Technology

CONICYT, apart from its role of policy making in science and technology, also develops activities in the field of information. Some of these are:

(a) The coordination and development of a national system of information and documentation. The system presently comprises 405 specialist units.
(b) Acting as a focal point for *CARIS* (Current Agricultural Research Information System), *INFOTERRA* (International Referal System Sources Environmental Information), *ISDS* (International Serials Data Systems), *ISORID* (International Information System Research Documentation) and *UNISIST* (Inter-governmental Programme of Cooperation in Science and Technology).
(c) The provision of services of cataloguing; translating, etc., of national and foreign literature and maintenance of a library of specialist collections in politics and science.

There are also the following information networks:

Instituto de Investigaciones Tecnológicas (INTEC) (Institute of Technological Research), which has a contract with the National Firm of Telecommunications *(ENTEL)* and is connected directly with the data bases of Lockheed and Systems Development Corporation.

Empresa Nacional de Computación e Información (ECOM) (National Computing and Information Enterprise), which uses the national network of telecommunications and is connected via satellite to other world systems. The firm has access to the TELENET network of the United States of America, through which it is enabled to use the data banks of Lockheed, *New York Times*, Systems Development Corporation and others.

Universidad de Chile (Chile University)
Subscribes to the COMPENDEX data system of the United States of America.

Instituto de Recursos Naturales (IREN) (Institute of Natural Resources)
Maintains a cartographical data base of the natural resources of the country and a national bibliographical data base of natural resources.

Information services are mainly to be found in the specialized libraries and centres. Some institutes, like the National Institute of Statistics, provide, through their publications, a large amount of information which forms the basic material for many studies.

In Chile there are thirteen large libraries, among which are the National Library, Santiago, with 1 200 000 volumes; University of Chile Library, with 1 000 000 volumes; and Library of the National Congress, with 700 000 volumes.

The information centres, many of which form part of the network of the Information Service of the United Nations, are:

Comité de Investigaciones Tecnológicas (INTEC–UNIDO) (Committee of Technological Research)

Comisión Nacional de Investigaciones Científicas y Tecnológicas (UNEP) (National Commission of Scientific and Technological Research)

Instituto de Higiene del Trabajo y Contaminación Atmosférica (ILO) (Institute of Occupational Hygiene and Environmental Contamination)

Biblioteca Central, Instituto de Investigaciones Agropecuarias (FAO) (Central Library of the Institute of Land and Livestock Research)

Centro Nacional de Información y Documentación (CENID) (UNESCO) (National Centre of Information and Documentation)

Centro de Documentación Pedagógica (UNESCO) (Centre of Pedagogic Documentation)

Comisión Chilena de Energía Nuclear (IAEA) (Chilean Commission of Nuclear Energy)

Meteorology and Astrophysics

Observatorio Astrofísico 'Manuel Foster' (Manuel Foster Astrophysics Observatory)
Founded in 1904, this is part of the Catholic University of Chile and carries out photometric studies and stellar spectrographic surveys. It has a technical staff of eight and a library of 4500 specialized volumes.

Observatorio Astronomicó Nacional (National Astronomic Observatory)
Established in 1853, it belongs to the University of Chile and maintains an astronomical station on the Cerro el Roble and a radio-astronomical observatory in Maipu. Its specialized library contains 6000 volumes.

Observatorio Europeo Austral (European Observatory for the Southern Hemisphere)
Founded in 1962, this is dependent on the European Organization for the Astronomical Research of the Southern Hemisphere. It is situated in the Cerro la Silla in the Coquimbo Province and its main field of research is the photometric and spectrographic study of the southern hemisphere, especially the central region of the galaxy and the Magellanic clouds.

Observatorio Interamericano de Cerro Tololo (Cerro Tololo Interamerican Observatory)
Established in 1963, the observatory dedicates its work mainly to the observation of stars visible only in the southern hemisphere. It has a library of 10 900 specialized volumes.

Oficina Meteorológica de Chile (Meteorological Office of Chile)
Founded in 1884, the office has a staff of 170 and a library of 8000 volumes. Its work is mainly dedicated to hydrology, climatology and meteorology, for which purpose it possesses 106 meteorological stations, sixty-six synoptic stations, 616 rainfall and thermic recording stations, four radiosonde stations, one geomagnetic observatory (Easter Island) and twelve atmospheric radioactivity observatories.

International Cooperation in Science and Technology

The main international programmes in Chile are:

(a) *With multi-lateral organizations*
With the UNDP: developing activities in bacteriology, forestry technology, agricultural engineering, quality control, applied technology in civil aviation, nuclear science and technology (uranium prospecting), biological sciences and production technology of meat and its derivatives.
With OAS: projects in mathematics, physics, chemistry, oceanography, agricultural and livestock technology.
With UNESCO: regional projects on biosphere reserves, ecosystems of the prairies, ecosystems of the mountains, ecosystems of the lakes and waterways.
With FAO: preventing food loss, evaluation of poor harvests and loss of grain, agricultural protection and control, forestry development of the hot and arid regions of Chile, and controls against rare animal diseases.

(b) *In bilateral cooperative agreements*
Japan: programmes of mineral exploration; applied technology in smelting and refining copper; introduction of geothermic energy and the introduction of Pacific salmon.
France: forestry and agricultural development; exact sciences; nuclear technology; telecommunications; computing and medicine.

(c) *With scientific organizations*
National Science Foundation: cooperation in geochemistry of volcanic rocks; chemical reactions in natural products; geological studies of the southern Andes.
Royal Society: joint studies in ergonomy and occupational health; clinical pathology; animal production; biochemistry, etc.
CONICET of Argentina; studies in basic sciences and pharmaceuticals; optical research; astronomy and physics; and research into arid zones.

Consejo Superior de Investigaciones Científicas (CSIC) (Higher Council of Scientific Research, Spain): petroleum studies; civil construction; fisheries research; corrosion of iron; anthropology, etc.

Deutsche Forschungsgemeinschaft (DFG) (German Research Society): Antarctic oceanic geological studies; biology; and bivalves for commercial exploitation.

With *CNPq* of Brazil: joint research projects in zoology; dynamics; applied mathematics and energy.

Societies and Professional Associations in the General Field

There are in Chile eighty-five other organizations dedicated to the development of science and technology, among which are the six academies affiliated to the *Instituto Chile* (Chile Institute); *Academia Chilena de la Historia* (Chilean Academy of History); *Academia Chilena de la Lengua* (Chilean Academy of Languages); *Academia Chilena de Ciencias Naturales* (Chilean Academy of Natural Sciences); and *Academia Chilena de Ciencias* (Chilean Academy of Sciences). Societies include: *Sociedad Chilena de Química* (Chilean Chemistry Society); *Sociedad Científica de Chile* (Chilean Scientific Society); *Sociedad de Biología de Chile* (Chilean Biology Society); *Sociedad Científica Chilena 'Claudio Gay'* (Claudio Gay Chilean Scientific Society); *Asociación Chilena de Sismología e Ingeniería Antisísmica* (Chilean Association of Seismology and Anti-seismological Engineering). There are also thirty specialized museums.

Colombia

Geographical, Demographical, Political and Economic Features

Colombia is located in the semi-tropical equatorial zone of South America with borders to the east with Venezuela (2219 kilometres) and Brazil (1645 kilometres); to the south with Ecuador and Peru; to the west with the Pacific Ocean and to the north with the Antilles Sea, with 1600 kilometres of coastline.

The area is 1 138 914 square kilometres (456 535 square miles) and has territorial waters totalling 2900 kilometres in length, 1600 kilometres extending into the Antilles Sea and 1300 kilometres into the Pacific Ocean.

The population, according to estimates in 1980, is 27 326 463 with a projection for 1985 of 30 684 000. The density, from figures dating from 1976, is 23.9 per square kilometre. Fifty-eight per cent of the population is urban, with 16 per cent of the total population living in the capital city, Bogotá. The remaining 42 per cent reside in rural areas. The ethnic composition is 48 per cent mestizos (mixture of white and Indian); 24 per cent mulattoes (mixture of white and black); 20 per cent white; 6 per cent negroes and 2 per cent Indians.

The capital of the republic is Bogotá (4 424 000 inhabitants), which lies at a height of 8661 feet above sea level. The other main commercial towns are Medellín, with 1 589 000 inhabitants, which is the principal area for mining and coffee; Cali, with 1 380 035 inhabitants, an important centre for sugar and general industry; and Baranquilla (891 000), which has an international airport as well as a river and seaport. Cartagena (452 411 inhabitants) is another important port whose major industrial activities are centred around shipping and the oil pipe-line terminal. Manizales (249 618) and Bucaramanga (417 414) are major industrial centres of coffee and tobacco and, finally, Santa Marta (151 690), located on the Caribbean coast, is the terminating point of the *Ferrocarril del Atlántico* (Atlantic Railway).

Government

Colombia is a unitarian and democratic republic governed by the Constitution of 4 August 1886 and subsequent amendments. The country is divided into 23 departments; 4 intendancies; 4 commissariats and 1016 municipalities. The public power comprises 3 bodies – the executive power, represented by the President of the Republic (elected every 4 years by citizens over 18 years of age), who has the authority to nominate the governors of the departments, intendancies and commissariats and is advised by a cabinet of 13 ministers. The legislative power comprises a senate of 112 members and a chamber of representatives with 199 parliamentarians. The judicial power is in the hands of the high court (20 magistrates), the district courts and the various tribunals.

Natural Resources

AGRICULTURE

Although the soils are essentially fertile, their full potential has not been exploited, the main reason being the limitations imposed by the inadequate road system. Colombia experiences a highly variable climate caused by the wide range of altitudes, with the result that diverse agricultural techniques must be employed according to the climatic zone. The country possesses about 6 million acres of cultivable land, 96 million acres of pasture and 148 million acres of forestry.

Colombia is the second world producer of coffee; estimates in 1980 indicated a production of 738 000 tons. Other important products include cotton (101 000 tons); rice (1 797 000 tons); barley (125 000 tons); maize (831 000 tons); soy bean (155 000 tons); wheat (55 260 tons); bananas (1337 tons); cacao (41 000 tons) and sugar cane (1 060 060 tons).

LIVESTOCK

Livestock figures in 1976 show 23 860 head of cattle; 1.9 million pigs; 2.04 million sheep and 4.7 million poultry.

FORESTRY

Colombia has various forests for commercial production and in 1977 figures show 934 000 cubic metres of cut wood.

MINERALS

In May 1980 it was officially announced that there were oil deposits in the Casanare region in the south of the country and these could make Colombia a world exporter of crude oil. The country is also rich in mineral resources – gold is located in Cauca, Caldas, Tolima, Navino and Choco, and in 1977 production reached 256 951 troy ounces, the highest in South America. Other mineral resources include sulphur, chromium, mercury, silver, salt, coal, zinc, natural gas, iron, baryta, kaolin, and emeralds, in the mines of Muzo, Coscuez, Sonomdoco, Chivor and Guateque. There is also magnesium, manganese, nickel, platinum, mica and gypsum.

INDUSTRY

Main industries include iron and steel, cars, sugar, cement, cigarettes,

fish meal, petroleum derivatives, caustic soda, edible oils and fats, hydro-
chloric acid, electric batteries, tinned food, alcoholic and soft drinks, shoes
and other leather articles, detergents and soaps, nitrogenous fertilizers, non-
cellulose fibres, earthenware and china, cut wood, sewing machines, type-
writers, construction materials, railway materials, paper and cartons, paints
and chemical products and pharmaceuticals.

Education

According to figures in 1975, 87 per cent of the population is literate and
in the same year there were 1837 establishments of primary education to
furnish the needs of 96 566 students and 3855 teachers. In 1977 figures
showed 35 420 schools with 4 160 527 students and 128 494 teachers at
primary level, while at secondary level there were 5017 colleges for
1 616 111 students and 79 742 teachers. Regarding higher education, in
1978 there were 290 624 students and 53 092 teachers.

International Relations

Colombia is a member of the United Nations Organization, Organization
of American States, Andean Group, and the Latin American Free Trade
Association.

Organization of Science and Technology

In 1968 *Consejo Nacional de Investigaciones Científicas y Tecnológicas
(CONICYT)* was established to formulate policies for the development of
science and technology and *Fondo Colombiano para Investigaciones Científicas
(COLCIENCIAS)* was formed to coordinate and promote these policies.

The government sector is made up of the decentralized institutes affiliated
to the various ministries, and will be covered in the relevant sections.

The university sector holds the larger concentration of research and
development institutes. *COLCIENCIAS* has endeavoured to orient
research towards the productive sector and priority areas beneficial to the
country as a whole. With this objective in mind the universities have been
integrated into the *Comités Técnicos* (Technical Committees) who act as
advisory bodies to special programmes, and the government and productive
sectors are actively involved in the committees.

The private sector plays a limited role in research and development,
although certain important innovations have been detected in the productive
sector.

The census of 1978 reveals the following information on organizations
carrying out research and development activities: in the government sector
there is a total of 31 research organizations; in the university sector, 43
universities (of which 26 are public and 17 private); research centres and
institutes number 23 (of which 9 are public and 12 private, 1 of mixed
economy and 1 international); in the sector of scientific and technological

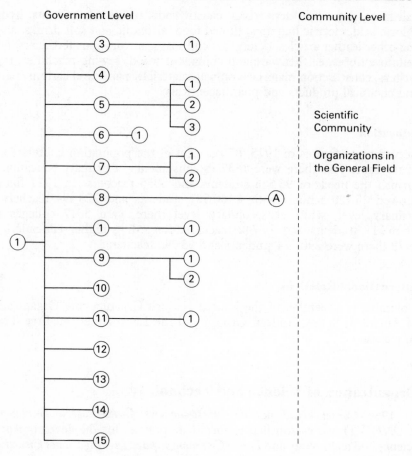

Fig. 5   Organization of science and technology

1	Presidencia (Presidency)
1.1	CONICYT
1.1.1	COLCIENCIAS
1.3	Ministerio de Obras Publicas (Ministry of Public Works)
1.3.1	Centro Interamericano de Fotointerpretación (Inter-American Centre of Photo-interpretation)
1.4	Ministerio del Trabajo (Ministry of Labour)
1.5	Ministerio de Minas y Energía (Ministry of Mines and Energy)
1.5.1	Instituto de Asuntos Nucleares (Nuclear Institute)
1.5.2	Instituto Nacional de Investigaciones Geologico-mineras (National Institute of Geological and Mineral Research)
1.5.3	Instituto Colombiano de Energía Eléctrica (Colombian Institute of Electrical Energy)
1.6	Ministerio de Salud Publica (Ministry of Public Health)
1.6.1	Instituto Nacional de Cancerología (National Cancer Institute)
1.7	Ministerio de Agricultura (Ministry of Agriculture)
1.7.1	Instituto Colombiano Agropecuario (Colombian Institute of Agriculture and Live-stock)
1.7.2	Instituto Nacional de Recursos Naturales (National Institute of Natural Resources)
1.8	Ministerio de Educación (Ministry of Education)
A	Universidades (Universities)

1.9   Ministerio de Desarrollo Económico (Ministry of Economic Development)
1.9.1   Instituto de Fomento Industrial (Institute of Industrial Promotion)
1.9.2   Instituto de Investigaciones Tecnológicas (Institute of Technological Research)
1.10   Ministerio de Justicia (Ministry of Justice)
1.11   Ministerio de Hacienda y Credito (Treasury and Credit Ministry)
1.11.1   Instituto Geográfico 'A. Codazzi' (A. Codazzi Geographical Institute)
1.12   Ministerio de Gobierno (Ministry of Government)
1.13   Ministerio de Comunicaciones (Ministry of Communications)
1.14   Ministerio de Defensa (Ministry of Defence)
1.15   Ministerio de Relaciones Exteriores (Ministry of Foreign Relations)

services there are 29 bodies of which 18 are public, 9 private and 1 belongs to an international organization; and finally in the productive sector there are 13 establishments of which 12 are public and 1 of mixed economy. Of the last sector the private productive establishments did not appear in the census.

Financing and Policies in Science and Technology

There is a consensus among the policy-making bodies that research and development activities in science and technology must adhere strictly to the general guide-lines for social and economic development of the country.

Three major objectives in science and technology are:

(1) To strengthen the scientific and technological structure of the country.
(2) To generate machinery by which science and technology can contribute more effectively to the social and economic development of the country, stimulating a greater capacity for innovation.
(3) To decrease Colombia's dependency on more advanced countries.

For each of these objectives certain guide-lines have been established:

For (1)
To obtain the maximum utility from scientific laboratories and experimental stations, etc.
To encourage the productive sector to channel funds into research and development.
To strengthen and create educational programmes for human resources.
To improve teaching and research in the basic sciences at university level, etc.

For (2)
To promote the coordination of research programmes in state, university and private bodies.
To achieve equal development in all areas of science in accordance with national needs.

For (3)
To coordinate national expenditure in science and technology in accordance with national scientific and technological policy.

To ensure that adequate funds are channelled into experimental science and technology.

To establish a programme of technical assistance for small and light industry.

Besides these propositions, positive actions have already been implemented in several fields: ie measures destined to control imports and exchange of technology.

In 1981 a *Programa Nacional de Desarrollo Científicó y Tecnológico* (National Programme of Scientific and Technological Development) was established. This programme has been structured around specific programmes in priority areas. Six programmes have been drawn up comprising more specific sub-programmes. (A list of these programmes and sub-programmes follows.)

1. *Programme on Agriculture and Livestock, Forestry and Marine Resources.*
1.1 Agricultural and livestock sub-programme.
1.2 Marine science and technology sub-programme.
1.3 Conservation, improvement and recuperation of ecosystems sub-programme.

2. *Energy Resources Programme.*
2.1 Basic energy resources sub-programme.
2.2 Conventional energy resources sub-programme.
2.3 Non-conventional energy resources sub-programme.

3. *Social Development Programme.*
3.1 Health sub-programme.
3.2 Housing sub-programme.
3.3 Employment sub-programme.
3.4 Education sub-programme.
3.5 Social sciences sub-programme.

4. *Industrial Programme.*
4.1 Sub-programme on the promotion of development capacity in the productive sector.
4.2 Food and nutrition sub-programme.

5. *Basic Sciences Programme.*
5.1 Physics sub-programme.
5.2 Chemistry sub-programme.
5.3 Mathematics sub-programme.
5.4 Biology sub-programme.
5.5 Earth sciences sub-programme.

6. *Programme to Develop a National Information System.*
6.1 National information system for agriculture and livestock; forestry and marine sectors.
6.2 National information system for the energy resources sector.
6.3 National information system for the social development sector.
6.4 National information system for the industrial development sector.

In accordance with the policy guide-lines contained in the National Programme for Scientific and Technological Development, *COLCIENCIAS*, in its capacity as a coordinating body, is presently being restructured. The measures that the policy has adopted are aimed at redressing deficiencies in the present system which have occurred due to the following factors:

(a) Lack of clear short-term and long-term policies;

(b) insufficient funds to ensure execution of programmes. While in Colombia 0.15 per cent of the GNP is destined to science and technology, in other countries of similar socio-economic development, between 0.7 per cent and 1.0 per cent of the GNP is dedicated to activities in science and technology, and in developed countries investment is in the region of 3 per cent of the GNP or more;

(c) poor relations between the various scientific and technological bodies;

(d) lack of adequate machinery to facilitate the transfer of technology to the productive sector;

(e) inadequate distribution of resources in teaching and administration in higher education, and general decay in research and education of scientists and technicians.

Table 4.1 entitled 'Scientific and technological manpower' uses data from the National Census of Scientific and Technological Activities in 1972–78.

Table 4.1 Scientific and technological manpower

Year	Population (million)	Total number (units)	Scientists and engineers					Technicians
			Those working in research and development					
			Basic sciences	Engineering	Agriculture and livestock	Health sciences	Social sciences	In research and development (units)
1972	22.0	1.140	188	154	348	127	323	
1978	26.1	3.404	995	627	527	647	608	704

Source: National Information on the *Sixth Meeting of the Permanent Conference of National Scientific and Technological Organizations in Latin America and the Caribbean.* La Paz, Bolivia, October 1981.

Table 4.2 entitled 'Financing in research and development' was drawn up using material from the *II National Census of Scientific and Technological Activities* (1978). The figures for research and development in higher education indicate expenditure only in specific projects and not in research and development as a whole, which would show a much higher level. Similarly, the table does not include expenditure in the private productive sector.

Table 4.2 Financing in research and development (by source and sector of activity)

Source / Sector of activity	National Government funds	Other funds*	Foreign	Total
General governmental services	384.714	86.003	46.665	517.382
Higher education	141.642	50.197	23.441	215.280
Firms of the productive sector	2.687	434	–	3.121
Total	529.043	136.634	70.106	735.783

*Including funds from state production firms and special funds (according to UNESCO classification).
Source: COLCIENCIAS – Division of Scientific Statistics. Prepared by the 'Statistical questionnaire on personnel and expenditure in experimental research and development' of UNESCO using base material from the *II National Census of Scientific and Technological Activities*.

Science and Technology at Government Level

Apart from the organizations already mentioned, the ministries have affiliated institutions that carry out research and development activities in certain areas of science and technology:

Departamento Administrativo Nacional de Estadísticas (Administrative Department of National Statistics)
Founded in 1953; publishes *Boletín Mensual de Estadisticas* (Monthly Statistics Bulletin) and *Anuario de Comercio Exterior* (Yearbook of Foreign Trade), etc.

Instituto Geográfico 'Agustín Codazzi' (Agustín Codazzi Geographical Institute)
Established in 1935; carries out basic geographical research and publishes geographical, topographical, cadastral and agricultural maps of Colombia. Much of the material produced by the institute is used as basic research material for studies into natural resources, agriculture, mining, road planning, etc.

Instituto Nacional de Investigaciones Geológico Mineraş (National Institute of Geological and Mineral Research)
Founded in 1968; formerly existed as a scientific commission. The staff, numbering 480, draw up official geological maps and issue specialized reports.

Centro Interamericano de Fotointerpretación (Inter-American Centre of Photo-Interpretation)

Instituto Nacional de Recursos Naturales (National Institute of Natural Resources)

Science and Technology at Academic Level

Universidad de Antioquia (University of Antioquia)
Located in Medellín; founded in 1822. The student population numbers

25 000 with 2000 teachers. Publications include various reviews, *Universidad de Antioquia* (Antioquia University) and *Noticiero Universitario de la Secretaria General* (University News of the General Secretariat). Apart from the humanities, faculties include engineering, medicine, veterinary medicine and zootechny, dentistry, chemistry and pharmacy, and public health. There is also an affiliated institute, the *Escuela Interamericana de Bibliotecnologia* (Inter-American School of Librarianship).

Universidad del Atlántico (Atlantic University)
Established in 1941 in Barranquilla, it has 8784 students with 592 lecturers. Faculties include chemistry and pharmacy, chemical engineering, architecture, nutrition and dietetics, and arts and humanities.

Universidad de Caldas (Caldas University)
Founded in Manizales, Caldas in 1943; has an academic year between February and December. Students number 3802 with 393 teachers.
  Faculties include agriculture, medicine, nursing, veterinary medicine and stockbreeding, plus the faculties of arts and humanities.

Universidad de Cartagena (University of Cartagena)
Located in Cartagena; was established in 1827. The academic year runs between February and December with 4202 students and 684 teachers. Publications include *Revista de la Facultad de Medicina* (Medicine Faculty Review). Apart from the faculties of humanities there are faculties of medicine, dentistry, pharmacy, engineering and nursing.

Universidad del Cauca (Cauca University)
Located in Popayan, it was founded in 1827. The academic year runs from January to June and from July to December. Two publications are issues – *Revista* (Review), every four months, and *Boletín Informativo* (Informative Bulletin) monthly. The student population numbers 3880 with a staff of 470 teachers. Faculties include civil engineering, electrical engineering and telecommunications, medicine and nursing, as well as the arts and humanities.

Universidad Francisco de Paula Santander (Francisco de Paula Santander University)
Founded in 1962 in Cucuta; affiliated to the National University. The library contains 4500 volumes and students number 2314 with 243 lecturers. Faculties include engineering, technology, humanities and education.

Universidad Nacional de Colombia (National University of Colombia)
Located in Bogotá and has an academic year running from January to July and August to December. There is a staff of 3061 lecturers for its 25 000 students in the faculties of sciences, agriculture, medicine, engineering, nursing, dentistry and veterinary sciences, plus the arts and humanities.

Universidad de Córdoba (Córdoba University)
Founded in 1966 in Monteria and has an academic year between April and March. Students number 2600 with a staff of 221 lecturers in the faculties of veterinary medicine and animal husbandry, agronomy, nursing and education.

Universidad Distrital 'Francisco José de Caldas' (Francisco José de Caldas University)
Founded in 1960 in Bogotá and has a staff of 350 teachers for its 4500 students. The library holds 35 000 volumes to serve the faculties of electrical engineering, forestry and land resources and a School of Topography.

Universidad de Nariño (Nariño University)
Established in 1827 as a provincial college by General Francisco de Paula Santander, it later became known as an academic college and finally obtained university status in 1964. The central library holds 10 000 volumes, and a library for agronomy with 15 000 volumes. Among its publications are *Anales* (Annals); *Foro Universitario* (University Forum) and *Numen Universitario 'A'* (University Talent). Faculties include law, education, languages and agronomy.

Universidad de Pamplona (Pamplona University)
The university has a staff of 100 teachers for its 1300 students in the faculties of sciences, humanities and education. Located in Pamplona.

Universidad Pedagógica Nacional (National Pedagogic University)
Established in 1955 in Bogotá, the university has 4778 students and 746 lecturers and an academic year between January and December divided into two semesters. Two publications are issued: *Revista Colombiana de Educación* and *Documentación Educativa* (Colombian Review of Education and Educational Documentation). The university is divided into departments of biology, chemistry, physics, mathematics, industrial education and the humanities. There are two affiliated institutes: *Instituto Pedagógico Nacional* (National Pedagogic Institute) and *Instituto Técnico Industrial* (Industrial Technical Institute).

Universidad de Quindío (Quindío University)
Located in Armenia and founded in 1962, it has a staff of 165 lecturers for its 3300 students. Courses of four months' duration are offered in mathematics, physics, biology, chemistry, topography and engineering, apart from humanities and sciences.

Universidad Industrial de Santander (Santander Industrial University)
Located in Bucaramanga, Santander, it has a staff of 380 lecturers and 5500 students. Publications include *Revista UIS Investigaciones* (University Research Review); *Landa*; *ION* and a Geology Bulletin. Apart from the humanities, faculties include sciences, physics, mechanics, chemistry, health sciences, biology, electricity and electronics, mathematics and chemistry, mechanical engineering, industrial production, computing systems, metallurgy, chemical engineering, petro-chemical engineering, morphopathology, physiological sciences, microbiology, nursing, internal medicine, paediatrics, psychiatry, surgery, physical rehabilitation, nutrition, public health and preventive medicine, industrial consultancy, biological medicine, chemical engineering, and postgraduate studies in physics and metallurgical engineering.

Universidad de Tolima (Tolima University)
Although originally founded in 1945 it only received university status in
1954. Located in Ibague, it has an academic year from February to June
and August to December. The student population numbers 3500, with 250
teachers for the faculties of agricultural engineering, forestry, commerce,
and education sciences, and an institute of technology.

Universidad del Valle (Valle University)
Located in Cali; was founded in 1945. The library contains 110 000 volumes for its 884 lecturers and 8056 students. Publications include *Boletín
del Departamento de Biología* (Bulletin of the Biology Department); *Revista
Division de Ingeniería* (Engineering Review); *Acta Médica del Valle* (Medical
Notes of Valle), among others. The university is divided into Divisions,
the Division of Engineering has departments in agricultural engineering,
civil and electrical engineering, computing systems, fluid and heat sciences,
and chemical and biological processes. The Division of Health has departments in morphology, physiology, pathology, radiology, anaesthesiology,
internal medicine, surgery, paediatrics, obstetrics and gynaecology,
psychiatry, social medicine, microbiology, stomatology, nursing and clinical
laboratories, and a school of physiotherapy. The Division of Architecture
has departments of construction, environmental planning and design. The
Division of Sciences has departments of mathematics, physics, chemistry,
and biology. There is also a Humanities Division.

Universidad Pedágogica y Tecnológica de Colombia (Pedagogic and Tech-
nological University of Colombia)
Founded in 1953 in Tunja, Boyaca with 4592 students and 415 lecturers.
Apart from the humanities, faculties include agronomy and engineering.

Universidad Tecnológica del Magdalena (Magdalena Technological University)
Established in 1958 in Santa Marta, it started functioning in 1966 and
presently has a total of 500 students and sixty teachers. Faculties are agricultural engineering, farm management and sciences.

Universidad Tecnológica de Pereira (Pereira Technological University)
Located in Pereira, it was founded in 1958. Students number 3891 with
269 lecturers. Publications include *Revista UTP* (Pereira University
Review); *Noticiero Bibliografico e Informativo* (Bibliographical and Informative Review) and the *Serie Arte y Cultura* (Art and Culture Series).
The library houses 14 393 volumes. There are faculties of electrical, mechanical and industrial engineering, besides the department of medicine,
polytechnic institute and a computing centre.

Escuela Superior de Administración Pública (Higher School of Public Administration)
Located in Bogotá, it was founded in 1958 with an academic year between
February and November. The library holds 17 300 volumes. There are 753
students and seventy-two teachers. The university is divided into centres of
research and information, political sciences and administrative sciences.

Private Universities of Colombia

Fundación Universidad de Bogotá 'Jorge Tadeo Lozano' (Jorge Tadeo Lozano University Foundation of Bogotá)
Located in Bogota, it has an academic year divided into two periods between February and December. Students number 6100 and staff 650. Apart from the faculties of humanities there are faculties of food technology, geographical engineering, agronomy and industrial design.

Universidad Autónoma Latinoamericana (Autonomous University of Latin America)
Founded in 1966 in Medellín, it has a total of 6100 students and 650 teachers and an academic year running from January to June and from July to December. Faculties include arts and humanities and industrial engineering.

Universidad Pontificia Bolivariana (Bolivariana Pontificial University)
Located in Medellín; was established in 1936. The student population numbers 9867 with 560 lecturers. There are faculties of chemical engineering, architecture and urban planning, electrical engineering and electronics, mechanical engineering, medicine and humanities.

Universidad Externado de Colombia (Externado University of Colombia)
Founded in 1886 in Bogotá, it does not possess faculties in the field of science and technology.

Universidad de Gran Colombia (Gran Colombia University)
Situated in Bogotá, it was founded in 1951 and has a staff of 650 lecturers and 7251 students. Faculties are civil engineering and the humanities.

Pontificia Universidad Javieriana (Javieriana Pontificial University)
Located in Bogotá, it was founded in 1622 by the Jesuits and was reestablished in 1931 and reached its present status in 1937.

Universidad Libre de Colombia (Free University of Colombia)
Located in Bogotá, it has 2000 students and 220 lecturers. Founded in 1923 and has a faculty of metallurgy as well as the humanities and arts. The academic year extends between January and November in two semesters. Among its publications is *Universitas Médica* (University Medical Notes), quarterly. The teaching staff numbers 1146 and there are 12 512 students in the faculties of medicine, engineering, sciences, electrical engineering, architecture, psychology, dentistry, and nursing. The *Instituto Geofísico de los Andes Colombianos* (Geophysical Institute of the Colombian Andes) is affiliated to the university.

Universidad de los Andes (University of the Andes)
Founded in 1948 in Bogotá, it has an academic year between January to December. The library houses 90 000 volumes. The student population is 4000 with 150 teachers. Technical publications are issued by the faculty of engineering. Affiliated institutes are the Centre of Electronics and Computing *(CCE)*, the Centre of Hydraulic Studies and Research and the Genetic Institute.

Universidad de Medellín (Medellín University)
Established in 1950 in Medellín, it has 5036 students and 420 teachers divided among the seven faculties of law, industrial economy, business studies, statistics, civil engineering, education and public accounting. Affiliated to the university are the *Instituto de Estudios de Postgrado e Investigaciones* (Institute of Post-graduate Studies and Research) and the *Instituto de Derecho Penal y Criminología* (Institute of Penal Law and Criminology). Publications include: *Revista U. de Medellín* (University Review) (quarterly); *Prospecto Facultad de Derecho* (Prospectus of the Faculty of Law) (annually); and *Universidad y Pueblo* (University and People) (twenty per year).

Universidad del Norte (University of the North)
Founded in 1966 in Barranquilla, with an academic year divided into two, from January to May and August to December. Faculties include psychology, engineering, health sciences and company administration. The library contains 11 000 volumes for the use of its 2180 students and 296 lecturers.

Universidad Santiago de Calí (Santiago de Calí University)
Founded in 1958 in Calí. Students number 3800 for the faculties of law, business studies, education, and accounting and book-keeping.

Universidad de San Buenaventura (San Buenaventura University)
Established in Bogotá in 1715, but only reached university status in 1964. There are two main branches in Cali and Medellín. The student population numbers 1600 with 110 teachers. The faculties are philosophy, theology, sociology, psychology, law, business studies and education. The academic year runs between February and November. Publications include – *Franciscanum* (three times yearly). Most of the lecturers are priests.

Universidad de Santo Tomás (Santo Tomás University)
Originally founded in 1580 but was restored and reorganized in 1965. Located in Bogotá with a student population of 3250 and 800 lecturers. *Universidad Santo Tomás* is published. There are six faculties – law and political sciences; public accounting; civil engineering; philosophy, economics and company administration; and sociology.

Universidad Social Católica de la Salle (La Salle Catholic University)
Founded in 1965 in Bogotá, it has 300 teachers and 3450 students in the seven faculties of architecture; civil engineering; economics; education; philosophy and letters; information sciences and optometry. There are also eleven departments: social work; statistics; business administration; agricultural administration; biology and chemistry; mathematics and physics; languages; religious studies; education; audio-visual aids; and accounting.

Colegio Mayor de Nuestra Señora del Rosario (College of Our Lady of the Rosary)
Established in 1653 in Bogotá. The academic year runs between February and November. Faculties include: law; economics; philosophy; letters and history; medicine and business administration, with a School of Translators and Interpreters. Students number 3007, with 607 teachers. *Revista* (Review) is published.

Escuela de Administración y Finanzas y Tecnologías (School of Administration, Finance and Technology)
Established in 1960 in Medellín, it has an academic year divided into two periods between February to June and August to December. Student population numbers 1735 with 154 lecturers.

Industrial Research and Technology

Instituto Colombiano de Normas Técnicas (ICONTEC) (Colombian Institute of Technical Standards)
Established in 1963, it publishes *Boletín Informativo* (Informative Bulletin) and *Estandars* (Standards) once a month.

Instituto de Investigaciones Tecnológicas (Institute of Technological Research)
Founded in 1958 as an autonomous organization, it carries out applied research and acts as an information and advisory service to industry. The institute specializes in food, metallurgy, metal mechanics, agricultural industry and industrial processes in the chemical industry. *Tecnología* (Technology) is published twice monthly.

Instituto de Fomento Industrial (Institute of Industrial Promotion)

Agricultural and Marine Science and Technology

Instituto Colombiano Agropecuario (ICA) (Colombian Agricultural and Livestock Institute)
Founded in 1962, it promotes and coordinates research, teaching and development in agriculture and livestock. Various publications are issued: *Revista ICA* (Institute Review), *ICA Informa* (ICA Information), etc.

Instituto de Investigaciones Marinas de Punta Betín (Punta Betín Institute of Marine Research)
Founded in 1965; created by *COLCIENCIAS* in cooperation with the Universidad Nacional de Colombia and the German university Justus Liebig of Giessen. Research is carried out into preservation of flora and fauna.

Medical Science and Technology

There are fourteen organizations that carry out research activities in the field of medicine, among which are:

Instituto Nacional de Medicina Legal (National Institute of Forensic Medicine)

Instituto Nacional de Cancerología (National Institute of Cancer)

Specialized societies include:

Sociedad Colombiana de Cancerología (Colombian Cancer Society)

Sociedad Colombiana de Psiquiatría (Colombian Psychiatry Society)

Sociedad Colombiana de Patología (Colombian Pathology Society)

The *Instituto Nacional de Salud* (National Health Institute) also carries out research and development activities although this is not one of its main functions.

Nuclear Science and Technology

Instituto de Asuntos Nucleares (Institute of Nuclear Affairs)
Established in 1959 to study the peaceful use of atomic energy. There is a staff of 120 researchers and technicians and a library of 3000 specialized volumes; 30 000 documents and 60 000 micro-films.

Meteorological and Astrophysical Research

Observatorio Astronómico Nacional (National Astronomical Observatory)

Professional Associations and Societies in the Field of Science and Technology

There are about thirty organizations that fall into this category, developing research and development in various areas. Among these are:

Academia Colombiana de Ciencias Exactas, Físicas y Naturales (Colombian Academy of Exact, Physical and Natural Sciences)

Academia Nacional de Medicina (National Academy of Medicine)

Sociedad Colombiana de Químicos e Ingenieros Químicos (Colombian Society of Chemists and Chemical Engineers)

Sociedad Colombiana de Ingenieros (Colombian Society of Engineers)

Centro de Estudios sobre Desarrollo Economico (Centre of Economic Development Studies)
Affiliated to the Universidad de los Andes, Bogotá.

Information Services in Science and Technology

There is, in Colombia, a *Sistema Nacional de Información (SNI)* (National Information System), which provides information services through its subsystems – which are: agricultural and livestock *(SNICA)* (National Information System for Agricultural and Livestock Sciences); health sciences *(SNICS)* (National Information System for Health Sciences); education

(SNIE) (National Information for Education); economics and business studies *(SNICEA)* (National Information System for Economics and Business Studies); marine sciences *(SNIM)* (National Information System for Marine Sciences); energy resources *(SNIRE)* (National Information System for Energy Resources); environment and natural resources *(SNIMA)* (National Information System for the Environment and Natural Resources) and industrial information *(SNII)* (National Information System for the Industrial Sector). Information is obtainable through reproduction of documents; inter-library loaning system, etc. Bibliographical information on the different areas of science and technology is available from the *Centro de Documentación e Información del Instituto Colombiano para el Fomento de la Educación Superior* (Documentation and Information Centre of the Colombian Institute for the Development of Higher Education).

There are at least thirty-two important centres of documentation including libraries and museums:

Archivo Nacional de Colombia (National Archives of Colombia)
Founded in 1868 and has a total of 40 600 volumes and other documents.

Biblioteca Nacional de Colombia (National Library of Colombia)
Established in 1777 and has 400 000 volumes.

Biblioteca Central de la Universidad Nacional de Colombia (Central Library of the National University of Colombia)
Founded in 1867, has 125 000 volumes and issues various specialized publications.

Biblioteca Central de la Pontificia Universidad Javeriana (Central Library of the Javeriana Pontifical University)
Founded in 1931 and possesses 145 000 volumes among other documents.

International Cooperation in Science and Technology

International cooperation is promoted in activities currently being developed and which are of national interest. Through *COLCIENCIAS*, the following areas are benefiting from international cooperation: metallurgy; food and nutrition; natural resources; information; energy resources; environmental conservation; housing and construction materials; health and teaching; and training in the sciences.

In 1977, Colombia received, from the Organization of American States, US$669 315, to develop the areas mentioned above. In 1978 the contributions were US$304 780, as well as the regular budget of US$266 560, for three special projects: coal; technological information services; and the use of tropical wood.

In 1979 Colombia received US$250 000 for eleven research projects, among which are: *Programa Colombiano de Ciencias del Mar* (Colombian Programme of Marine Sciences); *Mecanismos Institutionales de Política*

Científicas y Tecnológicos (Institutional Mechanisms for Scientific and Technological Policy); the evaluation of cellulose pulp.

In 1980 US$337 207 were channelled into projects on energy resources; food and nutrition; natural plant resources; metallurgy and marine sciences.

In 1981 another US$260 000 was allocated to the projects already mentioned.

Costa Rica

Geographical, Demographical, Political and Economic Features

The Republic of Costa Rica shares its border to the north with Nicaragua, with 300 kilometres of frontier; to the north-east with the Antilles Sea, with 212 kilometres of coastline; a frontier of 363 kilometres separates it from Panama to the east, and to the south and west the Pacific Ocean forms a natural border with 1016 kilometres of coast.

The area of Costa Rica is 50 700 square kilometres. Estimates for 1979 showed the population to be 2 183 625 with a projection for 1985 of 2 631 000. The density per square kilometre is 42.7. The capital city, San José, had a population of 249 074 in 1979 and the other principal towns are Cartago (75 458 inhabitants); Limón (48 217) and Alajuela (38 481).

The climate in the coastal regions and lowlands up to 900 metres is hot and humid with temperatures fluctuating between 22 °C and 28 °C. The temperate region falls between the altitude of 900 and 1500 metres where temperatures vary between 14 °C and 20 °C, and it is cold in the mountainous region with an average temperature of 14 °C.

Costa Rica is a free democratic republic governed by the Constitution of 7 November 1949 (the eleventh of the republic). The executive power is in the hands of the President of the Republic who is elected by universal suffrage for a period of four years; the president may not be re-elected for a consecutive period of office. The president exercises his duties with his cabinet of thirteen ministers. The legislative power has one chamber – the Legislative Assembly – with fifty-seven deputies, elected by popular vote, from the seven provinces of the country. The provinces are divided into eighty-one cantons and these, in turn, into 415 districts. The judicial power lies with the High Court of Justice, whose seventeen members are elected by the Legislative Assembly for a minimum period of eight years. The High Court of Justice nominates the magistrates of the lower courts of justice. Voting is obligatory for all adults over eighteen years of age.

The constitution prohibits the organization of an army; law and order are preserved by the police force – Guardia Civil – with its 3000 members.

Natural Resources

AGRICULTURE

Costa Rica is an essentially agricultural country, the main products being cacao, coffee, rice, bananas, almonds, citrus fruit, cotton, sugar cane, beans, maize, peanuts, potatoes, tobacco, tomatoes, millet and sorghum.

Land utilization can be broken down as follows:

Agriculture and livestock	15 000 square kilometres	30.0 per cent
Forestry protection	12 000 square kilometres	23.3 per cent
Forestry production	24 000 square kilometres	46.7 per cent

More than 35 per cent of the total area of the country is covered in forest and shrubland and 60 per cent or 70 per cent of these reserves are commercially viable species.

LIVESTOCK

Figures for 1973 show 1.7 million head of cattle and 215 792 pigs as well as sheep, horses, mules, asses and poultry.

FISH

The main species fished commercially are tuna, other sea fish in general and shrimps. Costa Rica is in a favourable position geographically, having access to two oceans.

MINERAL RESOURCES

Costa Rica is a volcanic country, rich in mineral resources. However, geological exploration has not been carried out on a large scale; therefore the exact value of reserves is not known. Nevertheless, there are mines of salt, sulphur, calcium carbonate, zinc, copper, iron, manganese, mercury, nickel, gold, silver and lead.

Economic Information

Monetary Unit: Colon (8.57 = US$1 in November 1980).

Gross National Product per person (1978): US$1540.

Distribution of GNP (1977): agriculture 22 per cent; industry 27 per cent; services 51 per cent.

In 1979 the GNP decreased for the second consecutive year; GNP in 1979 was 4.6 per cent compared to 5.7 per cent in 1978 and 8.9 per cent in 1977. The GNP in the agricultural and livestock sector showed only a 1 per cent increase (excess rain and lack of manpower seriously affected the coffee crop, although banana production increased by 1.1 per cent. Industry achieved a 5 per cent growth, commercial activities 5.6 per cent and governmental services 3 per cent.

Education

Costa Rica has a very low percentage of illiteracy. Table 6.1 relates to education in 1977.

The general budget for education in 1979 was 28.4 per cent of the national budget.

Table 5

Level	Schools	Teachers	Students
Primary	416	598	18 971
First	2 865	12 500	367 026
Second	n/a	5 915*	121 202
Third	n/a	617†	38 629

* 1976.
† 1965.

International Relations

Costa Rica is a member of the United Nations and the Organization of American States.

Organization of Science and Technology

As in most Latin American countries, scientific and technological research is concentrated in university institutes and central Government organizations. The system of science and technology has as its central axis the *Consejo Nacional de Investigaciones Científicas y Tecnólogicas (CONICIT)* (National Council for Scientific and Technological Research), which was created in 1972 to advise the Government in matters of science and technology. This body, together with the *Oficina de Planificación Nacional* (National Planning Office) outlines national policies in science and technology. With the forming of *CONICIT*, many proposed scientific research programmes, whether at community level, or in university or governmental organizations, have been able to reach a conclusion, and recommendations from *ad hoc* commissions in different fields and agreements with international organizations have been acted upon positively. Examples of the work carried out by *CONICIT* are: the establishing of the *Instituto de Investigaciones en Salud* (Institute of Health Research) in the University of Costa Rica and in collaboration with the Ministry of Health; and the founding of the *Centro de Investigaciones Marinas* (Centre of Marine Research) in conjunction with the *Comisión de Ciencias del Mar y Pesquería* (Commission of Sea and Fisheries).

Financing and Policies in Science and Technology

Since the formation of *CONICIT*, policy-making for research activities has been in the hands of this body, which, in conjunction with the specialized institutes and organizations, elaborates and plans the development of activities in science and technology.

Special emphasis is placed on the teaching and instruction of staff in the areas of planning and management, whether these be in foreign or national concerns. The improvement of the standard of human resources in research is one of the central issues, and with this objective, *CONICIT* has granted aid for: post-graduate studies abroad; financial assistance for scientific events; the organization of seminars and symposiums; reinforcing the scientific capacity; integrating foreign scientists into the system; and stimulating incentives to avoid the 'exodus of the intelligentsia'.

Priority areas of research are listed as being: public health; sea and fisheries; veterinary medicine; low-cost housing; conservation and exploitation of natural resources; cultivation of basic food materials; education; agricultural sciences; social sciences and energy.

CONICIT also studies and analyses the work carried out by scientists and technicians in the different organizations with the objective of promoting and coordinating these activities. Financial resources are channelled through *CONICIT* to the various areas of research, and the budget of this institute has been growing steadily since its formation. In 1975 funds were US$264 000 and in 1977 US$583 000. Thirty per cent of these amounts were allocated to the direct financing of research projects. As well as the funds provided by *CONICIT*, one must also consider the financing of various projects carried out in governmental institutions and the universities; as yet these figures have not been evaluated, but in 1979 an expert from the Organization of American States was working in this field.

The *Instituto Interamericano de la OEA* (Inter-American Institute of the Organization of American States) promotes international cooperation among the Central American states in research and development, especially in the field of livestock and agriculture. In Costa Rica, the institute, together with the government, has formed the *Centro Agronómico Tropical de Investigación y Enseñanza* (Centre for Research and Teaching of Tropical Agronomy).

Science and Technology at Government Level

Apart from *CONICIT*, which has already been covered, the following governmental bodies also fill an important role in science and technology:

Dirección General de Estadísticas y Censos (Head Office for Statistics and Census)
This body provides a major part of basic material for various research projects. Its library contains 3000 volumes and publishes several periodicals.

Dirección General de Geología, Minas y Petróleo (Directorate-General of Geology, Mining and Petroleum)
Founded in 1951, besides carrying out research programmes this organization acts as the central body for policy-making and advising in matters related to geology, mining and petroleum. *Informes Técnicos* (Technical Bulletin) and *Notas Geológicas* (Geological Notes) are published.

Instituto Geográfico Nacional (National Geographical Institute)
Established in 1944, its main function is the drawing-up of official maps

and charts of Costa Rica. Its specialized library contains 3000 volumes and *Informe Semestral* is published.

Centro para el Mejoramiento de la Enseñanza de las Ciencias (CEMEC) (Centre for the Improvement of Science Teaching)
Founded in 1977, the centre studies and applies teaching methods in science and to this end organizes refresher courses for teachers.

Instituto Centroamericano de Investigación y Tecnología Industrial (ICAITI) (Central American Institute of Industrial Research and Technology)
Although the headquarters are in Guatemala, Costa Rica was one of the founder countries of this institute in 1956, and through the Ministry of the Economy plays an active role in developing research programmes for the institute. Research is concentrated in the following areas: food technology; utilization of agricultural waste; special research into coffee; sugar cane; cacao and oil seeds; textile fibres and hides.

Science and Technology at an Academic Level

Universidad de Costa Rica (Costa Rica University)
Founded in 1843 and reformed in 1940 with its campus in the capital, San José de Costa Rica. Teachers and researchers number 2662 and the student population is 28 378. Faculties in agronomy; fine arts; economics; science; arts; law; education; pharmacy; engineering; medicine; microbiology; dentistry and social sciences.
    The following research institutes are affiliated to the university:

Organización de Estudios Tropicales (Organization of Tropical Studies)

Centro Latinoamericano de Demografía (Latin American Centre of Demography)

Centro de Estudios Sociales y de Población (Centre of Population and Social Studies)

Escuela Centroamericana de Geografía (Central American School of Geography)

Universidad Estatal a Distancia (Open University)
Founded in 1977 with its headquarters in San José. Study programmes on various subjects are organized through the twenty-one regional centres where students can register, receive tuition, sit exams and use the libraries. The staff numbers 150 teachers for its 6095 students.

Universidad Nacional Autónoma de Heredia (National Autonomous University of Heredia)
Founded in 1973 with its seat in Heredia. The student population numbers 8000, distributed in the centres in Guanacaste, San Ramon, and San Isidro de el General.
    Faculties: agriculture; education; geography; religion; veterinary medicine; philosophy and arts. The *Instituto Tecnológico de Costa Rica* (Tech-

nological Institute of Costa Rica) is affiliated to the university; it was established in 1971 with 2500 students in its departments of engineering, agriculture and forestry. Its library houses 25 000 volumes. *Tecnología en Marcha* (Technology in Progress) and *Comunicación* (Communication) are published.

Industrial Technology and Research

A major part of research in this field is executed through agreements with international institutions and in particular with *ICAITI*, described in the section on 'Science and Technology at Government Level'.

Agricultural and Marine Science and Technology

Specialist government institutes carry out a large part of research in this field. In 1966 there were 6 agricultural and livestock experimental stations, of which 4 were state-run; 1 was affiliated to a university and 1 was a private concern. There were 124 technicians and specialists employed in the experimental stations. In 1979 a *Centro de Investigaciones Marinas* (Centre of Marine Research) was created as a dependant of the *Comisión de Ciencias del Mar y Pesquería* (Commission of Sea and Fisheries).

Centro de Ciencia Tropical (Tropical Science Centre)
Established in 1962, it is a private institute. It acts as an advisory service on the use of soils and carries out ecological surveys. There is a field centre at Monteverde Cloud Forest. Publications are issued irregularly.

Instituto Interamericano de Ciencias Agrícolas de la OEA (IICA) (Inter-American Institute of Agricultural Sciences of the Organization of American States)

Centro Agronómico Tropical de Investigación y Enseñanza (CATIE) (Centre for Research and Teaching of Tropical Agronomy)
Founded in 1973, it is a liaising association between the Costa Rican Government and the *IICA* and carries out research in Central America and the Caribbean. *Memoria Anual* (Annual Memorandum) and *Actividades en Turrialba* (Activities in Turrialba) are published.

Medical Science and Technology

Instituto de Investigación en Salud (Institute of Health Research)
This body was established by the Ministry of Health in conjunction with the University of Costa Rica.

A large part of research in this field is undertaken in the specialized hospital units, particularly in the San Juan de Dios Hospital in San José, and the medical associations of cardiology, internal medicine and surgery, obstetrics and gynaecology.

Centro de Estudios Médicos 'Ricardo Moreno Canas' (Ricardo Moreno Canas Centre for Medical Studies)

Asociación Costarricense de Cirugía (Costa Rican Surgery Association)

Asociación Costarricense de Pediatría (Costa Rican Paediatrics Association)

Academia Costarricense de Periodoncia (Costa Rican Academy of Periodontics)

Nuclear Science and Technology

Comisión de Energía Atómica de Costa Rica (Atomic Energy Commission of Costa Rica)
Conducts programmes in applied and basic research in agriculture, medicine, and physics.

Economic and Social Research

Instituto Centroamericano de Administración Pública (ICAP) (Central American Institute of Public Administration)
Founded in 1954, it undertakes research in the field of public administration, economic development and integration. Its specialized library contains 25 000 volumes.

Instituto Costarricense de Ciencias Políticas y Sociales (Costa Rican Institute of Political and Social Sciences)

Astrophysical and Meteorological Science and Technology

Instituto Meteorólogico Nacional (National Meteorological Office)
Founded in 1888, it develops climatic and hydro-meteorological studies. The staff numbers sixty.

Professional Associations and Societies in the General Field

Academia Costarricense de la Lengua (Costa Rican Language Academy)
Founded in 1923, it is run on similar lines to the *Real Academia Española* (Spanish Royal Academy) and its objective is to broaden awareness of the Spanish language. There are eighteen members and *Boletín de la Academia Costarricense de la Lengua* (Costa Rican Language Academy Bulletin) is published.

Academia de Geografía e Historia de Costa Rica (Costa Rican Academy of Geography and History)

Centro Cultural Costarricense-Norteamericano (Costa Rican–North American Cultural Centre)

Information Services in Science and Technology

Departamento de Información y Documentación de CONICIT (Information and Documentation Department of *CONICIT*)
This department was formed in 1976 and has been steadily building up its stock of information since that date. The department is at present perfecting an information and documentation network and training specialized staff.

Archivo Nacional de Costa Rica (National Archives of Costa Rica)
The archives contain 8500 volumes. *Revista* (Review) is published.

Biblioteca de la Asemblea Legislativa (Legislative Assembly Library)
Founded in 1953, it has 30 000 volumes.

Biblioteca 'Carlos Monge Alfaro' de la Universidad de Costa Rica (Carlos Monge Alfaro Library of the University of Costa Rica)
Founded in 1946, it has 300 000 volumes.

Biblioteca 'Alvaro Castro J.' (Alvaro Castro J. Library)
Founded in 1950; has 30 163 volumes.

Biblioteca del Centro Cultural Costarricense-Norteamericano (Library of the Costa Rican-North American Cultural Centre)
Founded in 1945; has 7500 volumes.

Biblioteca del Ministerio de Relaciones Exteriores (Library of the Ministry of Foreign Relations)
Founded in 1960; has 12 500 volumes.

Biblioteca Nacional (National Library)
Founded in 1888; has 175 000 volumes.

Museo Nacional (National Museum)
Founded in 1887. Possesses collections of pre-Columbian art, colonial art and natural history and has a library of 10 000 volumes. *Brenesia* and *Vínculos* (Links) are published.

Museo de Historia Eclesiástica 'Anselmo Liorente y Lafuente' (Anselmo Liorente y Lafuente Ecclesiastical History Museum)
Founded in 1972, it houses important documents on the Church in Costa Rica.

International Cooperation in Science and Technology

Links that Costa Rica maintains with Central American and Caribbean countries have already been mentioned.

Through the *CONICIT*, cooperation agreements have been established with: *Consejo Nacional de Ciencia y Tecnología de Mexico* (Mexican National Council of Science and Technology); the Israeli Government; and the *Conselho Nacional de Desenvolvimento Científico e Tecnológico de Brasil* (Brazilian National Council for Scientific and Technological Development).

Cuba

Geographical, Demographical and Economic Features

The Republic of Cuba is the largest and most western island of the Greater Antilles. Cuba is situated at some 217 kilometres from the southernmost point of the State of Florida of the United States of America. The area is 114 524 square kilometres; its islands make up a total of 3496 square kilometres. From the census of 1972, the population reached a figure of 8 749 000 and estimates for 1978 show 9 780 000.

Government

1959 was the year in which the revolution, led by Fidel Castro, deposed Fulgencio Batista. The existing constitution was suspended and replaced by the first Socialist constitution of the continent, dated 24 February 1976. By this constitution, Dr Fidel Castro Ruz became President of the Council of State, President of the Council of Ministers and First Secretary of the Communist Party of Cuba.

The country is divided into fourteen provinces and 169 municipalities. In 1976 elections were held to nominate Delegates for the Municipal Assemblies and for the National Assembly.

Natural Resources

AGRICULTURE

From 1959 land was nationalized and all properties over 30 'caballerizas' (approximately 402 hectares) were expropriated and converted into state haciendas. In 1963 private ownership of land was reduced to a maximum of five caballerizas (approximately 69 hectares) and by the end of 1966 nearly 65 per cent of the land was state-owned, the rest being in private hands.

The main agricultural products of the country are tobacco, cane sugar, coffee, cotton, maize, rice and potatoes. Cane sugar is the principal export and production in 1978 reached 7661.5 metric tons, of which 7231.2 metric tons were exported. Cuba is the second largest sugar producer in the world.

The main fruits for export are pineapple, citrus fruits, tomatoes and peppers. Rice cultivation started in 1967 in the south of the province of La Habana, and is presently developing at a highly mechanized rate with two crops a year being produced.

LIVESTOCK

Figures for 1976 show 1.5 million head of pigs; 80 000 horses; 300 000 sheep; 92 000 goats and 5.5 million cattle.

FORESTRY

Cuba has exceptional forest resources, among which are eucalyptus, mahogany and cedar, which is used predominantly in the production of cigar boxes. Mahogany is exported in vast quantities. Between the years 1960 and 1963 reforestation plans were put into effect with the planting of 9941.81 hectares of pines; 52 699.43 hectares of majagua (a tree belonging to the lime family) and 9615.61 hectares of casuarinas.

FISH

The fishing industry is one of the sectors of the most rapid growth in the Cuban economy; this is basically as a result of the large subsidies poured into the industry. Fish exports represent the second largest source of wealth of the country. In 1959 fish production reached 27 000 tons. The subsequent years have seen an annual growth rate of 12.5 per cent. In 1970 fish production reached 100 000 tons and in 1978 213 100 tons, which is a 15 per cent increase on the previous year's figures.

MINERAL RESOURCES

Among the principal mineral resources are iron, copper, refractory chrome, nickel, cobalt, gold and silver. The development of nickel mining is one of the main economic objectives of the country, especially in the export market. According to the five-year plan (1980–85) it is estimated that the four existing plants could produce 100 000 tons of nickel per year.

Oil production, although a modest enterprise, reached 288 000 tons in 1978. Copper production in 1978 totalled 390 000 tons. Salt production has increased notably and production for 1974 was at 138 300 tons.

Education

Education is obligatory between the ages of six and fourteen and is free and available in all parts of the country. The census for 1953 revealed that 22.8 per cent of adults over the age of ten were illiterate. In 1961, called the '*Año de la Alfabetización*' (Year of Literacy), an intense literacy programme was launched using all educational forces available, many of which were students of higher education, which resulted in bringing the illiteracy rate down to 3.9 per cent and in 1964 illiteracy being 'officially eradicated'.

In 1975–76 there were 1 560 193 students attending primary schools and 70 000 in the secondary schools, while technical schools provided an education for 42 517 students during the same period. In 1978 there were about 40 000 students attending the four existing universities. Estimates made in 1979 show that 62 950 students were receiving an education in the field of science and technology (exact sciences, engineering and technology, medicine and agricultural sciences).

International Relations

Cuba is a member of the Organization of American States, *COMECON*, and the United Nations and all its affiliated organizations.

Organization of Science and Technology

In 1974 the *Consejo Nacional de Ciencia y Tecnología (CNCT)* (National Council of Science and Technology) was established with the objective of outlining a policy for science and technology and the drawing up of specific plans for the development of scientific and technological research.

During the first two years the *CNCT* carried out various activities, among which were:

Studying the network of the various research and development organizations and their scientific and technological potential.

Preliminary diagnostic research into the country's science and technology.

Determining the priorities for the development of research and development in science and technology.

The coordination of forces towards common objectives within the various research and development organizations.

The establishment of specific bodies in specialized areas, for example, the *Comision Nacional para el uso pacífico de la Energía Atómica* (National Commission for the Peaceful Use of Atomic Energy).

The *CNCT* has taken several important measures such as the establishment of laws determining scientific grades and standards, categories of lecturers and researchers in higher education and the establishment of the *Ministerio de Educación Superior* (Ministry of Higher Education).

In November 1976 law number 1323 came into effect on the Organization of the Central State Administration. By this law was founded the *Comité Estatal de Ciencia y Técnica (CECT)* (State Committee of Science and Technology), which has taken over the functions of the *CNCT*. Also affiliated to this committee is the *Oficina Nacional de Invenciones, Información Técnica y Marcas* (National Office of Inventions, Technical Information and Trade Marks). The *Comité Estatal de Normalización* (State Committee of Standards) and the *Instituto Nacional de Sistemas Automatizados y Técnicas de Computación* (National Institute of Automated Systems and Computation) were formed to work in parallel with the *CECT*, among other bodies.

The *CECT*, apart from consolidating and developing the work initiated by *CNCT*, has become the principal organization for policy-making in the field of science and technology. Its functions also include institutional organization and financing. The *CECT* has developed various activities, among which are:

The establishment of the only scientific and technological information system.

The control of the progress of scientific and technological research.

The creation and functioning of the *Comisión Nacional de Protección al*

Medio Ambiente y Conservación de los Recursos Naturales (National Commission for Environmental Protection and Conservation of Natural Resources).

The relation of *CECT* with the other organizations can be seen in Figure 6.

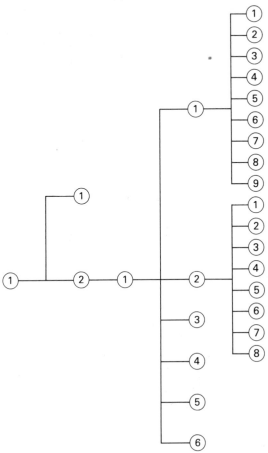

Fig. 6 Organization of science and technology

1	Consejo de Ministros (Council of Ministers)
1.1	Ministerios (Ministries)
1.2	Junta Central de Planificación (Central Planning Board)
1.2.1	Comité Estatal de Ciencia y Técnica (State Council of Science and Technology)
1.2.1.1.1	Direccion por ramas (Sectoral administration)
1.2.1.1.1	Industria básica (Basic industry)
1.2.1.1.2	Electrónica (Electronics)
1.2.1.1.3	Azucar y derivados (Sugar and by-products)
1.2.1.1.4	Industria ligera y alimenticia (Light industry and food industry)
1.2.1.1.5	Agricultura y pesca (Agriculture and fish)
1.2.1.1.6	Investigaciones económicas y sociales (Economic and social research)
1.2.1.1.7	Transporte (Transport)
1.2.1.1.8	Construcción (Construction)
1.2.1.1.9	Bio-médica (Bio-medicine)

Fig. 6 - *continued*

1.2.1.2	Direcciones funcionales (Functional administration)
1.2.1.2.1	Cuadros científico-técnicos (Scientific and technological cadres)
1.2.1.2.2	Información científica-técnica (Scientific and technological information)
1.2.1.2.3	Planificación de la Ciencia y la Técnica (Planning of Science and Technology)
1.2.1.2.4	Pronósticos científico-técnicos (Prognostications in Science and Technology)
1.2.1.2.5	Finanzas de la Ciencia y la Técnica (Economy in Science and Technology)
1.2.1.2.6	Economía de la Ciencia y la Técnica (Economy in Science and Technology)
1.2.1.2.7	Relaciones internacionales (International relations)
1.2.1.2.8	Evaluación de projectors tecnológicos (Evaluation of technological projects)
1.2.1.3	Comisión Nacional para le uso pacífico de la Energía Atómica (National Commission for the Peaceful Use of Atomic Energy)
1.2.1.4	Comisión Nacional para la proteccion del Medio Ambiente y Conservación de Recursos Naturales (National Commission for Environmental Protection and Conservation of Natural Resources)
1.2.1.5	Oficina Nacional de Invenciones, Información Técnica y Marcas (National Office for Inventions, Technical Information and Trade Marks)
1.2.1.6	Comisión Nacional para la Aprobación de Categorías de Investigación (National Commission for the Approvement of Research Categories)

Financing Policies in Science and Technology

Scientific and technological progress has been incorporated in the National Economic Plan. The following objectives fall within the National Economic Plan:

the increase of efficiency;
standardization and improvement of quality;
training of scientific and technological work forces;
the organization of scientific work;
the introduction of new technology (national or international);
the application of computer technology and automated systems.

The integration of science and technology into the economic and social development programme is executed through the scientific policy within the National Economic Plan. Special emphasis is placed on the 'rapport' between the teaching, research and production sectors. Among the objectives of the plans for the development of science and technology are the solution of key issues to the Cuban economy, such as: obtaining the maximum from natural resources; the introduction of new technology; increasing efficiency and the quality of production, etc.

Another function of *CECT* is to determine the financial needs of each research project. Since 1978 a strict control is maintained over the expenditure of each project and this enables the *CECT* to detect any financial difficulties which could result in the curtailment of any project.

Briefly, the principal aims of the national scientific policy are:

(a) the development of science and technology in parallel with social progress;
(b) the combination of basic and applied research with the emphasis on the latter;
(c) the transference and assimilation of technology;

(d) the rapid introduction of beneficial research results into the production and services sector;

(e) the strengthening of scientific and technical potential.

In 1977 expenditure and human resources in research and development in higher education were as follows:

A. Personnel

1.	Scientists and engineers	1 107
2.	Technicians	969
3.	Auxiliary staff	1 762

B. Expenditure (in Cuban pesos)

1.	Capital goods	12 423 507
2.	Overheads	1 739 212
	Total	14 162 719

According to data of 1977, the expenditure for research and development activities was 74 257 910 Cuban pesos, of which 93.9 per cent was from public funds and 6.1 per cent was from foreign sources.

Human resources in research and development for the same year were:

1.	Scientists and engineers	10 130
2.	Technicians	6 075
3.	Auxiliary staff	8 625

Science and Technology at Government Level

The principal body in this field *(CECT)* has already been dealt with. Other organizations in this section include:

Instituto de Geografía (Geography Institute)

Instituto de Investigación Técnica Fundamental (Institute of Fundamental Technical Research)
Carries out research into electronic circuits, laser and solar energy.

Instituto de Geología y Paleontología (Geology and Palaeontology Institute)

Industrial Technology and Research

Instituto Tecnológico 'Mártires de Girón' (Mártires de Girón Technology Institute)

Instituto Tecnológico de Electrónica 'Fernando Aguado Rico' (Fernando Aguado Rico Electronic Technology Institute)

Instituto Cubano de Investigaciones Mineras y Metalúrgicas (Cuban Institute of Mining and Metallurgical Research)

Centro de Investigaciones para la Industria Minero Metalúrgica (Centre of Research for the Mining and Metallurgical Industry)

Agricultural and Marine Science and Technology

There are at least eleven organizations in this field, among which are:

Instituto Nacional de la Reforma Agraria (National Institute of Agrarian Reform)
This institute is a complex of research centres, experimental stations and laboratories. The areas of research cover practically the whole range of agricultural products: coffee, cocoa, cereals and vegetables, citrus fruits and other fruits, tobacco, rice, etc. Studies are also made into irrigation systems, plant protection and cultivation, soils and fertilizers, agricultural machinery, etc. The *Centro de Información y Documentación Agropecuaria (CIDA)* (Centre of Agricultural and Livestock Information and Documentation) is affiliated to the institute, and acts as an advisory and agricultural information centre.

Instituto de Investigaciones de la Caña de Azúcar (Sugar Cane Research Institute) and the *Instituto de Investigaciones de los Derivados de la Caña de Azúcar* (Institute of Research into Sugar Cane By-products)
Both these institutes have experimental stations under their jurisdiction and the major part of research in this area is performed by them.

Instituto Nacional de Desarrollo y Aprovechamiento Forestales (INDAF) (National Institute of Forest Development and Exploitation)
Affiliated to this institute is the *Centro de Investigaciones y Capacitación Forestales* (Forestry Research Centre)

Centro de Investigaciones Pesqueras (Fisheries Research Centre)
Carries out research into fish resources, marine biology, oceanography and pollution control. The centre has a staff of 330 and a specialized library of 3551 volumes.

Science and Technology at an Academic Level

Universidad de la Habana (Havana University)
Originally founded in 1728, it has undergone several reorganizations in 1863, 1943 and 1962. Its students number 53 682 with 3066 teachers. *Universidad de la Habana* is published among other scientific and technological reviews. The faculties are technology, medicine, agricultural sciences, and the humanities. The following centres are affiliated to the university: *Centro Nacional de Investigaciones Científicas* (National Centre of Scientific Research); *Instituto de Ciencia Animal* (Institute of Animal Science); *Instituto de Ciencias Agrícolas* (Institute of Agricultural Sciences) and the *Estación Experimental de Pastos y Forrajes 'Indio Hatuey'* (the Indio Hatuey Pasture and Forage Experimental Station).

Universidad de Oriente (University of the East)
Founded in 1947, it has a student population of 16 000 and 800 teachers. *Mambi* is published and faculties include the humanities, technology, medicine, rural work, science, agriculture and animal husbandry.

Universidad de las Villas (Las Villas University)
Established in 1948, it has 8200 students and 400 teachers in its faculties of technology, agriculture, livestock studies, sciences, medicine and the humanities.

Universidad de Camaguey (Camaguey University)
Founded as a branch of the Universidad de la Habana in 1974, at present it has 5000 students. Its departments are agriculture, economics, medicine and technology.

Medical Science and Technology

In this field of research there are six important national centres and various regional units.

Instituto Nacional de Higiene (National Institute of Hygiene)
A dependant of the Ministry of Public Health.

Instituto de Oncología y Radiobiología de la Habana (Institute of Oncology and Radiobiology of Havana)
Founded in 1961, it carries out research into genetics, cancer, radiobiology, surgery, immunology and related subjects. There is a staff of 400 researchers and technicians and a specialized library of 2500 volumes.

Instituto de Investigaciones Fundamentales del Cerebro (Institute of Fundamental Research on the Brain)
Research is done into neurosis and neurophysiology in general.

Centro Nacional de Información de Ciencias Médicas (National Centre of Medical Science Information)
An information centre for biomedicine, odontology and pharmacy. *Revista Cubana de Medicina* (Cuban Medical Review) is published, among others.

Grupo Nacional de Radiología (National Radiology Group)
A dependant on the Ministry of Public Health.

Nuclear Science and Technology

Instituto de Investigaciones Nucleares (Nuclear Research Institute)

Meteorological and Astrophysical Research

Instituto de Geofísica y Astronomía (Geophysics and Astronomy Institute)

This body incorporates seismological and radio-astronomical stations throughout the country.

Instituto de Meteorología (Meteorology Institute)
The national centre for research in this field; it has more than sixty stations throughout the country

Associations and Societies

There are many organizations which could be classified in this sector, but only about seventeen of them have national importance. Among these are: are:

Academia de Ciencias de Cuba (Cuban Academy of Science)
Carries out research in various areas of science and technology and publishes research results and information through its ten bulletins and specialized reviews.

Casa de las Américas (Americas House)
Founded in 1959, it is involved with the interchange of knowledge between Latin American countries in all fields of art and culture.

Instituto de Matemática, Cibernética y Computación (Mathematics, Cybernetics and Computing Institute)
Affiliated to the *Academia de Ciencias*.

Instituto de Química y Biología Experimental (Experimental Chemistry and Biology Institute)

Instituto de Zoología (Zoology Institute)

Instituto de Botánica (Botany Institute)

Information Services in Science and Technology

There are more than thirty-six libraries, centres of information and exhibitions in Cuba, among which are:

Instituto de Documentación e Información Científica y Técnica (Institute of Scientific and Technical Documentation and Information)
National centre for scientific information and incorporating a number of provincial centres.

Biblioteca Nacional 'José Martí' (José Martí National Library)
Founded in 1901, it has 531 329 volumes and 56 680 discs, drawings and transparencies containing information. More than seven reviews are published, including *Bibliografía Técnico Científicas* (Technical and Scientific Bibliography), *Indice General de Publicaciones Cubanas* (General Index of Cuban Publications).

Biblioteca Central 'Rubén Martínez Villena' de la Universidad de la Habana

(Rubén Martínez Villena Central Library of the University of Havana) Founded in 1728; contains 202 881 volumes and other publications.

Biblioteca 'Rubén Martínez Villena' (Rubén Martínez Villena Library) The library has 34 213 volumes.

Museo Histórico de las Ciencias Médicas 'Carlos J. Finlay' (Carlos J. Finlay Historical Museum of Medical Sciences) Possesses a specialized library of 300 000 collections and 50 000 volumes.

International Scientific Cooperation

Cuba participates in various organizations in the sub-region of the Caribbean and Latin America. However, the interchange of scientific and technological information mainly occurs with Socialist countries, especially with the Soviet Union. With international collaboration in mind, Cuba is developing, in conjunction with the *Consejo de Ayuda Mutua Económica (CAME)* (Council for Mutual Economic Aid), a development plan for science and technology until 1990, with seventeen research programmes in view.

Dominican Republic

Geographical, Demographical, Political and Economic Features

The Dominican Republic is situated on the eastern side of Hispaniola and covers 74 per cent of the island, which is the second largest of the Antilles. Its borders to the west are shared with Haiti, with 388 kilometres of frontier; to the north with the Atlantic Ocean; to the east with the Mona Passage, and to the south is the Caribbean Sea. The total area is 48 442 square kilometres.

Official estimates in 1979 show the population to be 5 275 000 with a density of 108 per square kilometre. The capital, Santo Domingo, has a population of 1 170 463. Fifty-two per cent of the population reside in urban areas and the ethnic composition is 72 per cent 'mestizos' (mixed blood) and mulattoes; 16 per cent white; and 11 per cent negroes originating from Africa, Haiti and the United States of America. The white population by origin are mainly Spaniards, Lebanese and German Jews.

Government

The government is made up of three powers – Executive, Legislative and Judicial. The executive is represented by the President of the Republic, who performs this duty for a period of four years together with the Vice-President. A cabinet of twelve secretaries of state advises the president and vice-president. The legislative power is in the hands of two chambers, the Senate, which has a senator for each province and federal district, and the Chamber of Deputies, made up of a total of ninety-one members of parliament. All members of the legislative exercise their duties for a four-year period. Finally there is the judicial power, which is made up of the High Court and Appeal Courts.

Natural Resources

Agriculture is the main source of wealth for the country, sugar being the main export product. Of the total cultivable area of 9900 square miles, only 3700 square miles is being exploited effectively. The largest sugar states are in the south-east of the country. Estimates for sugar production in

1980 show 1 090 000 tons. Coffee is also produced, most being exported to the United States of America. Rice is cultivated for internal consumption as well as cacao, tobacco, molasses, yucca and beans. Tobacco is exported and production figures for 1981 reached 57 000 tons. Another export is bananas whose production figures for the same year totalled 315 000 tons.

LIVESTOCK
Figures for 1977 reveal 2 million head of cattle, 800 000 pigs and 51 000 head of goats.

FORESTRY
Although there are no statistics available for this sector, the principal woods are cedar and mahogany.

MINERALS
Bauxite is the main strength in this sector; almost all of it is exported to the United States of America. There are also deposits of silver, platinum and copper.

INDUSTRY
The most important industries are: glass, industrial textiles, cement, paper and paper products, and matches. There are also sugar refineries, oil refineries and chemical plants for the production of ethyl alcohol.

Education
Primary education is obligatory and free for children between the ages of seven and fourteen. There are 5245 primary schools as well as secondary schools, vocational training schools, training colleges and high schools. Estimates for 1968 show that 68.1 per cent of the population is literate.

International Relations
The Republic is a member of the United Nations Organization and the Organization of American States.

Organization of Science and Technology

Most research activities are carried out in government organizations (25 per cent) and in the universities (57.7 per cent) (fifty-three institutions in 1973).

Although there is no specific central body which deals solely with policy-making in the field of science and technology, the *Unidad de Ciencia y Tecnología (UNICYT)* (Confederation of Science and Technology), a dependant of the *Secretariado Técnico* (Technical Secretariat) of the Presidency, fills this role to a certain extent.

The development of the agricultural and livestock research is of prime economic importance to the Dominican Republic.

Figure 7 shows the organization of science and technology.

Financing and Policies for Science and Technology

In the document entitled *Posibilidades del Desarrollo Económico y Social de la República Dominicana 1976–1986* (Possibilities for the Social and Economic Development of the Dominican Republic 1976–1986), there is a chapter written for *UNICYT* in which science and technology at national level are analysed. The socio-economic requirements are studied and specific objectives are defined accordingly and priorities for the various sectors of science and technology are established.

For the agricultural and livestock sector three priorities were stated as:

(1) the development of agricultural and livestock research centres;
(2) the improvement of teaching and documentation; and
(3) the control of the transference of technology.

In the industrial sector:

(1) to endeavour to regulate the import of technology;
(2) to increase national capacity in science and technology; and
(3) to coordinate scientific and technological activities within the government, university and private sectors.

For the mining sector the priority was to carry out research into mineral resources.

There are no up-to-date statistics for financial and human resources but figures for 1972 show that 1 561 070 pesos were destined for research activities, out of which 61.1 per cent went to the agricultural and livestock sector; 5.2 per cent to the industrial sector and 1 per cent to mining.

Expenditure for the agricultural sector is being maintained at a steady level, while the industrial sector has increased considerably, and the *Instituto Dominicano de Tecnología Industrial (INDOTEC)* (Dominican Institute for Industrial Technology) has an estimated annual global budget of 2 000 000 pesos (US$1 = 1 peso according to the Economic Department of the Bank of America, 29 April 1981).

Regarding human resources, there is only an estimate made in 1974 based upon a time and manpower study which shows that 28.7 per cent of the total researchers work in the exact sciences; 8.4 per cent in engineering; 18.0 per cent in medicine; 18.5 per cent in agriculture and livestock; 13.9 per cent in the social sciences and 8.4 per cent in legal studies and humanities. In 1977 a *Diagnóstico de la Investigación Agropecuaria* (Diagnosis of Agricultural and Livestock Research) was carried out, which revealed that a total of ninety-four researchers with titles and recognized grades were employed in this sector. In 1978 *INDOTEC* employed a total of fifty-five people, among which were doctors, graduates, engineers and technicians.

Science and Technology at Government Level

At this level there are fourteen organizations which develop research in science and technology. Among them are *Universidad Autónoma de Santo*

Domingo (Autonomous University of Santo Domingo); *Centro de Investigaciones de Biología Marina (CIBIMA)* (Marine Biology Research Centre) and the *Centro de Estudios de la Realidad Social Dominicana (CERESD)* (Social Studies Centre). The most important organizations are those which carry out activities in agricultural and livestock research such as: *Centro Nacional de Investigaciones Agropecuarias (CNIA)* (National Centre for Agricultural and Livestock Research); the *Centro de Desarrollo Agropecuario (CENDA)* (Agricultural and Livestock Development Centre) and *Division de Ganadería y Boyada (CEAGANA)* (Livestock Division), a dependant of the *Consejo Estatal del Azúcar* (State Sugar Council).

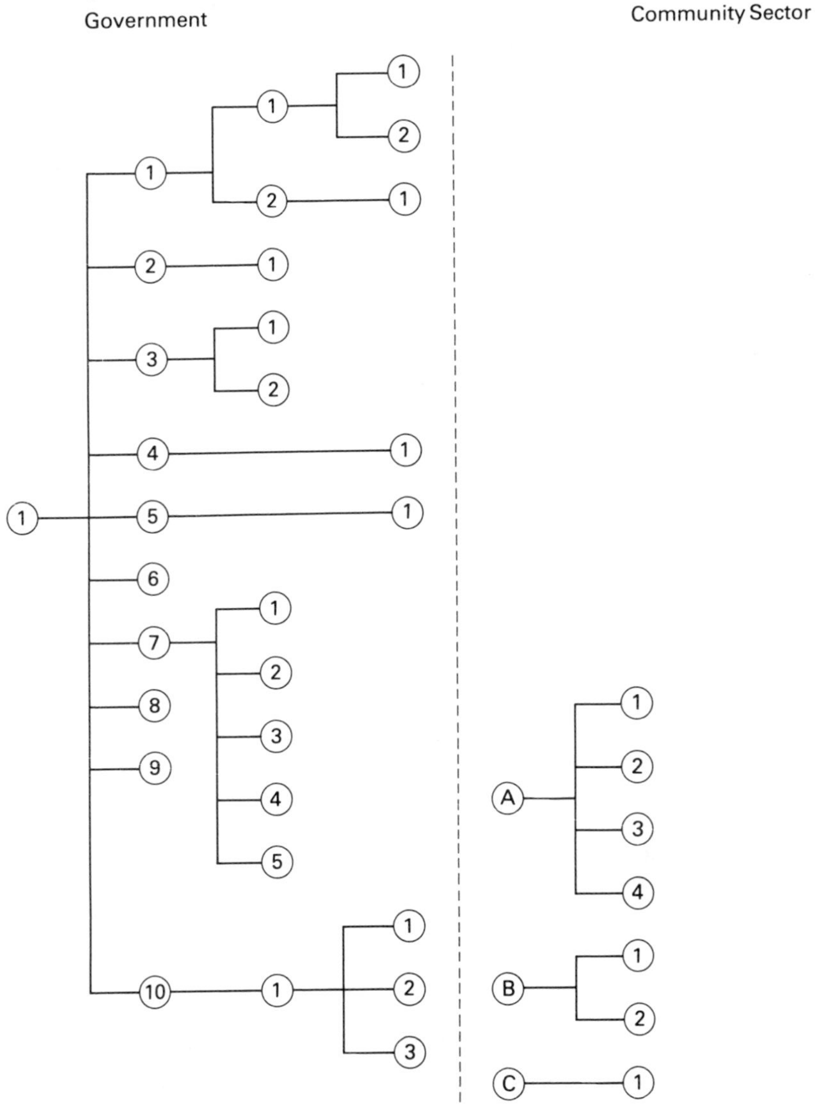

Fig. 7 Organization of science and technology

Fig. 7 – *continued*

1	Presidencia de la República (President of the Republic)
1.1	Secretaria de Estado de Agricultura (Secretary of State for Agriculture)
1.1.1	Subsecretaria de Estado de Investigación, Extensión y Capacitación Agropecuaria (Sub-secretary of State for Agricultural and Livestock Research, Development and Organization)
1.1.1.1	Centro de Desarrollo Agropecuario (Centre for Agricultural and Livestock Development)
1.1.1.2	Centro de Investigaciones Agropecuarias (Centre for Agricultural and Livestock Research)
1.1.2	Subsecretaria de Estado de Recursos Naturales (Sub-secretary of State for Natural Resources)
1.1.2.1	Departamento de Caza y Pesca (Department of Game and Fisheries)
2	Secretaría de Estado de Salud Publica (Secretary of State for Public Health)
2.1	Consejo Nacional de Población y Familia (National Council for the Family and Population)
3	Secretaría de Estado de Industria y Comercio (Secretary of State for Commerce and Industry)
3.1	Dirección General de Minería (General Mining Board)
3.2	Departamento de Control de Calidad (Department of Quality Control)
4	Consejo Estatal del Azúcar (State Sugar Council)
4.1	Division de Ganadería y Boyada (Livestock Division)
5	Banco Central de la República (Central Bank of the Republic)
5.1	Instituto Dominicano de Tecnología Industrial (Dominican Institute of Industrial Technology)
6	Museo del Hombre Dominicano (Dominican Museum of Man)
7	Secretariado Técnico de la Presidencia de la República (Technical Secretariat of the Presidency)
7.1	Unidad de Ciencia y Tecnología (Confederation of Science and Technology)
7.2	Unidad de Estudios Ambientales (Confederation of Environmental Studies)
7.3	Oficina Nacional de Planificación (National Planning Office)
7.4	Oficina Nacional de Presupuestos (National Budget Office)
7.5	Oficina Nacional de Estadísticas (National Statistics Office)
8	Museo de Geografía e Historia Natural (Museum of Geography and Natural History)
9	Biblioteca Nacional (National Library)
10	Universidad Autónoma de Santo Domingo (Autonomous University of Santo Domingo)
10.1	Dirección de Investigaciones Científicas (Board for Scientific Research)
10.1.1	Centro de Investigaciones de Biología Marina (Centre of Marine Biology Research)
10.1.2	Centro de Estudios de la Realidad Social Dominicana (Dominican Centre for Social Studies)
10.1.3	Finca Experimental de Egombe (Egombe Experimental Farm)
A	Universidad Nacional 'Pedro Henríquez Ureña' (National University of Pedro Henriquez Urena)
A.1	Dirección General de Investigaciones (General Board of Research)
A.2	Centro de Producción Ganadera (Centre for Livestock Production)
A.3	Instituto de Investigaciones Biomédicas (Institute of Bio-medical Research)
B	Universidad Católica Madre y Maestra (Madre y Maestra Catholic University)
B.1	Dirección General de Investigaciones (General Board of Research)
B.2	Consejo Técnico para el Fomento de las Ciencias Minero-metalurgicas (Technical Council for the Development of Mineral and Metallurgic Sciences)
C	Universidad Central del Este (Central University of the East)
C.1	Escuela Técnico – laboral (Technical School – labour)

Science and Technology at an Academic Level

Universidad Autónoma de Santo Domingo (Autonomous University of Santo Domingo)

Founded in 1538 by papal seal of Pablo III. The university was closed between 1801 and 1815 and was reorganized in 1914. It is the oldest university of America. The student population totals 28 628 and there are 1178 teachers. Apart from the faculties of humanities and arts, there are faculties of medicine, sciences, engineering and architecture, agronomy and veterinary medicine.

Universidad Católica Madre y Maestra (Madre y Maestra Catholic University)
Established in 1962 and run under private control. The academic year falls between August and June. With 5095 students and 465 lecturers, its faculties are arts, literature, sciences, engineering and general courses.

Universidad Nacional 'Pedro Henríquez Ureña' (Pedro Henríques Ureña National University)
Founded in 1966, it is a private university with an academic year between August and June. Students number 7000 with 500 teachers and a library of 30 000 volumes. The science faculties focus upon the areas of agronomy and veterinary medicine, engineering and technology, and health sciences. Also affiliated to the university are: *Instituto de Estudios Bio-médicos* (Institute of Biomedical Studies); *Centro de Investigaciones* (Research Centre) and the *Centro de Información de Drogas* (Centre of Drug Information).

Universidad Central del Este (Central University of the East)
Founded in 1970, it has 15 000 students and 370 teachers in its faculties of medicine, engineering, law and economics.

Universidad Tecnológica de Santiago (Technological University of Santiago)
Established in 1974, it has a library of 15 000 volumes for its 4000 students and ninety lecturers. Faculties are: arts and humanities, engineering and architecture.

Instituto Superior de Agricultura (ISA) (Higher Institute of Agriculture)
This is an independent institute which carries out agricultural programmes in conjunction with the *Universidad Católica Madre y Maestra*. Courses are offered in agricultural sciences at basic and advanced levels in areas such as horticulture, food technology, forestry, agricultural education, mechanization, irrigation, agrarian reform, agricultural economics and business studies. Occasionally publications are issued on work and research results of the institute.

Instituto Tecnológico de Santo Domingo (Technological Institute of Santo Domingo)
This institute offers courses at basic, advanced and post-graduate level and also research into science and technology, health sciences and the humanities. Students number 2700.

Centro de Asistencia Técnica (Technical Assistance Centre)
The centre functions as a liaising body between the universities and industry. Established in 1975, the year in which the Dominican Government together with the United Nations Development Programme (UNDP) launched a project to establish a *Unidad de Consultoria de la Dirección Industrial* (Consultative Organization for Industrial Development).

Industrial Research and Technology

Instituto Dominicano de Tecnología Industrial (INDOTEC) (Dominican Institute for Industrial Technology)
Founded in 1975, it has a research staff of fifty-five. The institute is dependent on the Central Bank, offers training courses in specialized areas and lends its services in areas of quality and standard control and the design and development of new products. *INDOTEC* carries out research and development activities for private industry, mainly in the area of food production.

Agricultural and Marine Science and Technology

In this area, the state sector has many organizations dedicated to research and development, and most of the country's researchers work for these bodies (61 out of the 94 previously mentioned as being the number of researchers, with titles and recognized grades working in the agriculture and livestock sector in 1976). Out of the 94 researchers, 18 work in the field of general agriculture; 2 in agricultural engineering; 8 in phytography; 14 in plant protection; 11 in soil research; 5 in plant nutrition; 1 in ecology; 1 in plant physiology; 1 in food technology; 1 in statistics and biometry and 32 in zootechnics. (This was the analysis in 1976 and is the most recent data.)

Centro de Desarrollo Agropecuario (CENDA) (Agricultural and Livestock Development Centre)

Centro Nacional de Investigaciones Agropecuarias (CNIA) (National Centre for Agricultural and Livestock Research)

Division de Ganadería y Boyada (Livestock Division)

University institutes carrying out research into agriculture and livestock are:

Centro de Investigaciones de Biología Marina (CIBIMA) (Marine Biology Research Centre)
Affiliated to the *Universidad Autónoma de Santo Domingo* (Autonomous University of Santo Domingo).

Finca Experimental de Engombe (Engombe Experimental Farm)
Affiliated to the Universidad Autónoma de Santo Domingo.

Centro de Producción Ganadera (Livestock Production Centre)
Dependant of the *Universidad Nacional 'Pedro Henríquez Ureña'* (Pedro Henríquez Ureña National University)

Instituto Azucarero Dominicano (Dominican Sugar Institute)

Medical Science and Technology

Instituto de Estudios Bio-médicos (Biomedical Research Institute)
Affiliated to the Universidad Nacional 'Pedro Henríquez Ureña'.

Asociación Médica Dominicana (Dominican Medical Association)

Asociación Médica de Santiago (Santiago Medical Association)
A major part of research in this field is developed in specialized hospital units.

Military Scientific and Technological Research

Instituto Cartográfico Militar de las Fuerzas Armadas (Institute of Military Cartography of the Armed Forces)

Professional Associations and Societies in the General Field

Among the seven most important organizations, the following three merit special mention:

Academia Dominicana de la Lengua (Dominican Language Academy)

Academia Dominicana de la Historia (Dominican History Academy)

Instituto de Cultura Dominicano (Dominican Cultural Institute)

Information Services in Science and Technology

In the Dominican Republic there are about twenty organizations whose function is documentation and information in science and technology. These organizations include museums and libraries; among them are:

Biblioteca de la Universidad Autónoma de Santo Domingo (Library of the Autonomous University of Santo Domingo)
Founded in 1458, it has a total of 104 441 volumes, also historical archives and maps.

Biblioteca Nacional (National Library)
Founded in 1971, it has 153 955 volumes.

Servicio de Documentación y Biblioteca (UNESCO) (Documentation and Library Service)

Biblioteca de la Sociedad Amantes de la Luz (Amantes de la Luz Society Library)
Situated in the town of Santiago de los Caballeros, it was founded in 1874 and has a total of 18 000 volumes.

International Cooperation in Science and Technology

Under the auspices of the Organization of American States, the Dominican Republic has established various agreements for the development of science and technology in the Caribbean region. *El Comité de Desarrollo y Cooperación del Caribe* (Caribbean Committee for Development and Cooperation) is one organization in which the Dominican Republic is an active member.

Ecuador

Geographical, Demographical, Political and Economic Features

The Republic of Ecuador shares its border to the north and north-west with Colombia, having 586 kilometres of frontier; to the east and south with Peru; and to the west lies the Pacific Ocean. The Galapagos Islands, lying 1120 kilometres from the coast, come under Ecuadorean jurisdiction (there are seventeen large islands and over 100 smaller). These islands have great scientific importance because of their unusual fauna.

In some areas of its border with Peru, Ecuador has reclaimed part of its territory which that country once held. This has led to continuous friction in the south of the country and for this reason it is difficult to determine the exact area of the country. Some estimate the area to be 268 178 square kilometres, excluding the disputed area that Ecuador reclaimed from Peru, which amounts to 190 807 square kilometres.

Ecuador can be divided into four main natural regions. Firstly the coastal region, which is an agricultural zone, lying between the Pacific Ocean and the spurs of the Andes. Secondly the Inter-Andean or Sierra region, which has at its eastern and western limits the rolling slopes of the Andes which forms the third region. Lastly come the Galapagos Islands.

The population of Ecuador, according to estimates made in 1980, is 8 372 193 with a projection for 1985 of 9 428 000. The density per square kilometre, according to the same figure, is 30.9. Forty-five per cent of the population is urban and 49.1 per cent reside in the ten mountainous provinces, while 48.9 per cent live in the low coastal region. The 'mestizo' is the predominant race, together with white, Indian and negro, but Ecuador has also some pure indigenous groups such as the 'Otavalos of Imbabura' and the 'Colorados' (Coloureds), of which only some 300 remain, and the 'Jibaros' of the east.

Government

In accordance with the Constitution, approved by referendum in 1978, Ecuador is a unitarian and democratic republic. The President and Vice-President are elected for a period of five years by secret and direct vote. The President of the Republic is advised by a Cabinet of fourteen Ministers. The legislative powers lie in the National Chamber of Representatives, a unicameral congress of twelve deputies elected by national vote, and fifty-nine elected from each province by proportional representation. The judicial power is represented by the Supreme Court, which has its seat in Quito, as well as by eight High Courts and the Courts of Justice of the provinces, districts and parishes.

From the point of view of administration, the country is divided into twenty provinces, and these, in turn, are divided into districts (123) and parishes (220 urban and 725 rural). The governors of the provinces are nominated by the President of the Republic. Voting is obligatory for all citizens over eighteen years of age.

Natural Resources

AGRICULTURE

Ecuador can be divided into two agricultural zones – the coastal zone, where the main products are bananas, cacao, coffee and cane sugar, and the Altiplano, where the main crops are cereals, potatoes and vegetables, all of which can easily adapt to the low temperatures of this region.

With the exception of the two agricultural zones and some arid areas along the coastal region, the country is largely covered with forest. The principal wood product for export is balsa wood. Wood production for 1977 reached 852 000 cubic metres of cut wood.

In 1980 exports of bananas represented 46 per cent of total exports while cacao stood at 12.5 per cent and coffee at 20 per cent.

FISH

The main fish product for export is shrimps. The total fish production for 1980 reached 475 000 tons.

LIVESTOCK

According to estimates made in 1976, livestock figures are as follows: 2.7 million head of cattle; 2.2 million of sheep and 2.7 million pigs.

MINERALS

Mineral production has been reduced somewhat and only a few firms still carry out mineral exploitation. There are mines of copper, gold, silver (2 tons were mined in 1972), lead and zinc.

The country is now turning to the exploitation of its oil reserves, which in 1978 were valued at $507.3 million. From specialized sources it has been confirmed that there are good prospects for the exploitation of new drillings in the area adjacent to the coast. Out of fifty-three exploratory drillings carried out in 1970, only six resulted in failure. Production in 1980 reached 1006 metric tonnes of petrol, 432 of kerosene, 788 of light oils and 2194 of heavy oil.

Education

Primary education is obligatory and free. State schools as well as private schools function under some supervision from the state. Figures for 1976–77 show a total of 1 318 475 primary students, 431 226 secondary, and 170 173 university students.

International Relations

The Republic of Ecuador is a member of the United Nations Organization and the Organization of American States.

Organization of Science and Technology

In 1980 an inventory was drafted of organizations that implemented activities in research and development during 1979. The results showed 533 organizations, of which 268 were public bodies, 48 private, 190 establishments of higher education, 25 private firms and 2 public firms. The inventory showed that 50 per cent of the organizations dedicated to research and development in science and technology were governmental institutions and 36 per cent were of higher education, indicating that institutions of the public sector are predominant in research and development. The importance of establishing a system of development in science and technology is relatively recent and to these ends the formation of the *Sección de Ciencia y Tecnología* (Section of Science and Technology) in 1973 as a dependant of the *Junta Nacional de Planificación y Coordinación Económica* (National Board for Economic Planning and Coordination) produced results at political level that were incorporated into the *Plan Integral de Transformación y Desarrollo 1973-1977* (Integral Plan for Transformation and Development 1973–1977). The section was later replaced by a division but no great differences were experienced in implementing policies. It was in 1978, as a result of the *Quinta Renunión de la Conferencia Permanente de Organismos Nacionales de Política Científica y Tecnológica en America Latina* (Fifth Meeting of the Permanent Conference of Scientific and Technological Politics within Latin American Organizations) which took place in Quito, that the criteria for founding a national body at policy-making level was established. In August 1978 the *Ley del Sistema Nacional de Ciencia y Tecnología* (Regulation of the National System for Science and Technology) was promulgated, and with this regulation, national organizations were integrated into a system. These organizations now form the *Consejo Nacional de Ciencia y Tecnología (CONACYT)* (National Council for Science and Technology), *Dirección Ejecutiva* (Executive Directorate) and the *Comisiones Sectoriales de Ciencia y Tecnología* (Sectorial Commissions in Science and Technology).

CONACYT is appointed to the *Consejo Nacional de Desarrollo (CONADE)* (National Development Council) and acts as an advisory body to the council. The main functions of *CONACYT* are advising, planning, programming and coordinating in matters of the development of science and technology and, additionally, managing the political machinery in science and technology in order to monitor progress, and to ensure that plans and programmes are fulfilled. The council supervises the growth of new organizations in science and technology; finances research; controls imports of foreign technology by administering licences and patents; establishes technological standards and quality control, etc.

CONACYT has two affiliated organizations which act as advisory bodies - *Comisión de Desarrollo Científico* (Commission for Scientific Development) and *Comisión de Desarrollo Tecnológico* (Commission of Technological Development). The first Commission comprises representatives of the basic sciences and the second representatives of the applied sciences.

Policies and Financing in Science and Technology

The objectives of scientific and technological development for 1980–84 concentrate on six major aspects: human resources; constitution; finance; technological management, international cooperation and conferences. The main aims in terms of human resources are to increase and diversify the number of researchers, to reach a high level of specialization and to promote the interchange of researchers and technology.

CONACYT is the main body responsible for implementing the policies and objectives in the development plan 1980–84 along with the other governmental institutions in their particular fields - for example, the Ministry of Industry, Commerce and Integration applies technology policies in matters of industry and supervises exchange of foreign technology in this area.

Human Resources

The government policy of increasing the number of highly qualified individuals in the national system of science and technology follows several aims: to create more opportunities to follow post-graduate studies; to enable researchers and professionals in specialized areas to complement their studies in foreign centres; to give financial aid to courses aimed at auxiliary personnel in science and technology, etc. Table 6.1 shows the most recent data on human resources available.

Table 6.1 Human resources in science and technology according to field of specialization

Years	Population (millions)	Scientists and engineers in science and technology					
		Public institutes	Higher education	Private institutes	Private firms	Others	Total
1970	6.0	454	292	—	35	4	785
1979*	8.1	1068	599	330	47	5	2049

* Figures are preliminary estimates of the *Inventario del Potencial Científicó y Tecnológico de 1979* (Inventory of Scientific and Technological Potential in 1979).

Source: Inventario del Potencial Científico y Tecnológico de 1979; Situación de las Actividades Científicas y Tecnológicas 1970 (Situation of Activities in Science and Technology, 1970).

Financial Resources

Public investment into research and development is indispensable to the execution of projects, most of which are priority areas as defined in the national plan. Thus it is important to stimulate public and national investment.

The goal for national investment is 0.4 per cent of GNP for 1984. This is seen as a significant increase and will be mainly orientated through *CONACYT* towards the national system.

Special attention is also being given to establishing a national fund for science and technology and to developing an adequate administration system.

Table 6.2 shows the financial resources in millions of sucres available to science and technology in 1979:

Table 6.2

Year	Public funds	Private funds	Foreign funds	Own funds	Total
1979*	627.1	49.2	142.1	90.4	908.8
Per cent	69.00	5.41	15.64	9.95	100.0

*Figures are preliminary estimates of the *Inventario del Potencial Científico y Tecnológico de 1979* (Inventory of Scientific and Technological Potential in 1979).
Source: Inventario del Potencial Científico y Tecnológico de 1979 by the *Dirección de Planificación del CONACYT.*

In 1979 the system of science and technology channelled into research and development a total of 908.8 million sucres, the most important source being public funds, contributing 69 per cent of the total, and 15.6 per cent coming from foreign sources.

The national development plan has also established sectorial programmes to complement the national plan and policies, such as:

Industry

Research is being done into the present state of quality control for the different branches of industry and on the impact that this control has on the improvement of technology in private and public companies.

With regard to the transfer of technology, emphasis is put on education and training of professionals and researchers.

Rural Development

Special aid is given to projects on replenishable natural resources, soil classification, climate and water systems, present and potential use of soil, etc., as well as research to determine land allocation at provincial and national level.

Foreign Trade

Decisions 24, 84 and 85 of the Cartagena Agreement, relating to technological policy and the transfer of technology, are being executed, studying the technical, administrative and legal implications.

Natural Resources

Special are projects orientated towards conservation and rational use of the country's natural resources, aiming to establish ecological areas of delimitation.

Energy

The main areas of research are into non-conventional energy and the potentialities of new combustibles. The introduction of the peaceful use of atomic energy is being studied jointly by the *Comisión Ecuatoriana de Energía Atómica* (Ecuadorean Commission for Atomic Energy), the universities and polytechnics, state institutes and health services, and inter-institutional work is being intensified in this field.

Housing

Socio-economic research highlights the needs and urgency of adequate housing, and projects are under way to develop technology based on traditional methods or new materials and designs using local materials to substitute or reduce imports.

Health

Research is being made into biological and environmental causes of the most common diseases, the development of new food sources, and nutrition, among other studies.

Science and Technology at Government Level

Dirección General de Geología y Minas (General Directorate of Geology and Mines)
This is a dependant of the Ministry of Natural Resources and Energy. It supervises the execution of laws relating to the mine works and establishes standards for the mining industry.

Dirección General de Hidrocarburos (General Directorate of Hydrocarbons)
This board oversees the implementation of laws relative to the exploration and development of the oil industry and sets out standards for this industry.

Instituto Nacional de Estadística y Censos (National Institute of Statistics and Census)

This institute publishes annually vital statistics on transport, hospital services, education, etc.

Instituto Oceanográfico de la Armada (Naval Oceanographic Institute)
The institute carries out oceanographic and hydrographic studies. It has a staff of 200 and a library of 2000 volumes.

Science and Technology on an Academic Level

Universidad Central del Ecuador (Central University of Ecuador)
Founded in 1769 from the *Seminario de San Luis* (Seminary of San Luis), established in 1594, and the *Universidad de San Gregorio Magno* which, in turn, was founded in 1622, and from the *Universidad Dominicana de Santo Tomás de Aquino*. The whole university was reorganized between 1822 and 1926. It is state controlled and has 60 000 students and 2500 teachers. *Anales* (Annals) is published by the university and there are faculties in human and medical sciences, chemistry, odontology, engineering, agricultural engineering and psychology.

Pontificia Universidad Católica del Ecuador (Pontifical Catholic University of Ecuador)
This is a private university founded in 1946. Its academic year runs between October and July. There are 14 924 students and a teaching staff of 573, and its faculties include the arts, engineering, human sciences and nursing.

Universidad Católica de Cuenca (Catholic University of Cuenca)
Founded in 1970 under private control, it has an academic year divided into two periods between October and July. Its library has 5000 volumes to serve its 2910 students and 300 teachers. Its faculties include the arts, medicine and health sciences, chemical engineering, and agricultural engineering. There is also a Centre of Communications and Television.

Universidad de Cuenca (Cuenca University)
Established in 1868 and currently has 14 212 students and 470 lecturers. The university publishes *Revista de la Facultad de Ciencias Médicas* (Faculty of Medical Sciences Review) among others. Its faculties are medicine, engineering, chemistry, dentistry and the arts.

Universidad Estatal de Guayaquil (State University of Guayaquil)
Founded in 1867, it has a staff of 400 teachers and 4500 students and faculties in agricultural and veterinary sciences, chemical engineering, dentistry, natural sciences, mathematics and medicine. There are also faculties in the art subjects.

Universidad Nacional de Loja (National University of Loja)
Established in 1869 as *Junta Universitaria*, it acquired the status of a university in 1943. Its academic year runs between October and July with 7536 students and 245 teachers. *Revista Universitaria* (University Review) is published among others from the various faculties. Faculties include the arts, agronomy and veterinary sciences, and medicine.

Universidad Católica de Santiago de Guayaquil (Catholic University of Santiago de Guayaquil)
This is a private university founded in 1962 and its academic year runs between April and December. There is a staff of 500 lecturers and 4290 students, and its faculties are in the arts, engineering, architecture, medicine, and technical education, with a bias towards development.

Universidad Técnica de Babahoyo (Technical University of Babahoyo)
Established in 1971 and at present has 1168 students and 80 teachers in its faculties of agronomy and education.

Universidad Técnica Particular de Loja (Private Technical University of Loja)
A private university, founded in 1971, with its academic year running from October to February and April to August. Its students total 4232 with a teaching staff of 120. *Revista Universitaria* (University Review) is published, and its faculties are civil engineering, agricultural engineering, economics and education. The following institutes are also affiliated to the university: human sciences, sciences, and computer sciences.

Universidad Técnica de Machala (Technical University of Machala)
Established in 1969, it functions under state control between the months of March and January and has a student population of 2743 and 182 teachers. Its library houses 3000 volumes and a review is published by the faculty of agronomy and veterinary sciences. There are faculties in agronomy and veterinary sciences, sciences, civil engineering and sociology.

Universidad Técnica de Manabí (Technical University of Manabí)
This is a state university whose academic year runs between May and January. With a total of 300 students and sixty lecturers, its faculties include mathematics, physics and chemistry, agriculture and veterinary medicine.

Universidad Técnica 'Luis Vargas Torres' (Luis Vargas Torres Technical University)
Founded in 1970, it has 757 students and 144 teachers. Its faculties are: agrarian sciences, education, business studies and sociology.

Universidad Laica 'Vicente Rocafuerte' de Guayaquil (Vicente Rocafuerte Lay University of Guayaquil)
Established in 1847, it has its own autonomy and attained university status in 1966. Its faculties are architecture, economics, civil engineering and education.

Escuela Politécnica Nacional (National Polytechnic)
Founded in 1869 and functions privately with an academic year between October and July. It employs a staff of 391 teachers for its 18 605 students. Its faculties are as follows: civil engineering; electrical engineering; geology, mining and oil; mechanical engineering and chemical engineering. Institutes of technical engineering and nuclear research are also affiliated to the university.

Escuela Superior Politécnica de Chimborazo (Higher Polytechnic of Chimborazo)
Founded in 1969 and has 1615 students and eighty-two lecturers. Its library houses 7000 volumes for its faculties in mechanical and agricultural engineering, nutrition and dietetics, and there is an institute of languages.

Escuela Superior Politécnica del Litoral (Higher Polytechnic of the Coast)
Founded in 1958 and is state run. Its students total 2500 with eighty-four full-time and fifty-three part-time teachers. It publishes *Revistas Polipesca* and *Revista Technológica* (Fish Review and Technological Review). Its faculties are in electrical and electronic engineering; oil and mine engineering and mechanical engineering, and its institutes comprise physics, chemistry, mathematics and general studies.

Colegio Nacional de Agricultura 'Luis A. Martínez' (Luis A. Martínez National College of Agriculture)
There are 500 students and the review *Germinación* (Germination) is published.

Agricultural and Marine Science and Technology

Instituto Interamericano de Agricultura Experimental (Inter-American Experimental Agriculture Institute)

Instituto Nacional de Investigaciones Agropecuarias (National Institute for Arable and Livestock Research)
This institute publishes scientific articles, bulletins and reports annually.

Instituto de Investigaciones Veterinarias del Litoral (Coastal Institute for Veterinary Research)

Instituto Nacional de Pesca (National Fisheries Institute)
This institute carries out research into fish resources, has a library of some 20 000 volumes and publishes scientific and technical bulletins.

Estación de Investigación Charles Darwin (Charles Darwin Research Station)
Founded in 1964 under the auspices of the Ecuadorean Government, UNESCO and the Charles Darwin Foundation. Research is carried out on the flora and fauna of the Galapagos.

Medical Science and Technology

A major part of research in this field is carried out in the specialist units of medical centres and hospitals.

Instituto Nacional de Nutrición (National Institute of Nutrition)
This institute runs teaching courses for specialized staff.

Instituto Nacional de Higiene 'Izquieta Pérez' (Izquieta Pérez National Institute of Hygiene)

Publishes *Revista Ecuatoriana de Higiene y Medicina Tropical* (Ecuadorean Review of Hygiene and Tropical Medicine).

Nuclear Science and Technology

Instituto de Ciencias Nucleares (Nuclear Sciences Institute)
This institute is a dependant of the *Escuela Politécnica Nacional* (National Polytechnic) and research is carried out mainly in the use of radio-isotopes in chemistry, agriculture and medicine and the control of radiation.

Comisión Ecuatoriana de Energía Atómica (Ecuadorean Commission for Atomic Energy)

Scientific and Technical Research in the Field of Meteorology and Astrophysics

Instituto Nacional de Meteorología e Hidrografía (National Institute of Meteorology and Hydrography)
Publishes *Boletín Climatológico* (Climatology Bulletin) and *Anuarios Meteorológico e Hidrológico* (Meteorological and Hydrologic Annual Report).

Observatorio Astronómico de Quito (Astronomical Observatory of Quito)
Research is carried out in astronomy, seismology and meteorology.

General Associations and Professional Societies in the Field of Science and Technology

There are fourteen organizations dedicated to the promotion of research in the field of science and technology. Among them are:

Instituto Ecuatoriano de Ciencias Naturales (Ecuadorean Institute of Natural Sciences)

Academia Ecuatoriana de Medicina (Ecuadorean Academy of Medicine)

Instituto Latinoamericano de Investigaciones Sociales (ILDIS) (Latin American Institute for Social Research)
The institute is affiliated to the Friedrich Ebert Foundation and carries out research in economics, sociology, political science and education.

Information Services in Science and Technology

There are in Ecuador twenty-eight documentation centres among museums and libraries:

Biblioteca Nacional del Ecuador (Ecuador National Library)
Founded in 1792 and has 55 000 volumes of which 7000 date from the sixteenth century.

Biblioteca de la Universidad Central del Ecuador (Central University of
Ecuador Library)
Founded in 1826 and has some 170 000 volumes.

Recent information shows that the 533 organizations of scientific research
and development have information resources available to them in one form
or another. Ninety-nine have specialized scientific and technological information and documentation services and eighteen have their own computers. There is still a long way to go in establishing an adequate information system in Ecuador.

CONACYT is structuring a *Sistema Nacional de Información Científica y
Tecnológica* (National Information System in Science and Technology).

Action is presently under way to develop the *Subsistemas Agropecuario e
Industrial* (Agricultural and Livestock and Industrial Sub-Systems). These
sectors already have an adequate information system on which to develop
a more extensive network.

Among the many units dedicated solely to documentation and information
is the *Biblioteca Nacional del Ecuador* (Ecuador National Library).

Summary of Data on Scientific and Technological Potential

Three research studies have been made on scientific and technological
potential, the first using information of 1970 for all sectors based on the
methodology of OAS. The second, for 1976 referred only to the agricultural
and livestock, fish and forestry sector, using internal methodology, and the
third, using information of 1979, referred to all sectors, using UNESCO
methodology.

The first two studies were carried out almost totally manually but the
third study was in collaboration with the Computer Centre of the National
Institute of Statistics and Census, *INEC*, and with *CONACYT*, who did
a lot of the groundwork in programming and statistics based upon direct
interviews in the various organizations of science and technology.

INEC is the institute that will in future be responsible for compiling
data on national scientific and technological potential together with
CONACYT.

Collating and analysing statistics is *INEC*'s main function.

International Cooperation in Science and Technology

The policy for international cooperation in science and technology is
structured and implemented through the participation of *CONACYT*
with the *Comité Nacional de Cooperación Técnica y Asistencia Económica*
(National Committee of Technical Cooperation and Economic Assistance).
The committee is the inter-institutional organization which programmes
and approves projects in which foreign countries and international organiza-

tions are involved. The committee considers the national repercussions and implications that research results could have in order to establish the most efficient ways of executing the projects.

Cooperation with International Organizations

Under the surveillance and guidance of the National Committee, Ecuador is in the process of carrying out projects in the field of science and technology in which support and cooperation of international organizations is involved.

The *Cartagena Pact* plays a major role in international cooperation and has strengthened scientific and technological capacity considerably through Decision 24, which monitors and controls the interchange of foreign personnel and technology. Other projects include *Proyectos Andinos de Desarrollo Tecnológico* (Andean Projects of Technological Development); the *Sistema Andino de Información Tecnológica* (Andean System of Technological Information); and debates and discussions are presently under way to form the *Consejo Andino de Ciencia y Tecnología* (Andean Council of Science and Technology). Research is also being carried out to determine technical standards, training of technical staff, etc.

With financial aid from UNDP a project was completed at the end of 1981, *Apoyo a la Estructuración de una Política y una Programa Científico y Tecnológico Nacional* (Aid to Structure National Policy and Programmes in Science and Technology) in collaboration with *CONACYT*. This project is to continue into 1982 under the title of *Desarrollo Científico y Tecnológico* (Scientific and Technological Development).

Organization of American States

Research is currently under way in collaboration with OAS in the form of ordinary projects financed by *Fondo Especial Multilateral de Consejo Interamericano para la Educación, la Ciencia y la Cultivo (FEMCIECC)* (Special Multilateral Fund of the Inter-American Council for Education, Science and Culture), and special projects financed by *Cuenta Mar del Plata* (Mar del Plata Fund).

Sistema Económico Latinamericano (SECLA) (Economic System of Latin America)
Is actively involved in establishing a *Red de Información Tecnológica Latinoamericana (RITLA)* (Latin American Network of Technological Information).

Agencia para el Desarrollo Internacional (AID) (International Development Agency)
Participates in projects such as: *Sistema de Transferencia de Tecnología Rural (STTR)* (System of Transference of Rural Technology), and is presently preparing a research study to identify priority areas in research and development and to develop activities accordingly. This should lead, ultimately, to the founding of a *Fondo Nacional de Ciencia y Tecnología* (National Fund for Science and Technology).

Agreements and Specific Projects

Negotiations are presently under way to establish an agreement between *Conselho Nacional de Desenvolvimento Científico y Tecnológico del Brasil (CNPq)* (National Council for Scientific and Technological Development of Brazil) and *CONACYT* of Ecuador, to develop activities beneficial to both countries.

A project has been initiated to strengthen the *Servicio de Información Técnica* (Technical Information System) in collaboration with *CONACYT* and the *Centro de Desarrollo Industrial del Ecuador (CENDES)* (Industrial Development Centre) and with the eventual participation of the Canadian Government. Similarly the Mexican Government has agreed to cooperate in the field of information and documentation, and in the training of personnel in public administration.

El Salvador

Geographical, Demographical, Political and Economic Features

The Republic of El Salvador has its borders to the north and east with Honduras, and to the south with the Pacific Ocean, having a coastline of 320 kilometres from the border with Honduras to the Gulf of Fonseca. To the west and the north El Salvador shares its frontier with Guatemala.

According to estimates made in 1979, El Salvador has a population of 4 436 000, with a projection for 1985 of 5 226 000. According to the same figures the population density in 1979 was 211 per square kilometre, which is one of the highest in Latin America. Of the population 40.2 per cent live in urban areas and the composition is 84 per cent mestizos (Ladinos) (mixed blood); 10 per cent white and 5.6 per cent Indians.

In accordance with the Constitution of 1962 the executive power is in the hands of the President of the Republic, who is elected by popular vote for a period of five years which cannot be prolonged. The president is advised by the council of Ministers and the General Secretariat of the Presidency. The legislative power is exercised by the Legislative Assembly, made up of fifty-two deputies elected for a period of two years (which is presently dissolved). The judicial power lies with the High Court, composed of thirteen magistrates nominated by the Assembly, and other regional courts. Administratively the country is divided into fourteen departments under the charge of governors designated by the President of the Republic, thirty-nine districts and 261 municipalities, which have municipal councils elected by universal suffrage, and 2547 hamlets.

Natural Resources

AGRICULTURE

The principal agricultural products of the country are: olives, 'agave', 'aguacates', cotton (fibre and seed) (of which 69 000 tons were produced in 1980), sugar cane, onions, citrus fruits, rice, bananas, coffee, beans and vegetables. Of the national territory 32.5 per cent is dedicated to crop cultivation while 30.2 per cent is for pasture. Coffee production covers approximately 308 000 acres. Almost 50 per cent of the population is involved in agricultural activities.

LIVESTOCK

In 1976 estimates showed a total of 1.1 million head of cattle, 4000 pigs, 400 000 sheep and 11 000 goats.

FORESTRY

The main species include cedar, mahogany, balsam and walnut.

MINERALS

The country possesses deposits of sulphur, copper, zinc, quartz, iron, lignite, mercury, gold, silver, platinum, lead, salt and gypsum. However, mining has run into neglect in recent years.

INDUSTRY

The main industrial products comprise: sugar; beer; cigarettes; electrical products; by-products of oil; fish meal; television sets; wine; fertilizer; vegetable oils; electric light bulbs; instant coffee; cement; sisal cord and sacks; car assembly; metallic furniture; pasta; radios and products pertaining to animals (leather; dairy goods, etc.).

Education

Schooling is free and obligatory. Basic education in El Salvador lasts nine years with three three-year cycles leading to two types of *bachillerato* (bachelorship), one varied (agriculture, industrial, pedagogic, commercial, etc.) while the other is academic. Figures for 1978 show that 63 per cent of the population are literate.

International Relations

The Republic of El Salvador is a member of the United Nations Organization and the Organization of American States.

Organization of Science and Technology

The organizations that carry out activities in the field of scientific and technological research are not integrated into a specific network, but the responsibility is distributed between the ministries and the universities. As can be seen in Figure 8 some areas of research, such as agriculture, have reached a higher level of development than the rest.

Policies and Financing in Science and Technology

Guidelines for scientific and technological policies have been included in the National Development Plan and the Ministry for Planning and Co-ordination of Economic and Social Development has the responsibility for implementing the policies.

Among the basic policies are:

adaptation, selection and control of imported technology,
promoting national technology,
developing basic and applied scientific activities.

The Development Plan outlines priority areas for scientific and technological development:

agriculture, livestock and agro-industry,
natural resources,
health and nutrition,
basic infrastructure.

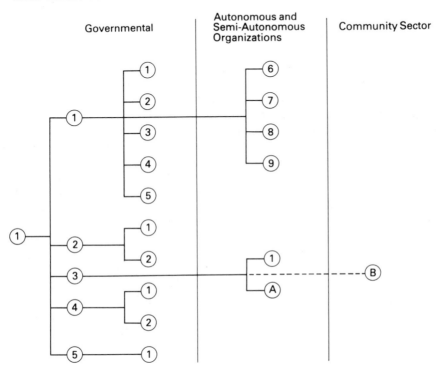

Fig. 8 Organization of science and technology

1	Presidencia (Presidency)
1.1	Ministerio de Agricultura (Ministry of Agriculture)
1.1.1	Dirección General de Economía y Planificación Agropecuaria (Head Office of Agriculture and Livestock Economics and Planning)
1.1.2	Dirección General de Recursos Naturales Renovables (Head Office for Replenishable Natural Resources)
1.1.3	Centro Nacional de Tecnología Agropecuaria (National Centre of Agriculture and Livestock Technology)
1.1.4	Dirección General de Riego y Drenaje (Head Office for Irrigation and Drainage)
1.1.5	Dirección General de Sanidad y Reproducción Animal (Head Office of Animal Health and Reproduction)
1.1.6	Banco de Fomento Agropecuario (Development Bank for Agriculture and Livestock)
1.1.7	Instituto Salvadoreño de Investigaciones de Café (Salvadorean Institute for Coffee Research)
1.1.8	Instituto Regulador de Abastecimientos (Institute for the Regulating of Supplies)

Fig. 8 - *continued*

1.1.9	Instituto de Colonización Rural (Institute for Rural Colonization)
1.2	Ministerio de Obras Públicas (Ministry of Public Works)
1.2.1	Centro de Investigaciones Geotécnicas (Centre for Geotechnical Research)
1.2.2	Instituto Geográfico Nacional (National Geographical Institute)
1.3	Ministerio de Educación (Ministry of Education)
1.3.1	Instituto Centroamericano de Telecomunicaciones (Central American Institute of Telecommunications)
1.3.A	Universidad Nacional de El Salvador (National University of El Salvador)
1.3.B	Universidades Privadas (Private Universities)
1.4	Ministerio de Economía (Ministry of the Economy)
1.4.1	Instituto Salvadoreño de Fomento Industrial (Salvadorean Institute of Industrial Development)
1.4.2	Centro Nacional de Productividad (National Centre of Productivity)
1.5	Ministerio de Salud Publica (Ministry of Public Health)
1.5.1	Hospitales (The Hospitals)

Special emphasis has also been given to establishing an efficient scientific and technological information system and the coordination of human and financial resources in science and technology.

Data on human resources in the field of research and development have not been evaluated, mainly because figures are available only for professional researchers and technicians. For example, in 1974, of a total of 533 professionals with recognized university qualifications, 29 per cent were in medicine; 26 per cent in engineering; 25 per cent in social sciences and law; 9 per cent in exact and natural sciences; 6 per cent in agricultural and livestock sciences, and 5 per cent in the humanities, education and others.

With regard to the 2370 technicians with average experience and qualifications, the distribution was as follows: 29 per cent in para-medicine; 25 per cent in engineering; 22 per cent in humanities, education and others; 20 per cent in agriculture and livestock; and 4 per cent in social sciences.

For the financial resources, the situation is similar. There is no breakdown of expenditure for specific research activities, however, there is some indication of financial resources for sectors as a whole:

Table 7   Estimates of funds for research and development in 1974

Total: 11.9 million colones	
Distribution:	*Per cent*
Agriculture and livestock	52.0
Exact and natural sciences	18.7
Social sciences	12.1
Construction	5.6
Water and electricity	6.1
Medicine	2.6
Industrial sector	2.9

It is estimated that for the same year 17 per cent of the funds was destined for experimental development; 65 per cent to applied research and 18 per cent to basic research.

Science and Technology at Government Level

Centro de Investigaciones Geotécnicas (Centre for Geotechnical Research) Affiliated to the Ministry of Public Works.

Instituto Geográfico Nacional (National Geographical Institute)

Scientific and Technological Research at an Academic Level

Universidad Nacional de El Salvador (National University of El Salvador) Founded in 1841, it has faculties in agronomy, engineering and architecture, medicine, economics, odontology, chemistry and pharmacy, sciences and humanities.

There is a private university with faculties in economics, engineering, the sciences of man and natural sciences.

Industrial Technology and Research

Instituto Salvadoreño de Fomento Industrial y el Centro Nacional de Productividad (Salvadorean Institute of Industrial Development and the National Centre for Productivity)
There are two bodies affiliated to the Ministry of the Economy which carry out research activities in this field.

Instituto Salvadoreño de Tecnología Industrial (Salvadorean Institute of Industrial Technology)
The main functions of this institute are industrial research and development and the coordination of these activities in state and private organizations and in education; the strengthening and extension of existing information systems and technological assistance in industry. The institute may also take over the activities and operations incorporated in the *Sistema Nacional de Metrología, Normalización y Control de Calidad* (National System of Metrology, Standards and Quality Control).

Agricultural and Marine Science and Technology

A major part of research in this field is executed by the *Instituto Interamericano de Ciencias Agrícolas* (Inter-American Institute of Agricultural Sciences) in Turrialba in Costa Rica; research in this field in El Salvador is somewhat restricted. However, the following organizations carry out a certain amount of research: *Centro Nacional de Tecnología Agropecuaria* (National Centre for Agricultural and Livestock Technology); *Dirección General de Recursos Naturales Renovables* (Institute of Renewable Natural Resources); *Dirección General de Sanidad y Reproducción Animal* (Institute of Animal Health and Reproduction); and the *Instituto Salvadoreño de Investigaciones del Café* (Salvadorean Institute of Coffee Research). All these are affiliated to the Ministry of Agriculture, and only the last one is a semi-autonomous organization. The ministry also has under its jurisdiction several agricultural experimental stations and units as well as the *Escuela Nacional Agrícola* (National School of Agriculture).

Medical Science and Technology

Most research in this field is carried out in the specialist institutes in hospitals. The *Liga Nacional contra Cancer* (National Alliance against Cancer) promotes research and development in this field.

Information Services in Science and Technology

Biblioteca Nacional (National Library)
Has about 100 000 volumes, with archives of Salvadorean publications.

Biblioteca de la Facultad de Ciencias Agronómicas (Library of the Agronomy Faculty) of the *Universidad Nacional*.

Centro de Información de las Naciones Unidas (United Nations Information Centre)

Oficina de Información y Documentación (Information and Documentation Office) of the *Centro Nacional de Tecnología Agropecuaria* (National Centre for Agricultural and Livestock Technology) in Santa Tecla.

French Guiana

French Guiana is one of the French Overseas Departments and is therefore an integral part of the French Republic, administered by a Prefect, with an elected general council and with elected representatives in the French National Assembly and in the Senate of the Republic in Paris. It is situated on the north-eastern coast of South America, with Surinam to the west and Brazil to the south and east. Its total area is 90 000 square kilometres and it has a population of approximately 66 600 (January 1981, estimate).

The economy of French Guiana is based on forestry and agriculture. Cassava, bananas, maize and other tropical crops are grown for local consumption. Sugar cane is the only cash crop of importance. In 1980, 7500 metric tons of sugar cane, 7650 metric tons of cassava, 1000 metric tons of bananas and 450 metric tons of rice were produced. In the same year the livestock count was: 8040 cattle, 420 sheep, 240 goats, 6500 pigs and 110 000 poultry. Fishing has been increasing in importance since 1965 and the total catch in 1980 was 4457 metric tons. Shrimps are the most important catch and these are mainly exported to the USA. There are extensive timber reserves and important mineral resources, particularly of gold, bauxite and tantalite, from which extractive industries are being developed.

With regard to education, there are sixty-one primary schools and nineteen secondary schools and technical colleges and schools. There is a teaching body of 1096 to serve a pupil population of about 13 000. There is a college of the *Centre Universitaire Antilles-Guyane*, situated in Cayenne, which conducts courses in law.

The international importance of French Guiana has increased significantly with the setting up of the *Centre Spatial Guyanais* (Guiana Space Centre) at Kourou. It is operated by the French Government agency, the *Centre National d'Etudes Spatiales (CNES)* (National Centre of Space Studies). The main objective of the Guiana Space Centre has been the development of Ariane, the European Space Agency's communication satellite launcher, and there was a successful test launching in December 1981.

The decision to build the launching site was taken in 1964. Work was begun in 1968 and the first rocket launched in that year. The site is particularly suitable as a launching base because of the consistent climate and the absence of the threat of hurricanes and earthquakes. The coastal strip on which it is situated faces north-east and provides a wide launching range of over 120 degrees. Its latitute is 5 degrees north of the equator, which increases the orbiting possibilities as both polar and equatorial orbits can be obtained. This low latitude offers optimum conditions for the launching of synchronous telecommunications.

All activities of the centre, including flights by aeroplanes equipped with telemetry emitters, radar responders and telecommand receivers; the launching of Eridan, Dauphin and Véronique probe rockets; and the launching of Super-Arcas rockets in the meteorological programme Exametnet, which has taken place regularly since 1972, are all undertaken with a view to realizing the Ariane programme. The *CNES* also maintains at Kourou a telemetry and telecommand station as part of its international network.

Other scientific establishments in the country include the *Office de la Recherche Scientifique et Technique Outre-Mer – Centre ORSTOM de Cayenne* (Office for Overseas Scientific and Technical Research (ORSTOM) Cayenne Centre), which conducts research in pedology, hydrology, botany and plant biology, medical entomology, agricultural entomology, and phytopharmacology; a Pasteur Institute, which conducts medical and biological research; a station of the *Institut de Recherches Agronomiques Tropicales et des Cultures Vivrières* (Research Institute for Tropical Agronomy and Cultivation of Foodstuffs), which researches into general agronomy and the cultivation of food crops, tobacco, sugar cane, forages, spices, etc.; the *Institut de Recherches sur les Fruits et Agrumes* (Fruit and Citrus Research Institute), affiliated to the institute of the same name in Paris; and the *Centre Technique Forestier Tropical* (Tropical Forestry Technical Centre), which is affiliated to the centre of the same name in Nogent-sur-Marne in France.

Guatemala

Geographical, Demographical, Political and Economic Features

The Republic of Guatemala formed the General Spanish Captaincy for all of Central America and gained its independence from Spain in 1821. The capital, Ciudad de Guatemala, is the most northern city of all the Central American republics, with the exception of Belize. The borders to the north and west are shared with Mexico; to the east with the Caribbean Sea and Belize; to the south-east with Honduras and El Salvador; and to the south with the Pacific Ocean.

The total area is 108 889 square kilometres. According to estimates made in 1980, the population was 7 006 020 with a projection for 1985 of 8 041 965. Population density per square kilometre in 1980 was 64.3.

The geographical position in the tropics with mountainous areas give Guatemala a varied climate, ranging from freezing areas to extreme heat. Average temperatures on the altiplano are 16 degrees to 20 degrees centigrade, and on the coast from 25 degrees to 30 degrees centigrade. The dry summer season falls between November and April and the rainy season is between May and October.

In accordance with the constitution which came into force in May 1966, the president and vice-president are elected for a period of four years without re-election. The executive power is in the hands of the president and a cabinet made up of ten ministers. The legislative power rests with a congress of two chambers of fifty-five members elected for four years. The judicial power is led by a High Court of Justice appointed by the Congress; it carries out its duties for a period of four years. Guatemala is divided into 22 departments where governors hold office, and 236 municipalities. The country has 28 cities, 29 towns, 266 villages, 2486 hamlets and 4400 settlements.

Natural Resources

AGRICULTURE

The main product for export is coffee, of which there are about 1200 plantations providing work for some 426 000 peasants. The two principal customers for Guatemalan coffee are the USA and West Germany.

Besides coffee the following are also produced:

In thousand tons

Cotton (fibre)	156	Beans	80
Cotton (seeds)	260	Maize	800
Bananas	566	Sorghum	60
Cacao	3	Tobacco	10
Coffee	151	Tomatoes	86

(1980 figures).

FORESTRY

Guatemala has 17 784 000 acres of forest land and the main species of wood for production is mahogany. In 1977 production of cut wood was 396 000 cubic metres.

LIVESTOCK

Figures for 1980 show 76 000 head of goats; 600 000 sheep and 14 096 poultry.

FISHERIES

The exploitation of fish resources has not been developed to its full potential and figures for 1977 reveal that exports were valued at 4.9 million quetzals.

Economic Information

Monetary unit: Quetzal (1 per US$ in December 1980).
 GNP per inhabitant (1978) US$910.
 Annual rate of increase of GNP per inhabitant (1960 to 1978) 2.9 per cent.
 Distribution of GNP (1979): agriculture 26 per cent; industry 20 per cent and services 54 per cent.
 Economically active population (1978) 54 per cent.

Education

According to figures in 1975, illiteracy was 47 per cent. Primary education is obligatory and free between the ages of seven and fourteen (three years in rural schools). There are five universities with 50 000 students.

International Relations

Guatemala is a member of the United Nations and the Organization of American States.

Organization of Science and Technology

According to the results of a study made by the *Departamento de Ciencia y Tecnología* (Department of Science and Technology) affiliated to the *Secretaría General del Consejo Nacional de Planificación Económica* (General

Secretariat of the National Council of Economic Planning), the national system of science and technology comprises 153 units, of which 140 are national institutions and thirteen come under the sub-region of Central America and the Caribbean.

Of the 140 national institutions, 102 belong to the public sector and 51 per cent of these are affiliated to central bodies (ministries) and 49 per cent to decentralized bodies. Of the remaining thirty-eight units, thirty-two belong to centres of higher education, mainly the state universities, and the other six units to private non-profit-making organizations.

Of the total units, 57.5 per cent are dedicated to complementary research; 23.3 per cent to technological research and 19.2 per cent to scientific research.

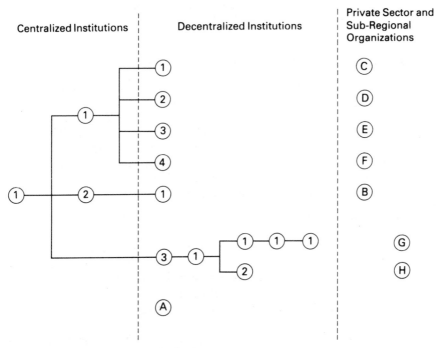

Fig. 9 Organization of science and technology

1	Presidencia (Presidency)
1.1	Ministerio de Agricultura (Ministry of Agriculture)
1.1.1	Instituto de Ciencia y Tecnología Agrícola (ICTA) (Institute of Agricultural Science and Technology)
1.1.2	Instituto Nacional de Comercialización Agrícola (INDECA) (National Institute of Agricultural Commercialization)
1.1.3	Banco Nacional de Desarrollo Agrícola (BANDESA) (National Bank for Agricultural Development)
1.1.4	Instituto Nacional Forestal (INAFOR) (National Forestry Institute)
1.2	Ministerio de Economía (Ministry of the Economy)
1.2.1	Instituto Nacional de Energía Nuclear (INEN) (National Institute of Nuclear Energy)
1.3	Consejo Nacional de Planificación Económica (National Council of Economic Planning)
1.3.1	Subsecretaria (Sub-secretariat)

Fig. 9 - *continued*

1.3.1.1   Planificación Intersectorial (Inter-sectorial Planning)
1.3.1 1.1   Departamento de Ciencia y Tecnológia (Department of Science and Technology)
A   Universidad de San Carlos de Guatemala (University of San Carlos de Guatemala)
B   Associación Nacional del Café (ANACAFE) (National Coffee Association)
C   Academia de Ciencias Médicas, Físicas y Naturales de Guatemala (Guatemala Academy of Medicine, Physics and Natural Sciences)
D   Centro Mesoamericano de Estudios de Tecnología Apropriada (CEMAT) (Middle-Americas Centre of Technological Study)
E   Universidad del Valle de Guatemala (University of Valle de Guatemala)
F   Instituto Centroamericano de Investigación y Tecnología Industrial (ICAITI) (Central American Institute of Research and Industrial Technology)
G   Instituto de Nutrición Centroamericano y Panama (INCAP) (Central American and Panamanian Institute of Nutrition)
H.   Secretaría Permanente del Tratado General de Integración Económica (SIECA) (Secretariat for Economic Integration)

Policies and Financing in Science and Technology

In 1977/78 a study was carried out called 'The description and analysis of scientific and technological activities in Guatemala'. This study, executed by the *Secretaría General del Consejo Nacional de Planificación Económica* (General Secretariat of the National Council of Economic Planning), provided a large part of the material used for policy-making in science and technology and was incorporated in the *Plan Nacional de Desarrollo 1979–1982* (National Development Plan 1979–1982).

Some long-term objectives of the plan are:

(a) to establish a liaison between science and technology and productive activities;
(b) to reduce the dependency on developed countries for science and technology. With this in mind, it is proposed to generate facilities for scientific and technological research and the assimilation and application of results according to national needs;
(c) to explore, evaluate, conserve and develop natural resources, conserving the environment;
(d) to contribute to the forming and functioning of the national system of science and technology and to coordinate international technical co-operation.

The improvement of the information system in science and technology plays a vital role in achieving the objectives in the development plan, as does an improvement in the liaison between the various organizations in matters of planning and coordination.

Human resources in research and development activities have risen to 1142; included in this figure are professionals at post-graduate level, licentiates, undergraduates and non-qualified. These scientists and technicians devote about 75 per cent of their time to basic and applied research and 57 per cent are employed in the government sector. Research is concentrated on agriculture and livestock, engineering, economics and technology in general.

In the sector of higher education, the emphasis is put on the social sciences, engineering and technology.

Financial resources in science and technology in 1977 rose to a total of 27 081 quetzals and were distributed as follows:

	Per cent
Science	6.32
Socio-economic diagnosis	12.38
Conservation of natural resources	10.88
Education	2.92
Health	11.37
Housing	2.11
Exploitation of natural resources	15.37
Industry	2.95
Infrastructure	29.78
Transport and telecommunications	—
Commerce and other services	0.82
Others	5.10
	100.00

Out of this total expenditure in science and technology, 13.37 per cent was destined for scientific research, 24.04 per cent to technological research and 62.59 per cent to supportive or complementary research.

Science and Technology at Government Level

Instituto Geográfico Nacional (National Geographic Institute)
Founded in 1964, it is affiliated to the Ministry of Communications and Public Works. One of its functions is the charting of national territory.

Dirección General de Estadística (Statistical Office)
A dependant of the Ministry of the Economy, this institute studies and publishes demographic and economic statistics.

The remaining government bodies are covered in the corresponding sections.

Science and Technology at Academic Level

Universidad de San Carlos de Guatemala (University of San Carlos de Guatemala)
Founded in 1676 by King Carlos II, the university was established in its present form in 1967. It is financed independently and has branches in Quezaltenango and Chiquimula. The academic year runs between January and November with a student population of 38 000 and 3007 teachers. Faculties are agronomy, architecture, chemistry and pharmacy, engineering, dentistry, veterinary medicine, zoology and biology. There are eight

regional centres in Huehuetenango; Coban; Quezaltenango; Chiquimula; Escuintla; Mazatenango; Jalapa and Monterico.

Universidad del Valle de Guatemala (University of Valle de Guatemala)
Established in 1966, it has an academic year between February and November, with 500 students and eighty-five teachers. There are schools of sciences and humanities; a research institute and a university college.

Universidad Francisco Marroquin (Francisco Marroquin University)
A private university founded in 1971, with an academic year between January and November. There is a student population of 1838 and 100 teachers for the faculties of medicine, and humanities and the school of computer studies.

Universidad Mariano Gálvez de Guatemala (Mariano Gálvez de Guatemala University)
Established in 1966, it has an academic year between February and November. There are 4000 students and 92 teachers. *Boletín Universitario* (University Bulletin) is published monthly. There are schools of engineering and structural engineering as well as the humanities.

Universidad Rafael Landívar (Rafael Landívar University)
Founded in 1961 with an academic year between the months of January and November and with a student population of 3000 and 250 teachers. Its faculties are in industrial engineering as well as arts and humanities.

Industrial Research and Technology

A large part of the research and development activities in this field are carried out by the sub-regional organization, *Instituto Centroamericano de Investigación y Tecnología Industrial* (Central American Institute of Industrial Research and Technology). There are four other centres of importance in this field, these are:

Departamento de Investigaciones Agropecuarias e Industriales (Department of Agricultural, Livestock and Industrial Research)
A dependant of the Bank of Guatemala.

Centro de Investigaciones de Ingeniería (Centre of Engineering Research)
The faculty of engineering of the University of San Carlos.

The remaining two are private organizations:

Instituto Técnico de Capacitación y Productividad (INTECAP) (Technical Institute of Training and Productivity), and

Oficina de Investigación Técnica (Office of Technical Research) and the *Asociación de Productores de Aceite Esenciales* (Association of Essential Oil Producers).

Agricultural and Marine Science and Technology

There is a sub-regional organization that carries out a large part of research in this field – *Instituto Interamericano de Ciencias Agrícolas* (Inter-American Institute of Agricultural Sciences). Various organizations affiliated to the Ministry of Agriculture develop activities in agricultural sciences. Among them are the *Instituto de Ciencia y Tecnología Agrícola* (Institute of Agricultural Science and Technology) and the divisions of Forestry and Soil of the *Dirección General de Recursos Naturales Renovables* (General Office of Renewable Natural Resources). At an academic level, the faculty of agronomy carries out research in this field, for example the *Departamento de Ingeniería Agrícola* (Department of Agricultural Engineering) and the *Departamento de Estaciones Experimentales* (Department of Experimental Stations). At private level there are the *Departamento de Café* (Department of Coffee) and the *Departamento de Diversificación de Cultivos* (Department for the Diversification of Cultivation), a department of *ANACAFE* that deals with agricultural advice and counselling.

Medical Science and Technology

The *División de Epidemiología* (Epidemiology Division), of the *Dirección General de Servicios de Salud* (General Board of Health Services) of the Ministry of Public Health and Social Assistance, and the university departments of pathological anatomy, microbiology and parasitology of the faculty of medicine are the principal research centres in medical science and technology.

Nuclear Science and Technology

Instituto Nacional de Energía Nuclear (National Institute of Nuclear Energy)
A dependant of the Ministry of the Economy.

Meteorological and Astrophysical Research

Instituto Nacional de Sismología, Vulcanología, Meteorología e Hidrología (National Institute of Seismology, Vulcanology, Meteorology and Hydrology)
Founded in 1976, it publishes annually *Anuario Hidrológico* (Hydrological Annual Report); monthly, *Boletín Meteorológico* (Meteorological Bulletin) and daily *Reporte Meteorológico* (Weather Report) and *Boletín Sismológico* (Seismological Bulletin).

Observatorio Meteorológico Nacional (National Meteorological Observatory)
A dependant of the Ministry of Agriculture.

Professional Associations and Societies in the General Field

There are thirteen organizations in the country that carry out research activities in different fields of science and technology. Among them are:

Academia de Ciencias Médicas, Físicas y Naturales de Guatemala (Guatemala Academy of Medicine, Physics and Natural Sciences)

Asociación Pediatríca de Guatemala (Guatemalan Paediatrics Association)

Colegio de Ingenieros de Guatemala (Guatemalan College of Engineers)

Information Services in Science and Technology

There are twenty organizations, among them libraries and museums, which offer documentation and information services. They include:

Biblioteca Nacional de Guatemala (Guatemalan National Library)
Founded in 1879; has a total of 352 000 volumes.

Museo de Artes e Industrias Populares (Museum of Art and Traditional and Ethnic Industry)
Founded in 1959; has a collection of working tools and instruments of indigenous cultures.

International Cooperation in Science and Technology

Guatemala is the seat of several important Central American and Caribbean research centres, some of which have already been mentioned, as well as the *Instituto de Nutrición de Centroamerica y Panama* (Central American and Panamanian Institute of Nutrition), patronized by the *Organización Panamericana de la Salud* (Pan American Health Organization) and the World Health Organization. This institute promotes and carries out research and post-graduate teaching in subjects relating to nutrition. It represents Costa Rica, El Salvador, Guatemala, Honduras, Nicaragua and Panama. It is administered by the Pan American Health Organization (PAHO)/World Health Organization (WHO). Its main activity is research into food utilization and nutrition. The library has over 70 000 volumes.

Oficina Sanitaria Panamericana (Pan American Sanitary Office) and regional offices of the FAO.

Guyana

Geographical, Demographical, Political and Economic Features

Guyana is situated in the northern part of South America with borders to the east with Surinam; to the south and south-east with Brazil; to the north with the Atlantic Ocean and the west with Brazil and Venezuela. There is a narrow coastal belt with a moderate climate and inland there are tropical forests and savannah. The total area of the country is 83 000 square miles.

According to estimates made in 1979 the population is 865 000, with a projection of 935 000 for 1985. The density is four people per square kilometre and 45.6 per cent of the population is urban. Ethnic composition is 50 per cent of Hindu descent; 31 per cent African negroes; 11.5 per cent mestizo; 5 per cent American Indians; 1 per cent Portuguese; 1 per cent Chinese and 0.5 per cent other Europeans.

The economy is based on agriculture and the production of bauxite. There are 364 000 hectares of arable land. The chief crops are sugar and rice. In 1980, 3 780 000 metric tons of sugar cane were produced and 313 000 metric tons of rice. Sugar accounts for about 30 per cent of export earnings. This is made possible because Guyana can harvest two sugar crops each year since there are two wet seasons and two dry seasons. Other products are citrus fruits, coconuts, groundnuts, oil palms and a variety of vegetables. In 1980 the count of livestock was: cattle 295 000, pigs 135 000, sheep 114 000, goats 70 000, chickens 12 500 000. The total catch of fish in 1980 was 21 300 metric tons. By 1979 the country had become self-sufficient in sugar, rice, vegetables, fish, meat, poultry and fruit.

Government

Guyana, formerly a British colony, has a president appointed by the National Assembly (made up of fifty-three members elected by proportional representation for five years). The executive consists of a cabinet (of the vice-prime minister and twenty-three ministers) which is presided over by the prime minister.

Natural Resources

AGRICULTURE

The main agricultural products include almonds, rice, bananas, citrus fruits, sugar cane, maize, tomatoes and pineapples.

FORESTRY

Only 20 per cent of forestry land is exploited, with production for 1977 reaching 82 000 cubic metres of cut wood.

MINERALS

Bauxite is Guyana's chief export earner and Guyana is a founder member of the International Bauxite Association. However, only 1.8 million tons was produced in 1980 as compared to 3.6 million tons in 1973–74, owing largely to industrial disputes. Gold and industrial diamonds and mica are also mined.

INDUSTRY

Sugar, beer, cigarettes, soap, wood, construction materials, furniture, food products and pharmaceuticals are among the most important sectors of industrial production. Industrial development depends on the expansion of energy sources and the 750 MW hydroelectric project on the Upper Mazaruni, which is due to be completed in 1985, should make possible the construction of a local aluminium smelter. In 1981 a 'year of energy' was announced and feasibility studies were undertaken on three hydro-electric projects.

Education

Education is free and compulsory between the ages of five and fourteen. The estimated literacy rate is 90 per cent. In 1980 Guyana had 374 nursery, 432 primary and eighty-four secondary schools. Children either receive secondary education in a general secondary school for five years or stay on at primary school for a further three years. The total number of pupils in all schools was 252 160 in 1980. There are also sixteen technical, vocational, special and higher education institutions. These include the University of Guyana in Georgetown and two colleges of education.

There are five technical and vocational institutions; Georgetown and New Amsterdam Technical Institutes, Carnegie and Fredericks Schools of Home Economics and the Guyana Industrial Training Centre. There are also thirty-six Home Economics and Industrial Arts Centres and many primary and secondary schools have departments attached to them. There are also a number of technical and vocational institutions not under the aegis of the Ministry of Education.

International Relations

Guyana is a member of the United Nations Organization; the Commonwealth, and *CARICOM*, and is an ACP state of the EEC.

Institutional Organization of Science and Technology

Figure 10 shows organizations at national level that carry out research and development activities in science and technology.

Financing and Policies in Science and Technology

In 1977 the National Science Research Council *(CNIC)* was established as an advisory body on scientific and technological policies. Recommendations from this council have been included in the Third Development

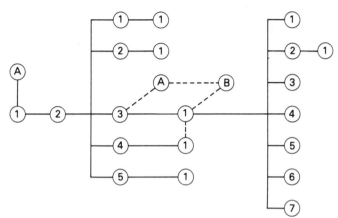

Fig. 10 Organization of science and technology

1	Primer Ministro (Prime Minister)
1.2	Gabinete (Cabinet)
1.2.1	Educación (Education)
1.2.1.1	Universidad de Guyana (University of Guyana)
1.2.2	Agricultura (Agriculture)
1.2.2.1	Estación Central Agricola (Central Agricultural Centre)
1.2.3	Desarrollo Social y Cooperatives (Social and Cooperative Development)
A	Actividades de UNESCO (UNESCO Activities)
1.2.3.1	Consejo Nacional de Investigación Científica (National Council of Scientific Research)
B	Consejo de Asuntos Cientifícos de la Commonwealth (Council of Commonwealth Scientific Agreements)
1.2.3.1.1	Comité Agrícola (Agricultural Council)
1.2.3.1.2	Comité Científico Industrial (Industrial Scientific Committee)
1.2.3.1.2.1	Instituto de Ciencias Aplicadas y Tecnológia (Institute of Applied Science and Technology)
1.2.3.1.3	Comité Forestal (Forestry Committee)
1.2.3.1.4	Comité de Investigaciones Médicas (Committee of Medical Research)
1.2.3.1.5	Comité de Estandardización y Control de Calidad (Committee for Standards and Quality Control)
1.2.3.1.6	Comité del Medio Ambiente (Committee for the Environment)
1.2.3.1.7	Comité Económico Social (Social Economic Committee)
1.2.4	Comercio y Protección de los Consumidores (Commerce and Consumer Protection)
1.2.4.1	Junta Nacional de Metricación (National Metrication Board)
1.2.5	Energía y Recursos Naturales (Energy and Natural Resources)
1.2.5.1	Cartografía de Recursos Naturales (Cartography of Natural Resources)

Plan 1978–1981. This plan lays out certain objectives for scientific and technological research such as:

(a) to achieve self-sufficiency in food, clothing and housing;
(b) to channel 1 per cent of the GNP into scientific and technological research;
(c) to develop research into areas where Guyana has the sole advantage, and
(d) to extend international cooperation.

The council is establishing a register of human and financial resources for research and development. This register will provide a breakdown of the available technical capacity; list the requirements of each research programme to meet its objectives; and provide a census of scientists and technicians in specialist areas, etc.

In 1977 thirty-five organizations represented 50 per cent of the potential work-force in research and development. The number of scientists and technicians was as follows:

(1)	Scientists and Engineers	611
	(a) Natural sciences	69
	(b) Engineering and technology	415
	(c) Medicine	89
	(d) Agriculture	38
	(e) Social sciences and the humanities	—
(2)	Technicians	968

There are no statistics for financial resources in science and technology.

Science and Technology at Government Level

Through various special committees, the council has promoted research programmes in agriculture, energy, natural products, etc. However, in spite of the efforts of the council and the universities the result has not really been significant. The council and its institutional network is the most important centre in Guyana for the development of science and technology.

Georgetown Hospital Compound, Georgetown
Has a Guyana Medical Science Library, which is attached to the Ministry of Health. It provides medical information to doctors, nurses, and health personnel.

Industrial Technological Research

Guyana Sugar Corporation Limited
A nationalized company, conducts agricultural research specifically in areas of breeding and selection of new cane varieties; methods of cultiva-

tion; control of disease and pests; efficient use of fertilizers and agricultural chemicals; and mechanization of sugar-cane harvesting and loading.

Scientific and Technological Research at Academic Level

University of Guyana
Founded in 1963, it runs courses in English. The staff numbers 183 researchers and teachers and 732 students. Faculties of the university are agriculture, arts, social sciences, natural sciences, education and technology. Affiliated to the university is the Institute of Development Studies.

Guyana School of Agriculture Corporation

Guybau Technical Training Complex
Established in 1958, it runs courses in mechanics, electricity, welding, plumbing, etc. The staff numbers seventeen for its 100 full-time and 200 part-time students.

New Amsterdam Technical Institute
Founded in 1971, it offers courses in car mechanics, electrical engineering, welding, secretarial studies, etc. Students number 1500 with forty-one teachers.

Professional Associations and Societies in the General Field

Guyanan Institute of International Affairs

Guyana Society
Formerly Royal Agricultural and Commercial Society.

Guyana Museum
Was founded by the Royal Agricultural and Commercial Society. Subjects covered include industry, art, history, anthropology, and zoology. The Guyana Zoo is affiliated to the Museum, and specializes in the display, care and management of South American fauna.

Information Services in Science and Technology

National Library
Founded in 1909, it has 160 000 volumes and is the official depository of all press material in Guyana.

Guyana Medical Science Library
Established in 1966, it is a dependant of the Ministry of Health. The library holds 5000 volumes, 300 specialist reviews and more than 1000 pamphlets.

Haiti

Geographical, Demographical, Political and Economic Features

The Republic of Haiti has an area of 27 750 square kilometres and is situated on the western side of Hispaniola, with its borders to the north with the Atlantic; to the east with the Dominican Republic; to the south with the Antilles Sea and to the west with the Windward Passage or the Maisi Straits, which separate the island from Cuba. Haiti is predominantly a mountainous region with a hot climate on the coast and a cold one in the high areas.

Estimates made in 1979 show the population to be 4 919 000 with a density of 177 people per square kilometre. The population can be divided into 90 per cent negroes and 10 per cent mulattoes. Twenty-four per cent live in the urban areas. French is the official language; however, 90 per cent speak creole (criollo) as a second language.

Government

The executive power is exercised by the President of the Republic, who is elected or nominated for life and this is ratified by the National Assembly. The President is advised by a council of fourteen secretaries of state. The legislative power is implemented by the National Assembly which comprises of fifty-eight members. The judicial power is in the hands of the Appeal Court (Corte de Casación), and the Departmental Civil Courts and Courts of Justice.

Natural Resources

Only one-third of the country has cultivable land and most of this land is taken up with smallholdings. This prevents large-scale development of agriculture and results in great poverty for the proprietors. The number of smallholdings is estimated to be in the region of 500 000. The principal product for export is coffee – the 1980 production figures show 35 000 tons. Cotton, cacao, bananas, sugar, sisal, maize, tobacco and rice are also cultivated.

LIVESTOCK
Figures for 1976 show 747 000 head of cattle; 81 000 sheep; 1.8 million pigs; 1.4 million goats; 3.4 million poultry and 387 000 horses.

MINERAL RESOURCES
Haiti is a producer of bauxite, which is exported by a United States consortium. It is thought that deposits of gold exist, while silver, antimony, salt, tin, lignite, iron, manganese, mercury, oil and gypsum are also exploited.

INDUSTRY
Haiti's relatively small productive sector includes cement, cigarettes, flour, plastics, essential oils, pharmaceuticals, cotton fabrics, glass, dairy products, meat, etc.

Education

The system in Haiti is a copy of the French model, and by law primary education is obligatory. Estimates in 1977 reveal that only 25 per cent of the population is literate.

International Relations

Haiti is a member of the United Nations Organization and the Organization of American States.

Science and Technology at an Academic Level

Université d'Etat d'Haiti (Haiti State University)
Established in 1920, it has a staff of 384 lecturers and 2926 students. Teaching is carried out in the French language. Its faculties comprise law and economics, medicine, odontology, ethnology, sciences, human and social sciences, agronomic and veterinary sciences. The *Escuela Normal Superior* (Teachers Training College) and the *Instituto Nacional de Administración, Gestión y Altos Estudios Internacionales* (The National Institute of Administration, Management and Higher International Studies) are also affiliated to the university.

There are four law schools in Haiti, situated at Cap Haitien, Cayes, Jeremie and Gonaives.

Ecole Polytechnique d'Haiti (Polytechnic of Haiti)

Institut Supérieur Technique d'Haiti (Higher Technical Institute of Haiti)

General Associations and Societies in the Field of Science and Technology

There are six important national organizations, one of which is:

Consejo Nacional de Investigaciones Científicas (National Council of Scientific Research)
Founded in 1963, this is a department of the Ministry of Public Health and Population. The Council coordinates research and scientific development, especially in the field of public health.

Information Services in the Field of Science and Technology

There are three libraries and three museums in Haiti. Among them is:

Bibliothèque Nationale d'Haiti (National Library of Haiti)
Founded in 1940, it has 12 000 volumes.

Honduras

Geographical, Demographical, Political and Economic Features

The Republic of Honduras has as its capital Tegucigalpa (444 749 inhabitants). To the north of Honduras lies the Antilles Sea (880 kilometres of coastline); to the south Nicaragua (783 kilometres border) and the Gulf of Fonseca (153 kilometres coastline); to the east Nicaragua and the Caribbean Sea, to the south-west El Salvador (341 kilometres border) and to the west Guatemala. Islands belonging to Honduras include: in the Atlantic, the Bay Islands (Utila, Roatan, Guanaja, etc.), the Cisne and the Cochinos keys; in the Gulf of Fonseca the island of Tigre (on which is situated the port of Amapala), Zacate Grande and others. The area of Honduras is 112 088 square kilometres.

The country is divided administratively into 18 departments and 282 municipalities. The population is made up of 91 per cent mestizos; 6 per cent indigenous; 2 per cent negroes and 1 per cent white.

Government

Complying with the Constitution of the Republic (1965), the executive power is exercised by the President of the Republic, who is elected by universal suffrage for a period of six years. The president is advised by a group of twelve secretaries of state. The legislative power is unicameral, with a National Congress, which has sixty-four deputies who give their services for a period of six years. The judicial power is made up of the Supreme Court with six magistrates nominated by the Congress for six years, which can be extended for an indefinite period; the Appeal Court is formed by four magistrates and justices of the peace.

Voting is obligatory for all citizens over eighteen years. The governors of the eighteen departments of the country are nominated by the President of the Republic.

Natural Resources

AGRICULTURE

Less than a quarter of the country is cultivable land and this is situated on the coastal plains of the Caribbean Sea and the Pacific Ocean. However this limitation does not prevent Honduras from being an essentially agricultural country. Agriculture employs 65 per cent of the active population and provides 80 per cent of exports. The principal agricultural products for export are bananas (an estimated 1 403 000 tons were produced in 1980); coffee (77 000 tons); maize (325 000 tons); cotton, tobacco (7 000 tons) and beans (20 000 tons).

FORESTRY
Forty-five per cent of the area of the country is covered in forests which produce hard and soft wood. Wood production for 1977 reached 628 000 cubic metres of cut wood. Woods such as walnut, rose, mahogany and pine are used predominantly in the production of finished items.

LIVESTOCK
Estimates for 1976 show 7.8 million poultry; 58 000 goats; 280 000 horses; 1.8 million cattle, 5000 sheep and 520 000 pigs.

FISH
The main fish product is shrimps, more than 2 million kilos of which are exported annually.

MINERALS
The main mineral resources exported are lead, antimony, silver, sulphur, coal, cinnabar, zinc, copper, iron, gold, pitchblende and salt. (See section on industrial production.)

INDUSTRIAL PRODUCTION
Main products include: oil by-products, petrol, kerosene, light oils, heavy oils. Manufactured products include beer, cigarettes, sugar, vegetable oils and soft drinks.

Education
Literacy for 1979 was 59.5 per cent. Primary education is obligatory and free between the ages of seven and fifteen. According to figures for 1979, the budget for education was 11.7 per cent of the country's total budget and with this a total of 6170 primary, secondary and specialized educational establishments function with the services of 20 843 teachers working to educate 767 178 students of both sexes.

Economic Information
Monetary unit: Lempira (2.00 per US$ in November 1980).
  Gross National Product per person: US$546.5 (1979).
  Annual growth of GNP per person: 0.4 per cent (1970–78).
  Annual rate of inflation: 8.0 per cent (1970–78).

Distribution of GNP (1978): 32 per cent agriculture; 26 per cent industry; 42 per cent services.
Economically active population: 29.3 per cent (1978).

International Relations

Honduras is a member of the United Nations Organization and the Organization of American States.

Science and Technology at Government Level

Instituto Geográfico Nacional (IGN) (National Geographic Institute)
This institute carries out studies on natural resources and minerals, their evaluation and exploitation. Several publications are produced: *Boletín de la Comisión Geográfica Especial de la Secretaría de Marina y Aviación* (Bulletin of the Geographical Commission for the Secretariat of Marine Affairs and Aviation); *Boletín de la Dirección General Cartografíca* (Bulletin of the General Directorate of Cartography); *Secretaría de Fomento* (Secretariat of Promotion); *Boletín de la Dirección General de Cartografía del Ministerio de Comunicaciones y Obras Publicas, etc.* (Bulletin of the General Directorate of Cartography of the Ministry of Communications and Public Works, etc.).

Scientific and Technical Research at an Academic Level

Universidad Nacional Autónoma de Honduras (Autonomous National University of Honduras)
Founded in 1847, it has 763 teachers and 15 404 students. Its faculties are juridical and social sciences; economics; medicine; chemistry and pharmacy; dentistry; and engineering. The three centres of *Costa Atlántica* (Atlantic Coast), *Estudios Generales* (General Studies), and *El Norte* (the North) are affiliated to the university.

Escuela Agrícola Panamericana (Pan American School of Agriculture)
Has thirty teachers and 250 students.

Escuela Nacional de Música (National School of Music)

Associations and Professional Societies in General

There are six such institutions, among which we can mention:

Academia Hondureña (Honduran Academy)

Instituto de Ingenieros y Arquitectos de Honduras (Institute of Engineers and Architects of Honduras)

Instituto Hondureño de Cultura Interamericana (Honduran Institute of Inter-American Culture)

Information Services in Science and Technology

There are eleven institutions involved in information and documentation. Among them are:

Biblioteca Nacional de Honduras (Honduras National Library) with 55 000 volumes.

Biblioteca de la Universidad Nacional (Library of the National University) which possesses 90 000 volumes.

Jamaica

Geographical, Demographical, Political and Economic Features

The island of Jamaica forms part of the Greater Antilles and is situated in the Caribbean Sea at some 128 kilometres to the south of Cuba and 160 kilometres to the west of Haiti. The island measures 231 kilometres from east to west and 80 kilometres from north to south. The climate is humid in the coastal region and moderately hot in the interior.

Official estimates made in 1979 show the population to be 2 162 000 with a projection for 1985 of 2 337 000 – the density is 197 per square kilometre. 65.7 per cent of the population is urban and the composition is 77 per cent negroes; 14.6 per cent white; 5 per cent Hindus and 3.4 per cent Chinese.

Jamaica is an independent state within the Commonwealth. The head of government is the Prime Minister, who is appointed by the Governor General from the sixty members of the Chamber of Representatives. The chamber is elected by universal suffrage (by citizens of eighteen years and over) for a period of five years. The Senate has twenty-one members designated by the Governor General (thirteen being recommended by the Prime Minister and eight by the leader of the opposition). The Prime Minister recommends the other Ministers to the Governor General.

Natural Resources

AGRICULTURE

Main products include bananas, citrus fruits, coffee, maize, tobacco, sugar, cocoa and potatoes.

LIVESTOCK

Estimates in 1976 show 280 000 head of cattle; 330 000 goats; 235 000 pigs; 3.77 million poultry.

MINERALS

Principal resources are bauxite, phosphates, marble, salt and gypsum.

INDUSTRY

Derivatives of oil are the main industrial products: petrol, kerosene, and light and heavy oils. Others are cement, sugar, cigarettes, tyres, electronics, textiles and food products.

Education

In 1979, 17.5 per cent of the national budget was channelled into education. From a survey made in 1975, 86 per cent of the population are literate.

International Relations

Jamaica is a member of the United Nations Organization, the Commonwealth, and the Organization of American States, and is an ACP Member of the EEC.

Science and Technology at Government Level

Consejo de Investigaciones Científicas (Scientific Research Council)
Founded in 1960, it promotes and coordinates scientific research on the island. Departments include mineral resources, agro-industry, food and nutrition, and industrial development.

Agencia Nacional de Planificación (National Planning Agency)
Comes directly under the jurisdiction of the Prime Minister and its objectives are to establish plans for scientific development.

Scientific and Technological Research at an Academic Level

University of the West Indies
Founded in 1948, it works in conjunction with the colonies and associated British states in the Caribbean and the Bahamas (Jamaica, Trinidad and Tobago, Barbados, San Vicente and Grenada, Santa Lucia and Dominica). The faculty of law is in Barbados; medicine in Jamaica and engineering and agriculture in Trinidad. Apart from the faculties already mentioned there are art and general studies, natural sciences, social sciences and a School of Education. The following institutes and centres are affiliated to the university.

Institute of Agricultural Research and Development of the Caribbean; Institute of International Relations; Institute of Economic and Social Research; Institute of Trade Union Education; Centre of Creative Art; and the *Department of Extra-mural studies.*

The university has a staff of 750 researchers and teachers and a student population of 8041.

College of Arts, Science and Technology
Founded in 1958, it has a staff of 144 full-time teachers and 106 part-time, and a student population of 1580 full time and 1100 part time. Courses

are offered at diploma and certificate level in mechanical, electrical and construction engineering; computing; business studies; medical technology; instrument technology; pharmacy; land surveying, etc.

School of Agriculture
Established in 1910 and has thirty teachers and 450 students. Courses of three years are run in agricultural economy; agronomy; agricultural engineering; animal husbandry; English; geology; physics; soil study; etc. The qualification received at the end of the course is Diploma in Agriculture.

Institute of Jamaica
Founded in 1879, it comprises various social and cultural study centres.

Caribbean Food and Nutrition Institute (CFNI)
Founded in 1967, it carries out research in these fields. The institute is a consultative body for English-speaking countries in the Caribbean.

Industrial Technological Research

Sugar Industry Research Institute
Founded in 1973, it was formerly the *Sugar Research Department of the Association of Sugar Producers*. The institute carries out research into sugar-cane production. The technical staff numbers sixty-five and there are several specialized laboratories. Associated to the Institute is the *Jamaican Association of Sugar Technologists*, founded in 1937, which conducts research into the technical problems of sugar production.

Medical Science and Technology

Medical Research Council Laboratories
Affiliated to the University of London and the Medical Research Council.

Information Services in Science and Technology

National Library
Associated with the Jamaican Institute and houses 30 000 volumes.

Library of the University of the West Indies
Established in 1948, it has 274 780 volumes and 10 000 periodicals.

Jamaica Library Service
The system offers 574 points of service including fourteen mobile libraries, and has a total of 1 079 025 volumes. Since 1952 a service has been in operation for primary and secondary schools with 1 036 684 volumes available.

Mexico

Geographical, Demographical, Political and Economic Features

Mexico is situated between the latitudes of 14 °N and 32 ° 43′ 35″ N, that is, to the north of the equator and in the western hemisphere with respect to the Greenwich Meridian. To the north lies the United States of America, with 3125 kilometres of frontier; to the west and north-west is the Pacific Ocean; to the south, Belize and Guatemala, and to the east is the Gulf of Mexico and the Antilles Sea. The climate is varied and moderate, with four seasons.

According to the census of 1980, the country has a population of 67 405 700, with a projection for 1985 of 85 177 000, and the population density is 34.4 per square kilometre. Although the official language is Spanish, there are several indigenous groups that speak Nahuatl, Otomi, Maya, Zapoteca, Tarasca, Totonaca and other languages.

Government

Mexico is a Federal Republic divided into thirty-one states and one Federal District (the capital) where the National Congress has its seat. The executive power is in the hands of the President, who is elected by universal suffrage for a period of six years and who can never be re-elected. The President carries out his duties in conjunction with seventeen ministers or secretaries of state. The legislative power is represented by two chambers: the Senate, with two members for each state, and the Chamber of Deputies, which has a representative for every 200 000 inhabitants. The senators remain in their posts for seven years and the deputies for three years, being elected by direct popular vote. The judicial power is administered by the High Court of Justice, consisting of twenty-one members appointed by the executive and ratified by the Congress. The states have the same divisions of power – the executive duties are carried out by the governor; the legislative by the Chamber of Deputies and the judicial power by the state High Court of Justice.

Natural Resources

AGRICULTURE

Of the total area of the country, 15.51 per cent is given over to agricultural activities; 14.20 per cent to forestry reserves; 53.26 per cent is pasture land; 4.70 per cent is untilled land and 11.29 per cent unproductive land from the point of view of agriculture. Main agricultural products are corn, beans, maize, wheat, sorghum and barley. The production figure for seed crops in 1980 reached 21 561 900 tons and the area sown was 11 415 100 hectares.

FORESTRY

Forestry reserves cover an area of 19 857 787.9 hectares, of which 32 per cent are woodland of commercial potential, with reserves of about 2 359 288.565 square kilometres.

LIVESTOCK

In 1980, figures showed about 80 153 725 head of livestock among which are sheep, goats, cattle, pigs and horses, giving a livestock production of 4 492 000 tons.

FISH

Fish production in 1975 reached 850 525 tons, of which 54 per cent was destined for human consumption and 46 per cent for industrial use.

MINERAL RESOURCES

Minerals are one of the most important resources of the country; of the forty-eight non-metallic minerals of the world, Mexico has twenty-three. Mineral production for 1980 was as follows: lead 134 026 tons; copper 154 105 tons; gold 5634 kilos; fluorite 844 000 tons; bismuth 663 tons; baryta 132 493 tons; iron 3 712 000 tons; manganese 144 752 tons; zinc 216 374 tons; cadmium 1647 tons; sulphur 1 912 000 tons; silver 1398 tons and coal 2 179 000 tons.

With regard to energy resources, Mexico holds fifth place in the world for reserves of hydrocarbons, producing the equivalent of 67 830 million barrels yearly. Mexico occupies fourth place among oil-producing countries.

Education

In accordance with figures for 1979, out of the total population over the age of ten years, 83.9 per cent are literate. Primary education is obligatory and free. In the Federal District education is controlled by the governmental authorities and in the states by the state educational bodies. Table 8.1 is based on 1979 statistics.

There are 507 institutes of higher education of which forty-eight are for post-graduate studies only.

International Relations

Mexico is a member of the United Nations Organization and the Organization of American States.

Table 8.1

Level	Schools	Students
Kindergarten	5 397	693 494
Primary	68 704	13 604 476
Secondary	5 778	2 141 127
Preparatory and vocational	1 145	607 961
Professional and specialized	645	189 884
Universities	531	543 112

Organization of Science and Technology

The network of organizations dedicated to research and development in science and technology has reached a highly developed level in Mexico. The scientific and technological system incorporates the private and public sector, industry, universities and individuals. Figures 11.1 and 11.2 are a simplified version of the network:

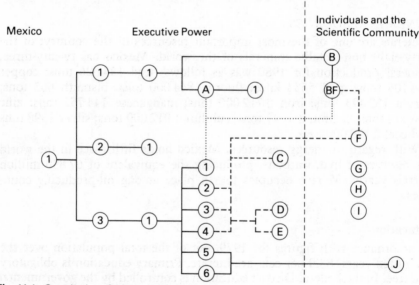

Fig. 11.1 Organization of science and technology

1	Presidencia (Presidency)
1.1	Secretaria de Educación (Secretariat of Education)
1.1.1	Subsecretaria de Educación Científica y Tecnológica (Subsecretariat of Scientific and Technological Education)
A	Universidades del Estado (State Universities)
A.1	Institutos de Investigación dependientes de las Universidades (Research Institutes affiliated to the Universities)
B	Universidades privadas (Private Universities)
BF	Institutos de Investigación dependientes de las Universidades y/o Empresas privadas (Research Institutes affiliated to the Universities and/or private firms)

1.2 Secretaria de Relaciones Exteriores (Secretariat for Foreign Relations)
1.2.1 Comisión Nacional para la UNESCO (National Commission for UNESCO)
1.3 Comisión sobre Ciencia y Tecnología (Commission for Science and Technology)
1.3.1 Consejo Nacional de Ciencia y Tecnológia (CONACYT) (National Council for Science and Technology)
1.3.1.1 Dirección Adjunta de Formación de Recursos Humanos (The Official Board for the Training of Human Resources)
1.3.1.2 Unidad de Planeación y Asesoría (Planning and Advisory Board)
1.3.1.3 Dirección Adjunta Técnica (Technical Board)
1.3.1.4 Dirección Adjunta de Servicos de Apoyo (Board of Support Services)
1.3.1.5 Unidad de Radio y Prensa (Radio and Press Board)
1.3.1.6 Dirección de Publicaciones (Board of Publications)
C Empresas y Organismos de Participación Publica (Public Firms and Organizations)
D Institutos de Investigación dependientes de las Empresas Públicas (Research Institutes affiliated to Public Firms)
E Institutos de Investigación Decentralizados (Decentralized Research Institutes)
F Empresa Privada (Private Firms)
G Asociaciones de Inventores (Association of Inventors)
H Academias de Investigación (Research Academies)
I Asociaciones de Investigadores (Association of Researchers)
J Programas de Divulgación Científica (Programmes of Scientific Media)

In the national system of science and technology, the central axis is formed by the *Consejo Nacional de Ciencias y Tecnología (CONACYT)* (National Council of Science and Technology). This was formed as a decentralized public body with autonomous characteristics. Its functions encompass the formulation, study, execution and evaluation of national policy in science and technology.

In Mexico nine research organizations employ 55 per cent of personnel in the field of research and development. There are several research centres with less than six members of staff in provincial areas of the country. The highest concentration of research organizations making up 70 per cent of the total research carried out in Mexico, is in Mexico City.

Policies and Financing in Science and Technology

The national policy for science and technology forms part of the National Plan for Economic and Social Development for 1976–82. This plan was set up with the objectives described as follows: 'science and technology form an active part in the economic development of the country. To this end the present policy is being considered in the fields of conventional and non-conventional energy; the need to reach self-sufficiency in food products; the urgency to improve public health; to combat unemployment and to reach self-determination in science and technology . . .' The plan underlines priority areas for the development of activities – basic research; agriculture, livestock and forestry; fisheries; nutrition and health; energy; industry; construction, transport and communications; social development and public administration.

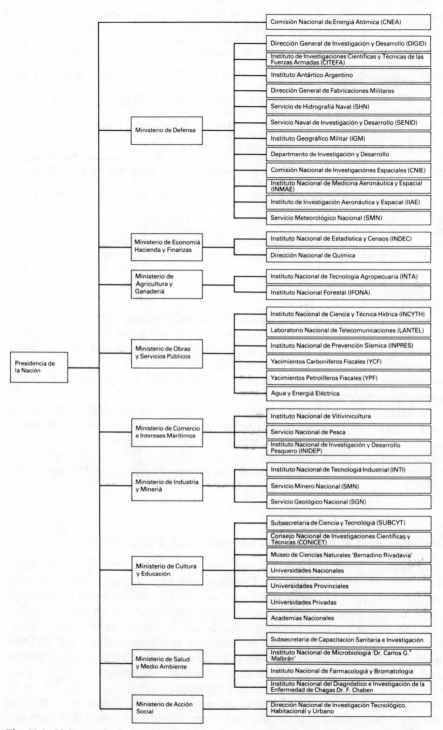

Fig. 11.2 Main organizations and institutes that carry out scientific and technological activities

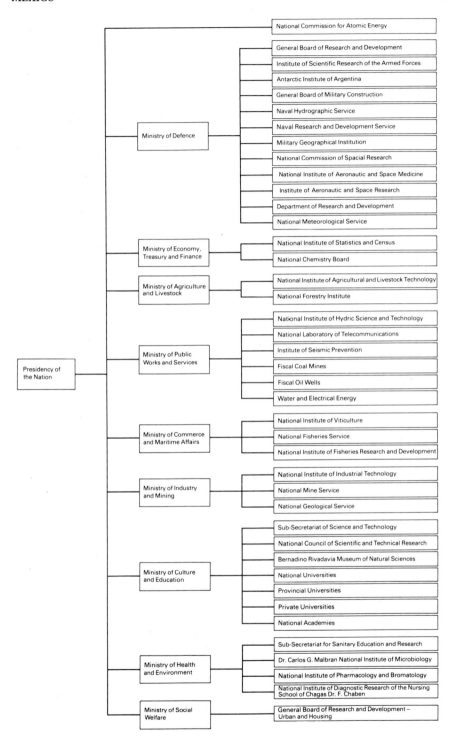

Presidency of the Nation

- Ministry of Defence
  - National Commission for Atomic Energy
  - General Board of Research and Development
  - Institute of Scientific Research of the Armed Forces
  - Antarctic Institute of Argentina
  - General Board of Military Construction
  - Naval Hydrographic Service
  - Naval Research and Development Service
  - Military Geographical Institution
  - National Commission of Spacial Research
  - National Institute of Aeronautic and Space Medicine
  - Institute of Aeronautic and Space Research
  - Department of Research and Development
  - National Meteorological Service

- Ministry of Economy, Treasury and Finance
  - National Institute of Statistics and Census
  - National Chemistry Board

- Ministry of Agriculture and Livestock
  - National Institute of Agricultural and Livestock Technology
  - National Forestry Institute

- Ministry of Public Works and Services
  - National Institute of Hydric Science and Technology
  - National Laboratory of Telecommunications
  - Institute of Seismic Prevention
  - Fiscal Coal Mines
  - Fiscal Oil Wells
  - Water and Electrical Energy

- Ministry of Commerce and Maritime Affairs
  - National Institute of Viticulture
  - National Fisheries Service
  - National Institute of Fisheries Research and Development

- Ministry of Industry and Mining
  - National Institute of Industrial Technology
  - National Mine Service
  - National Geological Service

- Ministry of Culture and Education
  - Sub-Secretariat of Science and Technology
  - National Council of Scientific and Technical Research
  - Bernadino Rivadavia Museum of Natural Sciences
  - National Universities
  - Provincial Universities
  - Private Universities
  - National Academies

- Ministry of Health and Environment
  - Sub-Secretariat for Sanitary Education and Research
  - Dr. Carlos G. Malbran National Institute of Microbiology
  - National Institute of Pharmacology and Bromatology
  - National Institute of Diagnostic Research of the Nursing School of Chagas Dr. F. Chaben

- Ministry of Social Welfare
  - General Board of Research and Development – Urban and Housing

The means to carry out the programmes in the National Plan are listed as follows: (1) indicative programmes; (2) science committees and specific consultative bodies; (3) consultative groups; (4) international agreements in science and technology; (5) inter-institutional groups involved in programming and sectorial budgeting; (6) regional research centres and technological assistance; (7) patent examination office for science and technology; (8) agreements between the various private and public productive sectors and the research institutes; (9) service centres for instruments and laboratory equipment; (10) inter-sectorial programming committees for science and technology.

The main body concerned with policy-making, as previously mentioned, is *CONACYT* whose main objectives are the gradual strengthening and rationalization of the scientific and technological system; to make headway towards solutions of national problems; to perpetuate research to useful ends; to gain a better knowledge of natural resources; to develop human resources, scientists and technicians to define the aims of science and technology in accordance with the National Development Plan.

The main fields of activity of *CONACYT* are the training of human resources and the development of scientific and technological institutions.

In Mexico there are 260 institutes of higher education, among which are sixty-nine universities, 101 institutes, sixty-seven schools, six colleges and seventeen centres, besides the decentralized bodies formed for research in specialist fields – all of which develop the objectives in science and technology previously mentioned.

There are eleven civil associations, two foundations and 210 private firms in the industrial sector that donate financial aid to research projects as well as training and educating personnel in science and technology. Of the total resources dedicated to research and development, 60 per cent are absorbed into the universities.

There are presently ninety-seven institutions that offer post-graduate activities covering 1232 specialized programmes. Table 8.2 shows the distribution of human resources in science and technology in the different fields of specialization.

Financial resources for science and technology are provided by various sources – the federal government, through the Secretariat of Public Education; *CONACYT*; state governments; national and foreign foundations and international bodies.

CONACYT, during its ten years of existence has channelled 7.8 per cent of the total national budget into science and technology from 1971–79.

The upward trend in expenditure indicates that the limit of 1 per cent of GNP will be achieved as stated in the *Plan Global de Desarrollo 1980–82* (Global Development Plan 1980–82).

The total expenditure in research and development is estimated to be distributed as follows – public sector 78 per cent; private sector 16 per cent and 6 per cent from foreign sources. Of the total budget for research and development, 25 per cent is destined for personnel – scientists and technicians – while the remaining 75 per cent covers expenses in general – equipment, facilities, etc.

Table 8.2 Human resources in science and technology

Year	Population (thousands)	Scientists and engineers		Those employed in research and development					Technicians	
		Total number of scientists and engineers (in thousands)	Total (in units)	Breakdown in fields of specialization (in units)					Total (in units)	Those employed in research and development (units)
				Natural sciences	Engineering and technology	Agricultural sciences	Medicine	Social sciences		
70	50 695		3 665	883	626	761	544	551		557*
74	58 526		7 582	1 638	1 095	1 149	1 038	2 662		1 013*
77	63 822	155 819	9 265	2 017	2 000	1 405	1 154	2 686	35 535	2 328
80	67 396	174 517	10 412	2 268	2 248	1 579	1 297	3 020	39 799	2 617
85	79 242†	310 640 (e)	18 598	4 051	4 015	2 820	2 316	5 396	70 842	4 675
90	87 489†	552 939 (e)	33 218	7 236	7 172	5 038	4 138	9 634	131 058	8 649

* Includes: Professionals, Technicians and Graduates.
† Estimate of the National Council of Population, Memorandum – Mexico 1979.
(e) Estimate of the Department of Control and Studies, DAAI–CONACYT.

Sources:

1970 – I.N.I.C. – National Policy and Programmes in Science and Technology.
1974 – CONACYT – DID Basic Statistics on the System of Science and Technology.
1977 – CONACYT – National Indicative Plan of Science and Technology.

Table 8.3 Expenditure in science and technology (1971–80) (millions of pesos)

Year	National budget	Government budget
1971	1 753	1 490
1972	2 229	1 895
1973	2 859	2 430
1974	3 653	3 105
1975	3 898	3 313
1976	4 732	4 022
1977	6 386	5 428
1978	9 519	8 091
1979	12 924	10 985
1980	13 644	11 597*

*Preliminary figure.
Source: *Informe Nacional México 1981* (National Mexican Brief 1981): 6th Meeting of the Permanent Conference of National Organizations for Policy in Science and Technology in Latin America and the Caribbean.

Table 8.4 Total expenditure on science and technology as envisaged on the global development plan

1978	8 844 490
1979	10 352 480
1980	12 251 120
1981	14 664 590
1982	17 626 840

Science and Technology at Government Level

The principal body at government level is the *CONACYT*, founded in 1970, whose main functions are the coordination of research and development and the drawing up of national policies in this field.

Dirección General de Relaciones Educativas, Científicas y Culturales (General Board of Educational, Scientific and Cultural Relations)
Formed in 1960 as a dependant of the Secretariat of Public Education. The Board offers technical assistance in education and in sciences in general. It also acts as a coordinating agency between UNESCO, OAS and the Mexican Government.

Dirección General de Estadística (General Board of Statistics)

Scientific and Technological Research at Academic Level

Asociación Nacional de Universidades e Institutós de Enseñanza Superior (National Association of Universities and Institutes of Higher Education)

Founded in 1950, it coordinates and represents the institutes of higher education in the country. The association studies and analyses the academic and administrative problems and promotes the interchange of staff, information and services between affiliated institutions (seventy-eight universities and institutions). Various publications are released – statistical, studies, etc.

Centro de Estudios Educativos (Centre of Educative Studies)
Founded in 1963; carries out scientific studies on educational problems in Mexico and Latin America. The staff numbers twenty-two researchers with a library of 15 000 volumes. *Revista* (Review) is published.

Instituto de Matemáticas (Institute of Mathematics)
Established in 1949; carries out research in mathematics. Publications include *Anales* (Annals) and monographs. The library houses 9432 volumes.

Centro de Investigación y de Estudios Avanzados del Instituto Politécnico Nacional (Research and Advanced Studies Centre of the National Polytechnic Institute)
Founded in 1961, the centre integrates the work of the departments of biochemistry, physics, mathematics, genetics, molecular biology, biotechnology, neuro-sciences and others. The centre offers research opportunities and teaching at post-graduate level.

Universities

Universidad Nacional Autónoma de México (UNAM) (National Autonomous University of Mexico)
Founded in 1551, it has 19 997 teachers for a student population of 271 266, not including post-graduate students.

FACULTIES
Business studies and accounting; chemistry; dentistry; economics; engineering; law; medicine; philosophy and the arts; social sciences and politics; psychology; veterinary sciences; and zoology.

RESEARCH CENTRES AND INSTITUTES
The following centre and institutes are affiliated to the university in the field of the sciences: Institutes of Astronomy; Institutes of Applied Mathematics and Computing; Institute of Biology; Institute of Biomedical Research; Institute of Chemistry; Institute of Engineering; Institute of Geography; Institute of Geology; Institute of Geophysics; Institute of Mathematics; Institute of Physics; Centre of Atmospheric Sciences; Centre of Computing Sciences; Centre of Scientific Instruments; Centre of Oceanography and Limnology; Centre of Material Research; Centre of Nuclear Studies and Centre of Scientific and Arts Research. Also affiliated to the university are the National Library, the Central Library and ninety-two specialized libraries attached to the faculties, schools and research centres.

PUBLICATIONS

The following publications are issued in connection with science and technology: *Anales del Instituto de Geofísica* (Annals of the Institute of Geophysics); *Anales de Veterinaria* (Veterinary Annals); *Tablas de Predicción de Mareas* (Tide Tables); *Anales del Instituto de Matemáticas* (Annals of the Institute of Mathematics); *Boletín de Estudios Médicos y Biológicos* (Bulletin of Medical and Biological Studies); *Anales Serie Botánico* (Botanical Annals); *Anales Serie Zoología* (Zoology Annals); *Biología Experimental* (Experimental Biology); *Anales Centro de Ciencias de Mar y Limnología* (Annals of the Centre of Oceanography and Limnology); *Datos Geofísicos* (Geophysical Data); *Serie A-Oceanografía* (Series A-Oceanography); *Anales de Geografía* (Geography Annals).

Universidad Autónoma del Estado de México (Autonomous University of the State of Mexico)
A private university founded in 1956. The following institutes are affiliated to the university: medicine, civil engineering, economics, commerce and business studies; and the schools of philosophy and arts, architecture, nursing and tourism.

The university employs 2000 teachers for its 20 000 students. The central library has 6000 volumes.

Universidad Femenina de México (Women's University of Mexico)
Founded in 1943; is a private university with an academic year between August and June. There are 170 teachers and 1100 students in the 12 schools of: law, education, chemistry, pharmacology, interior design, pedagogy, social work, psychology, commerce, business studies and tourism, journalism and communications, international relations, clinical laboratories, and secretarial. Its library holds 12 000 volumes.

Universidad Autónoma de Aguascalientes (Autonomous University of Aguascalientes)
Established in 1973, it is a private concern with an academic year beginning in September. The student population is 2229, with 423 teachers. Apart from the Centre of Humanities there are centres of biomedical studies, agriculture, technology and basic sciences. *Voz Universitaria* (University Voice) and *Correo Universitario* (University Post) are published.

Universidad de las Americas (University of the Americas)
Known as the College of Mexico City in 1940, when it was established, changed to its present name in 1963. A private university with teaching carried out in English and Spanish and an academic year from August to May. There are four faculties – social sciences; economics and business studies; technology; adult education. There is also a Department of Graduates. The university is affiliated to the *Centro de Estudios Universitarios* (Centre of University Studies). Students number 3500, with 242 teachers. Publications are *Meso-American Notes; Universitas Perspectivas* (University Perspectives); *Americas Alumnas* (American Students) and *Mexico Quarterly Review*.

Universidad Anáhuac (Anáhuac University)
Founded in 1963, it is run under private control with an academic year divided into two semesters (September to June). Its library holds 50 000 volumes for its 3500 students and 429 teachers. The university has ten schools including medicine and engineering, the other eight being in the humanities.

Universidad Autónoma de Baja California (Autonomous University of Baja California)
A private university established in 1957 with two seats in Ensenada and Tijuana. The students number 17 000 and there are 1200 teachers in the eleven schools – agronomy, practical agriculture, dentistry, engineering, medicine, nursing, architecture, law, business studies, education, and social sciences and politics.

Universidad Autónoma Chapingo (Autonomous University of Chapingo)
Known under its present name since 1978; previously was called *Escuela Nacional de Agricultura* (National School of Agriculture), dating back to 1854. A private concern with an academic year between the months of August and June. Courses are focused on agrarian studies: agricultural economics, agricultural industry, irrigation, parasitology, forestry studies, soil science, rural sociology.
There is a post-graduate college attached to the university dating from 1959, offering studies in genetics, entomology, botany, statistics, soil science, agricultural economics, irrigation, drainage, and higher studies in agriculture. Its library holds 100 000 volumes and 939 periodicals. The teaching staff, including the post-graduate college, is 475 for a student population of 2550. *Agrociencia* (Agro-science) is published.

Universidad Autónoma de Chiapas (Autonomous University of Chiapas)
Founded in 1975, it is a private university with an academic year between September and July with two semesters. Courses are offered in physics and mathematics; medicine; veterinary medicine; chemistry; agriculture; business studies; law and social sciences. *Criterio Universitario* (University Criterion) and *Vida Universitaria* (University Life) are published. Students number 2746 with 294 teachers.

Universidad Autónoma de Chihuahua (Autonomous University of Chihuahua)
Founded in 1954, it has 5000 students following courses in sciences; engineering; medicine; agriculture; law and philosophy. The university has a seat in the town in Juárez.

Universidad Autónoma de Coahuila (Autonomous University of Coahuila)
Re-established in 1957 but originally dates back to 1867. A state-run university with an academic year between August and June. Teachers number 900 for its 13 923 students. Faculties are education and law, with schools of chemistry, economics, civil engineering, architecture, psychology, nursing, obstetrics and medicine. There is also a school of medicine, nursing and dentistry in the seat of Torreon. The *Sección Norte* (northern

section) has a school of electrical and mechanical engineering, and one of mining studies and metallurgy.

Universidad de Colima (Colima University)
Founded in 1867 as the *Universidad Popular de Colima* and reorganized in 1962. A private university with an academic year between August and July. Faculties are law; social sciences; public administration; accounts; medicine; civil engineering; topography; agronomy and education. Students number 7822 with 632 teachers.

Universidad Juárez del Estado de Durango (Juárez University of the State of Durango)
Founded in 1856 as the *Colegio Civil* (Civil College) and became a university in 1957. A private university with an academic year divided into two periods, February to June and August to December. Integrated into the following schools are 7931 students: medicine; veterinary medicine; dentistry; agriculture; business studies and law. The *Instituto de Investigación Científica* (Institute of Scientific Research) is affiliated to the university. Teachers number 740.

Universidad de Guadalajara (Guadalajara University)
Controlled by the Consejo General de Universidades del Gobierno de Mexico (General Council of Universities of the Government of Mexico). Founded in 1792, it has an academic year between the months of September and August. The Central Library holds 450 000 volumes for the use of its 156 621 students and 6207 teachers. The university has the following faculties and schools in the field of the sciences: chemistry, engineering, medicine, veterinary medicine, dentistry, agriculture and nursing. Besides the faculties, there are also institutes in medical–biological sciences, botany, geography and statistics, experimental pathology and technology. The various faculties and departments publish reviews and reports and *Revista de la Universidad* (University Review) appears every three months.

Universidad Autónoma de Guadalajara (Autonomous University of Guadalajara)
Established in 1935, it is privately run with an academic year between August and May. Its library houses 80 265 volumes, 2729 maps and 2000 periodicals. The student population numbers 19 000, with 1430 teachers. Apart from the faculties in the humanities there is the faculty of bio-medicine, made up of the institute of biological sciences, school of nursing, schools of medicine and dentistry, school of community medicine, two university hospitals (Ramón Garibay and Angel Leaño), the office of health education and the office of audio-visual education. The faculty of technological sciences has institutes of exact sciences, and natural sciences, and schools of chemistry, engineering, mathematics, science and computing, and biology.

Universidad de Guanajuato (Guanajuato University)
Founded in 1732 under the name of *Colegio de la Purisima Concepción*, in 1928 it changed its name to *Colegio del Estado* and took its present name

in 1945. There are ten schools for its 7179 students and 1153 teachers. Apart from the humanities, courses are offered in mining engineering, civil engineering, topography and hydrology, medicine, chemistry and nursing. Affiliated to the university is the *Instituto de Investigaciones Científicas* (Institute of Scientific Research) and the *Instituto de Investigaciones Tecnológicas* (Institute of Technological Research).

Universidad Autónoma de Guerrero (Autonomous University of Guerrero)
Founded in 1867, it has schools of sciences, engineering, agronomy, and nursing and obstetrics.

Universidad Autónoma de Hidalgo (Autonomous University of Hidalgo)
Acquired its present status in 1961, but previously was named *Instituto Científico y Literario* (Scientific and Literary Institute) since 1869 when it was founded. Courses are offered in the social sciences, sciences, medicine, dentistry and social work. Scientific works are published irregularly. The teaching staff numbers 680 and there are 8293 students.

Universidad Iberoamericana (Ibero-American University)
A private university founded in 1943. There is a summer school with classes conducted in Spanish and English. The academic year consists of two semesters between the months of August and May. The student population numbers 6900 with 1200 professors, and the library houses 110 000 volumes. The following departments are affiliated to the university: anthropology; civil engineering; chemical engineering; mechanical and electrical engineering; mathematics and physics; and nutrition. *Catálogo General Anual* (General Annual Catalogue), *Boletín* (Bulletin, monthly), *Catálogo de Planes de Estudio* (Annual prospectus) and *Catálogo de Postgrado* (Catalogue of Post-Graduate Studies, annual) are all published by the university.

Universidad Intercontinental (Intercontinental University)
Founded in 1976. Besides the areas of social studies and the humanities, there are departments of computer studies and dentistry. There are 967 students and a staff of 209 teachers.

Universidad Autónoma de Ciudad Juárez (Autonomous University of Juarez)
The university was founded in 1973 and has 4000 students and 380 teachers. Teaching is carried out in English and Spanish. There is an Institute of Biomedical Sciences and an Institute of Engineering. *Tribuna Universitaria* (University Tribune) is published.

Universidad del Valle de México (University of Valle de Mexico)
Founded in 1960, it is run under private control. No courses are offered in the field of science and technology.

Universidad La Salle de México (La Salle de Mexico University)
Founded in 1962, it has an academic year between September and June and is a private concern.
Apart from the faculties in the humanities there are schools of medicine, chemistry, engineering and mathematics, computer studies and systems

analysis. Various publications are issued: *Gaceta* (Gazette); *Indicador* (Indicator); *10 Días* (10 Days); *Nomos; Ezeta; Logos;* and *M.A.S.* There is a teaching staff of 840 teachers and 8010 students.

Universidad Autónoma Metropolitana (Autonomous Metropolitan University)
The university was established in 1973 and has a student population of 20 170 and 1627 teachers. There are three campuses – Azcapotzalco, with the Centre of Basic Sciences and Engineering; Iztapalala, with the Centre of Biological Sciences and Health, and Xochimilco. Publications include: *Organo Informativo Semanal* (Weekly Informative Bulletin); *Boletín de Tablas de Contenido* (Bulletin of Contents, three per year); and a *Reporte de Investigación* (Research Report) published irregularly.

Universidad Michoacana de San Nicolás de Hidalgo (University of Michoacana de San Nicolás de Hidalgo)
Obtained the status of university in 1917, but had been functioning since 1540. A private university with an academic year between September and May, and a library of 25 000 volumes. Three informative bulletins are published. Teaching staff numbers 1282 for 29 167 students in its faculties of sciences and humanities.

Universidad de Montemorelos (Montemorelos University)
Founded in 1973, it offers courses in medicine, nursing, education and business studies. Students number 1531, with 100 teachers.

Universidad de Monterrey (Monterrey University)
A private university founded in 1969. *Foro* (*Boletín de Información de la Rectoría*) (University Information Bulletin) is published. Its institutes are: natural and exact sciences; health; education; law and the arts. There is also a post-graduate school and a department of extension studies. Students number 3335 with 379 teachers.

Universidad Autónoma del Estado de Morelos (State of Morelos Autonomous University)
Founded in 1939, it is run under private control and publishes, monthly, *Gazeta Universitaria* (University Gazette). The student population numbers 11 337 in the schools of biological sciences, chemistry, medicine, nursing and obstetrics, architecture, law and education.

Universidad Autónoma de Nayarit (Nayarit Autonomous University)
Established in 1930 as the *Instituto de Ciencias y Letras de Nayarit* (Nayarit Institute of Sciences and Humanities), it attained the status of university in 1969. Teachers number 230 for the 2480 students. Schools include agriculture, medicine, nursing, chemical engineering, zoology and veterinary medicine.

Universidad Autónoma del Noroeste (Autonomous University of the North-West)
Founded in 1974, it has 365 lecturers for its 2973 students. Faculties are psychology, biology, and arts and humanities. There are seats in Monclova, Torreon, Piedras Negras and Sabinas.

Universidad del Norte (University of the North)
Founded in 1973, it has a total of 681 students and fifty-eight teachers for its courses of engineering, business administration and commerce.

Universidad Autónoma de Nueva León (Nueva León Autonomous University)
Founded in 1933, it has at present 1400 teachers for 80 000 students amongst its nineteen faculties. In the field of science and technology the following disciplines are offered: medicine, chemistry, nursing and obstetrics, odontology, physics and mathematics, electrical and mechanical engineering, veterinary medicine and public health. Publications include: *Interfolia-biblioteca Universitaria Alfonzo Reyes* (University Library Review) (monthly) and *Humanitas* (Humanities) (annually).

Universidad Autónoma 'Benito Juárez' de Oaxaca (Benito Juárez University of Oaxaca)
Acquired the title of university in 1955, but had been an institute of education since 1827. A private university with an academic year between September and June and a library of 77 237 volumes. Students number 10 000, with 690 teachers. The faculties are: architecture; medicine; chemistry; nursing and obstetrics; odontology; veterinary studies; law; commerce and languages.

Universidad Autónoma de Puebla (Puebla Autonomous University)
A private university founded in 1937, with an academic year between February and September. There are thirteen schools with a total of 10 500 students and 1000 teachers. In the field of science and technology there are schools of physics and mathematics, chemistry, civil and chemical engineering, nursing and obstetrics, medicine and dentistry.

Universidad Popular Autónoma del Estado de Puebla (Autonomous University of the State of Puebla)
Established in 1973 and has departments in the humanities, technology and health sciences. *Revista Universidad* (University Review) is published. Students number 3580 with 500 lecturers.

Universidad Autónoma de Queretaro (Autonomous University of Queretaro)
Founded in 1618, it is run privately with an academic year between September and June. Students number 6000 with 300 teachers, of which only ninety-five are employed on a full-time basis. There are schools of engineering, chemistry, nursing, law, business studies and post-graduate studies. Publications include *Universidad* (University); *Diálogo Universitario* (University Dialogue) and *Nuestra Verdad* (Our Truth).

Universidad Regiomontana (Regiomontana University)
Founded in 1957, it is a private university with teaching carried out in English and Spanish. Faculties include engineering and exact sciences, research sciences and social sciences. There is also a research division and a practical division affiliated to the university. The student population is 7200 with 509 lecturers. The library houses 6500 volumes.

Universidad Autónoma de San Luis de Potosí (Autonomous University of San Luis de Potosí)
Recognized as a scientific and literary institute since 1826. There are at present 7800 students and 670 teachers, and a library of 65 000 volumes. The faculties are: medicine; chemical sciences; engineering; physics; nursing; law; business studies and economics. Also affiliated to the university is the *Instituto de Investigación de Zonas Desérticas* (Desert Region Research Institute) which was founded in 1954, has a staff of twenty-four and a library of 4283 volumes, and publishes *Acta Científica Potosina* (Potosí Scientific Record) and *Contribuciones* (Contributions).

Universidad Autónoma de Sinaloa (Autonomous University of Sinaloa)
Established in 1873, it has at present 6470 students and 403 teachers. Schools are chemistry, physics and mathematics, agriculture, nursing, social sciences and law.

Universidad de Sonora (Sonora University)
Officially inaugurated in 1942, it has an academic year between September and June. Apart from the schools of arts and humanities there are schools of nursing, chemistry, engineering, agriculture and agricultural business studies. *Revista de la Universidad* (University Review) and *Gaceta Uni-Son* (Uni-Son Gazette) are published every four months. The library holds 45 000 volumes and students number 8370 with 421 teachers.

Universidad del Sudeste (University of the South-East)
The university was re-established in 1935, but was originally founded in 1756. The schools include: medicine, nursing and dentistry, engineering, commerce and law. Students number 1959 with 178 lecturers.

Universidad Autónoma Juárez de Tabasco (Juárez Autonomous University of Tabasco)
With a university population of 2070 students and 171 lecturers, the university was founded in 1958. The library houses 23 000 volumes. There are schools of accounting and business studies, civil engineering, medicine, veterinary medicine and zoology, nursing and dentistry.

Universidad Autónoma de Tamaulipas (Tamaulipas Autonomous University)
Founded between 1950 and 1951, it is financed with private capital. The students number 14 243 with 1033 teachers. The library has a total of 11 055 volumes and *Boletín de Investigaciones* (Research Bulletin) is published monthly. Apart from the faculties of humanities there are faculties of dentistry, engineering, architecture, nursing, veterinary medicine, agriculture, medicine and chemistry.

Universidad Veracruzana (Veracruz University)
Founded in 1944, it has an academic year between August and June and a student population of 40 114 with 2520 lecturers. *Revistas Zeta* (ZETA Review) and *Cosmos* are published, among others. In the field of science and technology, faculties include medicine, dentistry, nutrition, veterinary medicine and animal husbandry, and there are schools of nursing and obstetrics and the interdisciplinary unit of engineering and chemistry. The

university has seats in Campus Veracruz, Cordoba-Orizaba, Posa Rica-Tuxpan, and Coatzacoalcos-Minatitlan.

Universidad de Yucatán (Yucatán University)
The university was originally founded in 1624 and was reorganized in 1922. Students number 8054, with 680 teachers, and the central library houses 35 453 volumes and 10 146 periodicals and reviews. There are also specialized libraries in the faculties of medicine, engineering, chemistry and dentistry. Affiliated to the faculty of medicine are the schools of nursing, chemistry and pharmacy, engineering and the School of Veterinary Studies.

Universidad Autónoma de Zacatecas (Zacatecas Autonomous University)
Founded in 1836, it has a student population of 8220 under the control of 640 teachers. The library holds 20 000 volumes for the faculties of veterinary medicine and animal nutrition, dentistry, chemistry, nursing, and engineering.

Instituto Politécnico Nacional – Universidad Técnica (National Polytechnic Institute – Technical University)
A state-run university with an academic year between July and September. Founded in 1936, it has at present a total of 110 000 students and 11 000 teachers. *Gazeta Politécnica* (Polytechnic Gazette), *Acta Politécnica* (Polytechnic Records), and *Actas Médicas* (Medical Records) are published as well as *Anales de la Escuela Nacional de Ciencias Biológicas* (Annals of the National School of Biological Sciences) and *Acta Mexicana de Ciencia y Tecnología* (Mexican Records of Science and Technology). There are schools of biology, physics and mathematics, chemical and electrical engineering, textiles, paediatrics and obstetrics, and specialist areas in technology. The *Centro de Investigaciones y de Estudios Avanzados* (Centre of Research and Advanced Studies) is an affiliated institute.

Instituto Tecnológico y de Estudios Superiores de Monterrey (Technological and Higher Studies Institute of Monterrey)
Founded in 1943, it is a private concern with an academic year between August and May. Students number 17 000 with 740 teachers and a library of 111 000 volumes and 21 000 periodicals. Publications include *Boletín de Agronomía* (Agronomy Bulletin) (bi-monthly) and *Anuario de la Escuela de Agricultura* (Annual of the School of Agriculture). There are departments of agricultural commerce, agronomy, biology, soil study, zootechny, physics, mathematics, computing systems, chemistry, civil, electrical, industrial, mechanical and chemical engineering, and departments of thermodynamics and natural resources.

Instituto Tecnológico y de Estudios Superiores de Occidente (Technological Institute and Higher Studies of the West)
Founded in 1957, it has a total of 1900 students and 300 teachers. The library houses 50 000 volumes for the courses of civil engineering, industrial and electrical engineering, chemistry, mathematics, computing systems, human development and industrial relations.

Instituto Tecnológico Regional de Chihuahua (Chihuahua Regional Institute of Technology)
Founded in 1948, it has a staff of forty-four full-time and 115 part-time teachers and 2600 students participating in the courses on industrial, mechanical, chemical, electrical and production engineering.

Instituto Tecnológico Regional de Durango (Durango Regional Institute of Technology)
A dependant of the *Dirección General de Institutos Tecnológicos Regionales SEP* (General Board of Regional Institutes of Technology)
The institute offers first-degree courses in engineering and post-graduate courses in industrial production.

Instituto Tecnológico Regional de Mérida (Mérida Regional Institute of Technology)
Founded in 1961, it has a student population of 3423 and a library holding 14 276 volumes. Courses are offered in industrial and civil engineering and biochemistry, and there is a special course in industrial technology.

Instituto Tecnológico Regional de Morelia (Morelia Regional Institute of Technology)
Founded in 1965, it has a student population of 3000 and offers courses in industrial engineering and technology.

Instituto Tecnológico Regional de Oaxaca (Oaxaca Regional Institute of Technology)
Established in 1968, it has 3000 students and 280 teachers. *Introsintesis* is published and courses are offered in mechanical, chemical, electrical and civil engineering.

Instituto Tecnológico Regional de Querétaro (Querétaro Regional Institute of Technology)
Founded in 1967; its courses of industrial engineering and technology are offered to 1800 students. Teachers number 200.

Instituto Tecnológico Regional de Saltillo (Saltillo Regional Institute of Technology)
Established in 1951, it has 3000 students and a library of 8750 volumes. Courses are offered in industrial and metallurgical engineering and technological engineering.

Instituto Tecnológico de Sonora (Sonora Institute of Technology)
Founded in 1955, it has 3000 students and 180 lecturers for its courses in civil, industrial and chemical engineering, agricultural technology and psychology.

Industrial Research and Technology

Research in this field is carried out in two main organizations:

Instituto Mexicano de Investigaciones Tecnológicas AC (IMIT) (Mexican Institute of Technological Research)

Founded in 1950, the institute carries out applied research into natural resources and the development of industrial processes, pre-investment studies and industrial engineering. There is a library specializing in chemical engineering with 6700 volumes.

Instituto Mexicano del Petróleo (Mexican Petroleum Institute)
Carries out research into oil by-products, staffing for the petrol industry, exploration for resources, and refining, and runs refresher courses and courses for specialization, etc. *Revista* (Review) is published.

Agricultural and Marine Science and Technology

A great deal of research is instigated by the government sector as well as by the academic and private sector. Much of this work is to develop new technology, production processes and basic research in sciences related to agriculture, etc. There are sixteen institutions of national importance, among which are:

Instituto Nacional de Investigaciones Agrícolas (National Institute of Agricultural Research)
Founded in 1960, it carries out research in all fields and aspects of agricultural development and production; however, the specialized area is the development of new species of grain. Five institutes are affiliated, with laboratories for soil analysis, entomology and phytopathology. *Agricultura Técnica en México* (Technical Agriculture in Mexico) is published twice a year.

Instituto Nacional de Investigaciones Forestales (National Institute of Forestry Research)
Established in 1958, it researches into forestry, forest botany, soil products and pathology, aero-photometric surveys and the conservation of forestry reserves.

Instituto Nacional de Investigaciones Pecuarias (National Institute of Livestock Research)
Founded in 1941, it carries out research and experiments into animal nutrition and veterinary sciences.

Centro Internacional del Mejoramiento de Maíz y Trigo (International Centre for the Improvement of Maize and Wheat)
Founded in 1966 through a bilateral agreement between the Rockefeller Foundation and the Ministry of Agriculture in Mexico. The centre researches and advises various countries on the improvement of grain. There are eight experimental stations in Mexico and their research results give assistance to neighbouring countries. The centre also offers programmes of specialization and improvement to scientists and technicians.

Instituto Mexicano de Recursos Renovables (Mexican Institute for the Conservation of Natural Resources)

Centro Nacional de Ciencias y Tecnologías Marinas (National Centre of Marine Science and Technology)

Instituto Nacional de Pesca (National Fisheries Institute)

Medical Science and Technology

In this field there are twenty-three institutions, including twelve professional and technical associations which, if they do not develop research directly, collaborate in the promotion and development of those activities.

Instituto de Salubridad y Enfermedades Tropicales (Institute of Public Health and Tropical Diseases)

Centro de Higiene y Estación de Adiestramiento en Enfermedades Tropicales (Centre of Hygiene and Educational Station for Tropical Diseases)
These two organizations, centred in Mexico City and Veracruz, carry out research into diagnosis and classification of infectious diseases of the tropical regions.

Instituto Nacional de Higiene (National Institute of Hygiene)

Instituto Miles de Terapéutica Experimental (Miles Institute of Experimental Therapy)
Carries out basic research and experiments into pharmacology.

Instituto Nacional de Cardiología (National Cardiology Institute)

Instituto Nacional de Neumología (National Institute of Pneumonology)

Among the professional societies the following are worth a mention:

Sociedad Mexicana de Parasitología AC (Mexican Society of Parasitology)

Sociedad Mexicana de Pediatría (Mexican Society of Paediatrics)

Sociedad Mexicana de Salud Pública (Mexican Society of Public Health)

Asociación de Medicas Mexicanas AC (Mexican Association of Women Doctors)

Nuclear Science and Technology

Instituto Nacional de Investigaciones Nucleares (National Institute of Nuclear Research)
Founded in 1979, it forms part of the *Instituto Nacional de Energía Nuclear* (National Institute of Nuclear Energy) and carries out research into nuclear technology, including the peaceful use of this energy. There is a specialized library of 30 000 volumes, 20 000 periodicals and about 290 000 scientific reports.

Meteorological and Astrophysical Research and Development

Dirección General del Servicio Meteorológico Nacional (General Board of the National Meteorological Service)
Publishes meteorological, hydrological and climatic reports.

Instituto de Astronomía (Institute of Astronomy)
Carries out research into astrophysics and astronomy and is affiliated to the Universidad Autónoma de Mexico.

Instituto Nacional de Astrofísica, Optica y Electrónica (National Institute of Astrophysics, Optics and Electronics)

Associations and Societies in the General Field of Science and Technology

There are about fifty such bodies in the country, among which are:

Academia Nacional de Ciencias (National Academy of Sciences)

Sociedad Matemática Mexicana (Mexican Society of Mathematics)

Academia de la Investigación Científica (Academy of Scientific Research)

Sociedad Química de México (Chemistry Society of Mexico)

Asociación Mexicana de Geólogos Petroleros (Mexican Association of Petroleum Geologists)

Information Services in Science and Technology

The body responsible for the revision, systematizing and disseminating of information is *CONACYT*. *CONACYT* accomplishes its objectives with the aid of bibliographies and computerized systems with access to data banks of national and international information, all of which forms back-up material for many research projects. *CONACYT* offers its services to light industry and small businesses.

There is also the *Centro de Investigación de Recursos Científicos y Tecnológicos (CIRCYT)* (Centre for Research into Scientific and Technological Resources), which plays a vital role in the field of information. The centre compiles data on institutions that carry out experimental research and development; on the scientific and technical personnel as well as on the projects presently under execution.

CONACYT publishes the *Repertorio Bibliográfico* (Bibliographical Index) which lists information on scientific and technical books in the country. Similar work is carried out into scientific and technical reviews published in the country.

For the purpose of issuing technical information of use to light industry and small businesses *Información, Innovación y Tecnología (INFOTEC)* (Technological Information and Innovation) has been in operation since

1973. There is also a multidisciplinary information service under the patronage of UNAM, the *Centro de Información Científica y Humanística de la Coordinación para la Investigación Científica (CICH)* (Scientific and Humanistic Information Centre for the Coordination of Scientific Research).

There are twenty-eight institutions in the country acting as documentation and data banks and thirty-four museums and similar centres with data banks of specialized information. Among them are:

Centro de Información Científica y Humanística (Centre of Scientific and Humanistic Information) of the Autonomous University of Mexico
The centre offers documentation and bibliographical services; a reference library and a data bank for specialized information. The centre receives publications from many different countries and houses 10 000 periodicals and publications.

Biblioteca Nacional de México (National Library of Mexico)
Founded in 1833, it now functions as one of the institutes of *UNAM*. The library holds 1 000 000 volumes and 150 000 documents.

Archivos Históricos y Bibliotecas (Historical Archives and Libraries)
Has 300 000 volumes and documents.

Biblioteca Central de la Universidad Nacional Autónoma de México (Central Library of the National Autonomous University of Mexico)
Has 300 000 volumes and 10 000 periodicals.

Biblioteca Nacional de Antropología e Historia 'Dr Eugenio Davalos' (Dr Eugenio Davalos National Library of Anthropology and History)
This is one of the specialized libraries in the field of archaeology, linguistics, history, ethnology and anthropology, and has the most complete material in the country. The library houses 500 000 volumes, 11 000 periodicals and other pamphlets.

Centro de Servicios de Información y Documentación, CONACYT-UNESCO (Centre of Information and Documentation Services)

International Cooperation in Science and Technology

Through *CONACYT* several agreements have been established for the interchange of technology with similar bodies in Latin America. Mexico participates in the specialist organizations of the United Nations such as UNESCO, FAO, UNICEF, UNEP, OIT, UNIDO, etc. The main objectives of international cooperation are: to incorporate specific projects to meet national objectives; to collaborate with developing countries, in particular with Latin America, and to establish joint projects which could lead to a high level of technical cooperation.

In recent years many agreements have been fulfilled in international cooperation and in particular special agreements have been promoted between countries of similar socio-economic development, in particular with Nicaragua.

There are presently thirty-four inter-governmental agreements; twenty-eight inter-institutional agreements and thirteen programmes of interchange of specialist information and technology under way.

Nicaragua

Geographical, Demographical, Political and Economic Features

The Republic of Nicaragua is the largest country in Central America. Its borders to the north are shared with Honduras; to the east with the Antilles Sea; to the south with Costa Rica and to the west with the Pacific Ocean. Islands belonging to Nicaragua are Little Corn Island and Great Corn Island in the Caribbean Sea facing Bluefields.

The population according to estimates in 1980 was 2 732 520, with a projection for 1985 of 3 347 000. The population density per square kilometre is twenty-two. The area of the country is 127 755 square kilometres, of which 9291 square kilometres is lakes. The capital, Managua, has a population of 615 000. Of the population 53.1 per cent, live in urban areas and the ethnic composition of the population is: 17 per cent white, 9 per cent negroes, 0.02 per cent 'Misquitos', 0.003 per cent Sumos, 0.0003 per cent Ramaquies and 73.9 per cent 'mestizos' (mixed blood).

Government

Until 10 July 1979 the country was governed by the Constitution of March 1974. During the period leading up to July 1979, Nicaragua was subject to an armed insurrection lead by the *Frente Sandinista de Liberación Nacional (FSLN)* (Sandinista Front for National Liberation) which eventually took power. Since the *FSLN* took power, the Republic has experienced radical changes. The president was replaced by a junta of three members who now lead the country; however, the real force of power lies with a body of nine members who represent the principal leaders of the revolution. There is also a cabinet represented by civilians, though important posts, such as the *Ministerio del Interior* (Home Office), are run by leading members of the *FSLN*.

The *Guardia Nacional* (National Guard), the former police guards of the country, has been outlawed and replaced by the *Ejército Sandinista* (Sandinista Army) and the *Milicias Populares* (People's Militia). With the objective of drafting a new constitution, a council of state has been formed with forty-seven members, the majority of whom are Sandinistas.

AGRICULTURE

Almost 50 per cent of the country is covered in virgin forests. Since the revolution of 1979 most of the potentially agricultural land has been put under state control. Among this is the land owned by the ex-president.

All expropriated land has been turned into state haciendas. The private sector still remains active in agriculture and their produce destined for export is sold through nationalized agencies. The main products for export are coffee, sugar, and cotton (although cotton production has deteriorated seriously as a result of the Civil War).

LIVESTOCK

The latest estimates (1978) show 2.8 million head of sheep and 710 000 pigs.

FORESTRY

Pine forests cover an area of 1.5 million acres. Mahogany, rosewood and cedar are also grown. Cut wood for export reached $7.6 million in 1978.

FISH

Although the potentialities have not been fully exploited, the *Instituto de Fomento Nacional (INFONAC)* (National Development Institute) has begun to develop programmes to reorganize the fishing industry. Fresh-water and salt-water crayfish and tuna fish are the main species being exploited. Lobsters and prawns are farmed and in 1980 these were valued at $25.9 million.

MINERALS

Although the real mineral wealth of the country is not known, in 1977 1904 kilos of gold were exported as well as 5 tons of silver and 5000 tons of copper, and there are also lead and zinc mines. Important reserves of tungsten and iron have been discovered but have not yet been exploited. Like the agricultural sector, mining proprieties were nationalized in 1979 and at present there are only programmes under way to increase gold production.

The potential exploitation of off-shore oil is faced with uncertainty at the moment and is a major preoccupation of the oil companies. The 'Union' held concessions (authorized by ex-President Somoza) in the north-west near the border with Honduras, and Chevron in the south and west, in the Prinzapolka area. These firms, as well as Texaco, Western Caribbean, General Crude Oil and Oceanic Exploration are all waiting for decisions on their contracts by the new Government.

Education

Before the Revolution it was estimated that 47.5 per cent of adults were illiterate; however, one of the first policies adopted by the junta govern-

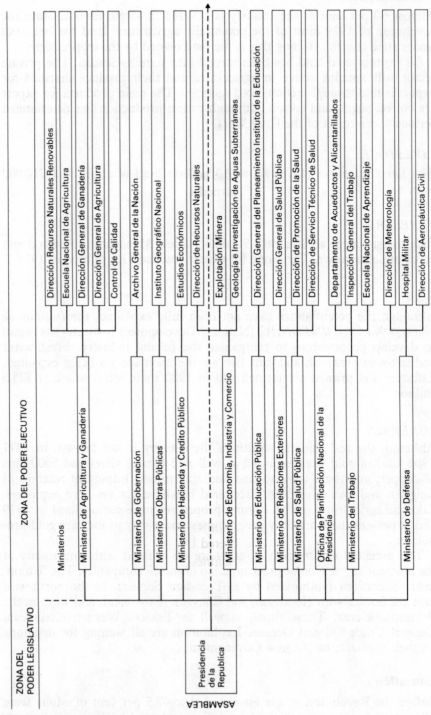

Fig. 12 Organization of science and technology

ZONA DEL CIUDADANO Y
COMUNIDAD CIENTIFICA

Colegio y Asociaciones
Profesionales

Cámara de Industrias
de la Construcción y
Comercio

Científicos y Técnicos
Individuales

Industrias Privadas

Cooperativas de
Productores

ZONA DEL PODER EJECUTIVO (continuation)

Entes Autónomos y Semiautónomos

Banco Central de Nicaragua

Banco Nacional de Nicaragua

Instituto Nacional de Comercio Exterior e Interior

Banco de la Vivienda

Instituto de Fomento Nacional

Empresa Nacional de Luz y Fuerza

Instituto Nacional de Energía Eléctrica

Empresa de Riego de Rivas

Instituto Nicaragüense del Café

Empresa Aguadora de Managua

Telecomunicaciones y Correos de Nicaragua

Empresa Aguadora Municipal

Ferrocarril del Pacífico de Nicaragua

Superintendencia de Bancos

Caja Nacional de Crédito Popular

Autoridad Portuaria de Corinto

Instituto Agrario de Nicaragua

Instituto Nacional de Seguridad Social

Universidad Nacional Autónoma de Nicaragua

Universidad Centroamericana

Instituto Politécnico

Junta Nacional de Asistencia Social

Instituto Técnico Vocacional

Instituto Tecnológico de Granada

Escuela de Pesca

Laboratorios

Instituto de Tecnogía de Alimentos

Instituto Mariano Fiallos

Oficina de Planeamiento Académico

Laboratorio de Análisis de Materiales

Centro de Investigación Científicas

Capacitación de Técnicos

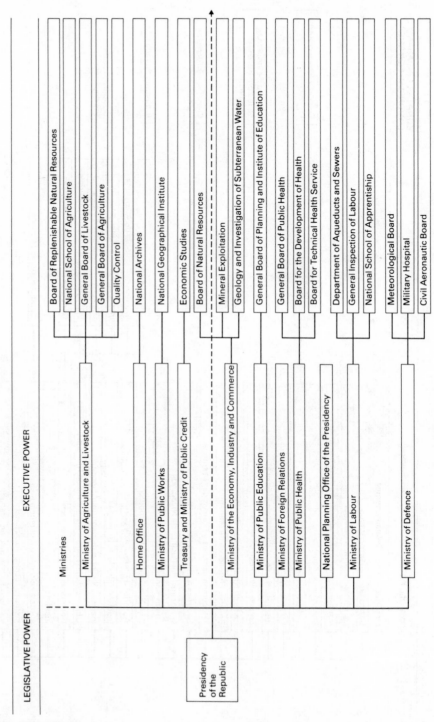

Fig. 12 – *continued*

EXECUTIVE POWER (continuation)

PUBLIC AND SCIENTIFIC COMMUNITY

Autonomous and Semi-Autonomous Bodies

- Central Bank of Nicaragua
- National Bank of Nicaragua
- National Institute of Trade, Foreign and Internal
- Building Society (Social Welfare)
- Institute of National Development
- National Firm of Light and Energy
- National Institute of Electrical Energy
- Rivas Irrigation Company
- Nicaraguan Coffee Institute
- Managua Water Board
- Post and Telecommunications
- Municipal Water Board
- Pacific Railway of Nicaragua
- Superintendancy of Banks
- National Cashiers' Office of the 'Credito Popular'
- Corinto Port Authority
- Nicaraguan Agrarian Institute
- National Institute of Social Security
- National Autonomous University of Nicaragua
- Central American University
- Polytechnic Institute
- National Board of Social Assistance
- Technical Vocational Institute
- Technological Institute of Granada

School of Fishing

Laboratories
- Institute of Food Technology
- Mariano Fiallos Institute
- Office of Academic Planning
- Material Analysis Laboratory
- Centre of Scientific Research

Technical Training

College and Professional Associations

Chamber of Construction and Commerce

Scientists and Technicians

Private Industry

Producers' Cooperatives

ment was a literacy campaign, launched in 1980, which after only one year is producing exciting results.

International Relations

Nicaragua is a member of the United Nations Organization, the Organization of American States and the Central American Common Market.

Scientific and Technological Research at Academic Level

Universidad Nacional Autónoma de Nicaragua (National Autonomous University of Nicaragua)
Founded in 1812, it has a staff of 756 teachers and 14 889 students. Its library contains 36 000 volumes and there are faculties of medicine, dentistry, humanities, economics, law and social studies, physics and mathematics, science and the arts. Affiliated to the university are the *Instituto de Investigaciones del Desarrollo* (Institute of Development Studies) and the *Instituto de Capacitación Sindical* (Institute of Union Education).

Universidad Centroamericana (Central American University)
Founded in 1961, it has a staff of 230 teachers and 4137 students. Faculties are: engineering; law; business studies; humanities and sciences; veterinary medicine and zootechics. Affiliated to the university are the *Instituto de Investigación Social 'Juan XXIII'* (Juan XXIII Institute of Social Research) and the *Instituto Histórico Centroamericano* (Central American Historical Institute).

Universidad Politécnica de Nicaragua (Polytechnic of Nicaragua)
Established in 1970, it has a staff of seventy teachers for its 600 students. There are schools of nursing and industrial technology and departments of commerce and decorative art.

Escuela Nacional de Agricultura y Ganadería (National School of Agriculture and Livestock)
A dependant of the *Instituto Nicaraguense de Tecnología Agropecuaria (INTA)* (Nicaraguan Institute of Agriculture and Livestock Technology).

Instituto Centroamericano de Administración de Empresas (INCAE) (Central American Institute of Commercial Administration)
In collaboration with the Harvard University, a course is offered in commercial business studies, as well as programmes of training for executives, advisory services, etc.

Instituto Nicaraguense de Cine (Nicaraguan Institute of the Cinema)

Industrial Technology and Research

Servicio Geológico Nacional (National Geological Service)

Mineral research and scientific research is carried out. Grants are awarded for specialist study.

Agricultural and Marine Science and Technology

Instituto Agrario de Nicaragua (Agrarian Institute of Nicaragua)

Instituto Nicaraguense de Tecnología Agropecuaria (INTA) (Nicaraguan Institute of Agriculture and Livestock Technology)

Professional Associations and Societies in the General Field

Colegio de Médicos y Cirujanos de Nicaragua (Nicaraguan College of Doctors and Surgeons)

Sociedad Nicaraguense de Psiquiatría y Psicología (Nicaraguan Society of Psychiatry and Psychology)

Information Services in Science and Technology

There are fourteen or more centres of documentation and data banks, among which are:

Biblioteca Nacional (National Library) with 70 000 volumes.

Biblioteca Central de la Universidad Nacional de Nicaragua (Central Library of the National University of Nicaragua) with 34 000 volumes.

Panama

Geographical, Demographical, Political and Economical Features

The Republic of Panama shares its borders to the north with the Antilles Sea; to the east with Colombia; to the south with the Pacific Ocean and to the west with Costa Rica. Panama has more than 600 islands in the Antilles Sea; among them is the Archipelago de las Mulatas or San Blas, which is made up of 332 islands, and there are more than 1000 islands in the Pacific Ocean, the islands of Coiba and the Rey being the largest.

The total area is 77 082 square kilometres including the 1432 square kilometres of the Canal Zone under the United States of America.

According to the census of 1980 the population is 1 830 175 with a projection for 1985 of 2 260 000, and the density per square kilometre is 23.7. The capital, Panama, has a population of 502 000, 54 per cent of the population live in urban areas.

Government

The country is governed by the Constitution of 1972 and has a democratic and representative government. The executive power is represented by the President of the Republic, the Vice-President and their ministers of state. The legislative power is in the hands of the *Asemblea Nacional de Representantes de Corregimientos* (National Assembly of Representatives of the Townships) and the *Consejo Nacional de Legislación* (National Council of Legislation). The judicial power is administered by the High Court of Justice, the Lower Courts of Justice and the judicatures established by law. The decisions taken by the powers are in cooperation with the *Fuerza Pública* (Public Force), which is represented by the *Guardia Nacional* (National Guard).

Natural Resources

AGRICULTURE

According to a study made in 1975, 18.5 per cent of the national territory is cultivated land; 57.1 per cent is pasture land and 9.5 per cent is fallow land. Of the remaining area, only a small part is cultivable land.

About 60 per cent of food products are imported. Of the land under cultivation, 26.4 per cent is privately owned property and 44.7 per cent is usufructuary. Important agricultural products are bananas (of which in 1980 1 071 000 tons were exported), cacao, coffee, cocoa and maize.

LIVESTOCK

Estimates made in 1978 show 1 379 000 head of cattle, 201 700 pigs and 4 422 000 poultry.

FORESTRY

The main species exploited commercially are oak, cativo, amarillo, balsam of tolu, mahogany and cedar.

FISH

Although Panama has a vast quantity of marine fauna, only anchovy, herring and *pez escama* are exploited for national consumption. Shrimps are also exploited, but solely for export.

MINERALS

According to various evaluations, Panama is potentially rich in mineral resources. Deposits of copper are to be found in the provinces of Chiriqui, Colón and Darien. The most important area for exploration is Cerro Colorado (Chiriqui) where studies were completed in 1978 by Texas Gulf, while in Pataquilla (Colón) exploitation licences have been granted to a Japanese group. There are deposits of gold (Central Zone), bauxite and manganese.

INDUSTRIAL SECTOR

Local industry includes cigarettes, textiles, food processing, shoes, soap, and cement, and there is an oil refinery in Colón.

Education

Primary education is obligatory for all children between the ages of seven and fifteen. Of the population 79.3 per cent is literate.

International Relations

The Republic of Panama is a member of the Organization of American States and the United Nations Organization.

Organization of Science and Technology

The governing bodies for the policy of science and technology at national level are the *Sección de Política Científica y Tecnológica* (Division of Scientific and Technological Policy), which is at government level and is a dependant of the Ministry of Planning and Economic Policy; and at academic level there is the *Comisión de Investigaciones* (Research Commission) of the National University.

The Ministry of Planning and Economic Policy is responsible for establishing social and economic policy, and this includes policy-making in the field of science and technology. The *Dirección de Presupuesto de la Nación* (National Budget Office) holds a key position, assuring that adequate funds are available to the ministry to meet its aims.

Within its organization, the ministry also calls upon the *Dirección de Planificación Económica y Social* (Board of Economic and Social Planning) and its dependant the *Departamento de Programación General* (Department of General Programming) to which belongs the *Sección de Ciencia y Tecnología* (Division of Science and Technology), all of which complement the ministry in implementing its policies and programmes.

Other important institutions in this area are the ministeries with their offices and departments; autonomous bodies and entities and the universities. Research and development activities are carried out in all these organizations and are classified in later sections.

Policies and Financing in Science and Technology

The plans for scientific and technological policy contained in the document *El Estado Actual y las Tendencias en el Campo de las Políticas Científicas y Tecnológicas* (The present state and tendencies in the field of scientific and technological policy), prepared by the Ministry of Planning and Economic Policy, are as follows:

(a) to establish the institutional machinery to analyse and evaluate technology in projects using public funds;
(b) to maintain a relation between policies in science and technology with the objectives in the socio-economic development of Panama;
(c) to create an infrastructure in research and development that corresponds to the problems in science and technology as determined in the national programmes and projects;
(d) to create a better liaison between the various national organizations in the field of science and technology;
(e) to promote national development of new technology with the aim of rationalizing the importation of technology;
(f) to develop human resources destined to the scientific sector.

The role that human resources play in the development of science and technology has been a major concern for the Panamanian Government and a concerted effort is being made to improve the situation with the aid of the Ministries of Planning and Economic Policy, Education, and Labour and Social Welfare; the *Instituto para la Formación y Aprovechamiento de Recursos Humanos* (Institute for Training and Improvement of Human Resources); *Contraloría General de la República* (General Controllership of the Republic) and the University of Panama.

Table 9.1 Human resources in science and technology

Year	Scientists and engineers						Technicians
	Total (in units)	Breakdown in field of specialization (in units)					Total (in units)
		Natural sciences	Engineering and technology	Agricultural sciences	Medicine	Social sciences	
1965	449	57	23	11	24	334	2
1970	559	121	20	12	47	313	42
1975	1112	265	72	23	73	679	505
1980	1489	215	121	109	156	888	710
1985*	2929	685	286	36	227	1755	822
1990*	3988	958	446	47	265	2272	1194

*Figure based on a projection.

The *Vice-Rectoría de Investigaciones y Post-grado* (Vice-Rectorship of Research and Post-Graduate Studies) in the University of Panama has been established as a further step towards remedying the present situation in scientific and technological personnel.

In Table 9.1 data from 1965 to 1980 can be studied with projections for 1985 and 1990.

In 1978, with the objective of instigating a development policy in science and technology, data was collected and analysed, under the auspices of the *Programa Regional de Desarrollo Científico y Tecnológico de la OEA* (Regional Programme of Scientific and Technological Development of the Organization of American States), the aim being to estimate costs and public expenditure in scientific and technological activities. At the same time the *Dirección de Presupuesto de la Nación* (National Budget Office) together with the General Controllership of the Republic were effecting a revision of the *Clasificaciones Presupuestarias del Gasto Público* (Budgetary Classification of Public Expenditure). All these efforts were made in order to classify the functions of public expenditure in science and technology.

The first such study was carried out in the budgetary year of 1981. The results can be seen in Table 9.2.

Table 9.2 Breakdown by source and sector of activity 1981 – in thousands of balboas

Sector of activity	Source National		Foreign	Total
	Government funds	Other funds*		
General governmental	28 776 063	—	—	28 776 063
Higher education	1 314 780	—	—	1 314 780
Companies of the productive sector	—	—	—	—
Total	30 090 843	—	—	30 090 843

*Principal source of funds: private funds; funds from productive firms; special funds; foundations, etc.

Table 9.3   Tendencies in expenditure in research and development

Year	Population	Gross National Product (GNP) (in thousands of balboas)	Total expenditure in research and development	Rate of exchange US$1 = B/1
1965	1 268 570	644.1	—	—
1970	1 464 392	1 019.4	—	—
1975	1 677 729	1 913.6	—	—
1980	1 896 385	3 206.4 (E)	—	—
1985*	2 116 773	4 551.0		
1990*	2 346 036	—		—

*Figure based on a projection (based in 1980).

Current and projected expenditure for the period 1965–90 can be seen in Table 9.3, and although the figures are only those of the central Government and certain decentralized institutions, this data is valuable for future programming of activities.

Science and Technology at Government Level

The major part of research activities is carried out by governmental organizations or in autonomous bodies closely related to government departments and in the universities.

These organizations will be analysed more fully in the following sections.

Research in Science and Technology at Academic Level

Universidad de Panamá (University of Panama)
Founded in 1935, it is located in the capital of Panama. There are 2300 researchers and teachers and 33 833 students. Faculties include agriculture, architecture, law and political sciences, medicine, natural sciences and pharmacy, dentistry, public administration and commerce, philosophy, art and education. There are six institutes affiliated to the university that carry out research activities, among them are: *Centro de Investigaciones Agropecuarias* (Centre of Agricultural and Livestock Research); *Centro de Ciencias del Mar y Limnología* (Centre of Marine Studies and Limnology); *Centro de Energía* (Centre of Energy); and the *Departamento de Investigaciones del Laboratorio Especializado de Análisis* (Department of Specialized Laboratory Research and Analysis). Recently the Vice-Rectorship of Research and Post-Graduate Studies has been formed and will act as the coordinating body at this level between the various university organizations.

The university publishes various works, among which are: *Boletín* (Bulletin); *Estadísticas Universitarias* (University Statistics); and *Boletín Informativo* (Informative Bulletin). The university library is the largest in the country with 206 436 volumes and maintains a bibliographical service with more than 200 organizations.

Universidad Santa María la Antigua (Santa Mariá la Antigua University)
Founded in 1965 in the capital of Panama, it has a staff of 178 researchers
and teachers and a student population of 1916. Faculties include law and
political science, business studies, technology and natural sciences,
humanities and religious studies, and social sciences.

Universidad Tecnológica de Panamá (Technical University of Panama)
Founded on 13 August 1981, it is being developed from the former *Instituto
Politécnico* (Polytechnic Institute) and the *Centro Experimental de Ingeniería*
(Experimental Engineering Centre) which were previously affiliated to the
University of Panama.

Other Academic Institutions

There are five specialist institutes including:

Escuela Naútica de Panamá (Nautical School of Panama)

Escuela Nacional de Artes Plasticas (National School of Plastic Arts)

Industrial Technology and Research

Ministerio de Comercio e Industrias (Ministry of Commerce and Industry)
Affiliated to this ministry are the *Dirección de Industrias* (Board of
Industry); the *Comisión Panameña de Normas Industriales y Técnicas*
(Panamanean Commission of Industrial and Technical Standards) and the
Departamento de Patentes (Department of Patents).

Corporación Azucarera la Victoria (Victoria Sugar Corporation)
An autonomous organization.

Empresa Estatal de Cemento Bayano (Bayano State Cement Company)
An autonomous body.

Agricultural and Marine Science and Technology

Ministerio de Desarrollo Agropecuario (Ministry of Agricultural and Livestock
Development)
Affiliated to this ministry are the *Dirección de Recursos Naturales* (Board of
Natural Resources); the *Dirección Nacional de Agroindustries* (National
Board of Agro-industry); the *Empresa Nacional de Semillas* (National Seed
Company) and the *Empresa Nacional de Maquinaria* (National Machinery
and Engineering Company).

Instituto de Investigaciones Agropecuarias (IDIAP) (Institute of Agricultural
and Livestock Research)
An autonomous body.

Cítricos de Chiriqui (Citrus Fruits of Chiriqui)
An autonomous body.

Dirección de Recursos Marinos (Board of Marine Resources)
A dependant of the Ministry of Commerce and Industry.

Specialist university institutes in this field include: *Centro de Ciencias del Mar y Limnología* (Centre of Marine Sciences and Limnology) and the *Centro de Investigaciones Agropecuarias* (Centre of Agricultural and Livestock Research).

Medical Science and Technology

Dependants of the Ministry of Health are: *Dirección de Nutrición* (Board of Nutrition) and the *Laboratorio Central de Salud* (Central Health Laboratory). The *Departamento de Farmacología Clínica* (Department of Clinical Pharmacology) is a dependant of the Social Security Department.

Gorgas Memorial Laboratory of Tropical and Preventative Medicine
Founded in 1928, it has a research staff of twenty-nine. An annual report is published.

Smithsonian Tropical Research Institute
Founded in 1949 and administered by the Smithsonian Institute, USA. There are ninety researchers and a library of 16 000 specialized volumes.

Professional Associations and Societies in the General Field

There are eight such institutions in Panama, among which are:

Academia Nacional de Ciencias (National Academy of Sciences)

Centro para el Desarrollo de la Capacidad Nacional de Investigación (Centre for the Development of National Research Capacity)
Founded in 1976; is a dependant of the University of Panama, whose main functions are to coordinate scientific and technological research in Panama.

Academia Panameña de la Historia (Panamanean Academy of History)

Information Services in Science and Technology

At the end of 1978 the Ministry of Planning and Economic Policy and the *Sindicato de Industriales de Panamá* (Industrial Trade Union of Panama) signed an agreement to form the *Servicio de Información y Extensión Tecnológica* (Information Service and Technological Extension) with the objective, among others, of orientating advisory services towards the improvement of industrial productivity. Also affiliated to the Ministry are the *Centro Nacional de Documentación sobre Población (CENDOP)*

(National Centre of Population Documentation), founded in April 1980, and the *Centro de Datos y Documentación (CEDARE)* (Centre of Data and Documentation), which is a dependant of the *Dirección de Planificación y Coordinación Regional* (Board of Regional Coordination).

Centro de Servicios de Informacion Agropecuarias (Service Centre for Agricultural and Livestock Information)
Affiliated to the *Instituto de Investigaciones Agropecuarias* (Institute of Agricultural and Livestock Research), founded in 1978 to aid research in this field. The Ministry of Agricultural and Livestock Development is also elaborating an information system in this sector.

Finally, the *Instituto para la Formación y Aprovechamiento de Recursos Humanos (IFARHU)* (Institute for Training and Improvement of Human Resources) has formed an information centre on human resources.

Apart from the services already mentioned, there are six important libraries and more than six museums containing specialized information. Of these institutions, it is worth mentioning the *Biblioteca Nacional* (National Library) with 200 000 volumes.

International Cooperation in Science and Technology

Panama has received US$12 million during 1979–80, which includes 129 projects, programmes, advice, donations, staff and teaching facilities to strengthen economic, social, technological, cultural and educational policies. Panama has scientific and cultural agreements with Mexico, Japan, Italy, Finland, Czechoslovakia, China, Holland, Great Britain, France, Spain, West Germany, Hungary, Bolivia and the Scandinavian countries.

Among the most important projects are:

CIID–IDRC of Canada: studies of development policies and interchange of technology in the agricultural sector.

OAS: basis for the formulation of a scientific and technological policy. A project initiated in 1976 with the principal objective of establishing criteria and guide-lines, through the analysis of completed projects, to aid the establishing of a scientific and technological policy in the productive sector.

OAS: exchange of technology and information. A project initiated in 1976 to study information systems in the country and to improve existing resources.

OAS: generic medicaments – initiated in 1980 to establish bio-physio-pharmaceutical standards for raw materials and manufacturing processes for drugs.

OAS: copper mining – project established in 1975; carries out geological surveys to be used in future exploitation and evaluation of the country's mining potential.

OAS: pharmaceutical analysis of the flora of Panama – researches into resources of raw material for the pharmaceutical and food industry and the most effective methods of production.

OAS: manufacture of paper pulp based on banana fibre – initiated in 1980, this is fundamentally a laboratory research project into the use of stems, peduncles and banana leaves for the manufacture of pulp for paper.

Paraguay

Geographical, Demographical, Political and Economic Features

Paraguay was a Spanish colony which obtained its independence on 14 May 1811. It is a landlocked country situated in the central zone of South America at 1448 kilometres from the Atlantic Ocean. Its borders to the north and east are shared with Brazil, with 1339 kilometres of frontier; to the south-east, south and west Argentina is the adjacent country, while to the north-west lies Bolivia.

The total area of Paraguay is 406 702 square kilometres. This figure can be divided into a western area of 246 925 square kilometres and an eastern area of 159 827 square kilometres.

The population, according to official estimates of 1979, reaches a figure of 2 973 000 with a projection for 1985 of 3 625 000. According to the same estimates the population density per square kilometre is 7.3. The population comprises mainly 'mestizos' resulting from inter-breeding of Spanish and Guaraní, with a small group of negroes. The official language of the country is Spanish, although about 50 per cent of the population can only speak Guaraní; 40 per cent are Spanish-speaking, while the remainder is bi-lingual. There are about 46 700 Indians who have not been integrated into the society and who reside mainly in the Chaco and in the forest in the east of the country. Since 1920, Paraguay has been receiving constant waves of immigrants, mainly from Canada, Japan and Germany.

Government

According to the Constitution of 1967, the Republic of Paraguay is a representative republic. The state comprises three powers: the executive power is exercised by the President of the Republic, who must be Catholic and over forty years, and is elected for a period of five years. The president is advised by a cabinet of eleven ministers. The legislative power is made up of two Chambers, the Senate with thirty members and the Chamber of Deputies with sixty members of parliament. The judicial power is represented by the Supreme Court of Justice (five members nominated by the President of the Republic); plus the Exchequer and various Courts of Justice. There is also a Council of State which is a consultative body of the executive power and is made up of ministers, the archbishop, the rector of the university and the armed forces.

Administratively speaking the country is divided into nineteen departments governed by delegates of the executive power. There are fourteen in the eastern zone and five in the Chaco zone which are at present under military administration. The departments are divided into municipalities and districts. The city of Asunción, the capital of the Republic, is a special municipality comprising the districts of Encarnación, Catedral, San Roque, Recoleta and Trinidad.

Natural Resources

AGRICULTURE

Agriculture is the main economic activity, employing 50 per cent of the economically active population, producing 35 per cent of the GNP and generating 95 per cent of the country's export earnings. According to figures of 1976, agriculture covers an area of approximately 1.5 million hectares of national territory. The main agricultural products can be broken down into the following groups according to 1980 figures:

Table 10

Cotton	150 000 tons	Beans	67 000 tons
Maize	600 000 tons	Cotton	77 000 tons
Tobacco	40 000 tons	Sweet potatoes	130 000 tons
Soya beans	550 000 tons	Coffee	8 000 tons

The nation has a vast amount of productive land which is unutilized. Land settlement programmes to take advantage of this potential have been carried out by public and private entities over the last two decades. As a result, 523 settlements with 104 236 family lots, averaging 52 hectares have been established.

FORESTRY

Paraguay has extensive forestry reserves in the western zone – especially hard wood and cedar – which have not been fully exploited. Wood production for 1977 reached 381 000 cubic metres of cut wood. In the last few years programmes for the planting of pines are being developed through projects financed by funds from the United Nations.

LIVESTOCK

Figures for 1977 show 5.4 million head of cattle, 315 000 horses, 800 000 pigs and 355 000 sheep. The same figures show exports of meat products at US$209 million, which represents 12 647 tons. In the same year the production of fresh meat reached 74 000 tons, and of processed meat 16 000 tons.

MINERALS
Paraguay possesses reserves of manganese and iron, but these are not being exploited because they are not considered to be commercially viable. Reserves of silver are being exploited on a small scale.

INDUSTRIAL PRODUCTION
The predominant industries are: beer; cigarettes; by-products of oil; sugar; cement; flour; almond oil; cocoa; castor oil; tung oil; alcohol and paper bags.

Education
Primary education is obligatory and free, although there is not full availability of educational facilities. Illiteracy is estimated to be 22 per cent in urban areas and 30 per cent in rural areas. Paraguay has about 2288 primary schools and 421 private schools to give and education to some 459 453 students. Besides this there are 652 establishments of secondary education for 66 746 pupils.

International Relations
The Republic of Paraguay is a member of the United Nations, the Organization of American States and LAFTA (Latin American Free Trade Association).

Research in Science and Technology at Academic Level

Universidad Católica 'Nuestra Señora de la Asunción' (Catholic University of Our Lady of the Assumption)
Founded in 1960, with 434 lecturers and a student population of 7180. There are faculties of philosophy and human sciences, business studies and accounts, law and diplomatic sciences, and the university also possesses the Institute of Theology and Religion. It has seats in Villa Rica, Concepción, Encarnación and Pedro Juan Caballero. The *Centro de Estudios Antropológicos* (Centre for Anthropological Studies) and the *Centro de Investigaciones* (Research Centre) are also affiliated to the university.

Universidad Nacional de Asunción (National University of Asunción)
Established in 1890 it has a staff of 500 lecturers and 8000 students. Its faculties are medicine, business studies and accounts, physics and mathematics, economics, odontology, chemistry and pharmacy, philosophy, agriculture and veterinary medicine, and architecture. The national schools of agriculture and higher philosophy, sciences and education belong to the university, as does also the National Institute of Parasitology.

Instituto Nacional de Investigaciones Científicas (National Institute of Scientific Research)
Carries out research and teaching in physics, chemistry, mathematics, psychology and education.

Agricultural and Marine Science and Technology

Servicio Técnico Interamericano de Cooperación Agrícola (Inter-American Technical Service for Agricultural Cooperation)

Medical Science and Technology

Instituto Nacional de Parasitología (National Institute of Parasitology)
This institute is a dependant of the Institute of Microbiology of the Faculty of Medicine of the National University.

Sociedad de Pediatría y Puericultura del Paraguay (Paediatrics and Child Welfare Society of Paraguay)

Scientific and Technical Research in the Military Field

Instituto Geográfico Militar (Military Geographical Institute)

Socio-Economic Research

Centro Paraguayo de Estudios de Desarrollo Económico y Social (Paraguayan Centre of Economic and Social Development Studies)

Associations and Professional Societies in General

There are twelve institutions that promote research in different fields of science – among them are:

Centro Paraguayo de Ingenieros (Paraguayan Centre for Engineers)

Sociedad Científica del Paraguay (Scientific Society of Paraguay)

Information Services in the Field of Science and Technology

Biblioteca de la Sociedad Científica del Paraguay (Library of the Scientific Society of Paraguay)
Has 30 000 volumes.

Biblioteca y Archivo Nacional (National Library and Archives)
Has 44 000 volumes.

Besides these there are thirteen other centres in the country.

Peru

Geographical, Demographical, Political and Economic Features

Peru is situated in the central western part of South America. Its borders to the north are shared with Ecuador; to the north-west with Colombia; to the east with Brazil; to the south-east with Bolivia; to the south with Chile; and to the west lies the Pacific Ocean.

The total area of the country is 1 285 216 square kilometres. Peru has a varied climate, with the coastal area being temperate; the south and centre are humid; the north is tropical; the sierra is variable, and the puna experiences extreme cold while the selvas are tropical and humid.

Peru can be divided into three geographical regions – a fringe of arid desert region that stretches along the entire coastline; a mountainous region in the centre; and an extensive region of wooded foothills and plains in the east.

The population, according to estimates made in 1979, is 17 779 500, with a projection for 1985 of 20 426 000. The density is 13.8 inhabitants per square kilometre. The capital, Lima, has a population of 5 100 000.

Estimates in 1980 show an urban population of 67 per cent, and the majority of the indigenous population reside in the rural Andean zones and on the selva. Mestizos of European ancestry are predominant in the coastal towns as well as a small cross-section of Asian and African immigrants, while the negroes, Chinese and Japanese are showing a downward trend. Lima now holds a quarter of the population of Peru.

Government

The government of Peru is defined as unitarian, representative and decentralized. The highest authority, the executive power, is represented by the President of the Republic. The other main bodies are the legislative power, the judicial and the Electoral. The administrative structure is made up of the ministers who are directed by their respective organizations, the municipal councils. Politically the country is divided into twenty-four departments and a constitutional province (Callao), 152 provinces and 1703 districts.

Natural Resources

AGRICULTURE

Almost half of the Peruvian population live from the benefits of agriculture. According to the 3rd Agrarian Reform in June 1969, the government decreed that the large farms and sugar mills in the north were to be turned into cooperatives. This also applied to properties on the sierra of at least 150 hectares of irrigated land capable of pasturing 5000 sheep. The aims are to obtain the maximum efficient exploitation of the land.

Fertile soil in Peru is somewhat scarce and the arid coastal land is presently under forestry development. The potentialities, from the point of view of agriculture, can be expressed as follows:

Table 11.1

	Potentially cultivable land	*Cultivated land (per cent)*
Coast	7 861 600 km^2	84
Sierra	17 674 010 km^2	73
Selva	5 877 830 km^2	90
Total	31 413 440 km^2	

Principal agricultural products are sugar, cotton and coffee.

LIVESTOCK

Estimates in 1977 show about 4.12 million head of cattle; 643 000 horses; 2.6 million goats; 14.5 million sheep; 19 million pigs and 30 million poultry.

FORESTRY

The total area of forestry land is 80 356 000 hectares, which is divided into productive forestry land 74 per cent, forestry development 8.6 per cent and protected land 17.4 per cent.

FISH

Peru is one of the most advanced countries in the world in terms of value and volume of fish, the main species being anchovies, which are made into fish meal for animal feed. Peru produces about 45 per cent of the world's production in fish – that is to say 2 million tons per year. There are more than 100 fish-processing plants in the twenty-two ports along the 1400 miles of coastline, providing work for 3000 workers. Almost 30 per cent of production is from Chimbote. Apart from anchovy, other important species are tuna, bonito (similar to tunny), hake, barrilete (crab covered in prickles), and swordfish. The tidal conditions on the coast of Peru are ideal for the anchovy – hence their economic viability.

MINERALS

Minerals make up 55 per cent of exports. The main mineral products are zinc, iron, lead and silver.

Oil deposits, according to official estimates, produce 1850 million barrels per annum of which 1000 million are heavy oil.

Education

Primary education is free and obligatory. In 1972 the Educational Reform Law was promulgated which restructured the national education system with the trend towards more technical courses. According to estimates of 1978, 79.7 per cent of the population are literate.

International Relations

Peru is a member of the United Nations Organization and the Organization of American States (OAS).

Organization of Science and Technology

In November 1968 the *Consejo Nacional de Investigación (CONI or CNI)* (National Research Council) was established with the objectives of formulating a development policy for science and technology; coordinating and stimulating scientific and technological development in general; and operating research institutes in the branches of science and technology that are beneficial to the national interest. However, for various political reasons *CNI* has not produced the expected results, since the present organization of scientific and technological institutions is not in any form of integrated system. In August 1980 the *CNI* was transformed into the *Consejo Nacional de Ciencia y Tecnología (CONCYTEC)* (National Council of Science and Technology).

The functions of this organization are:

To draw up an exclusive national policy for science and technology.

To coordinate this policy with the policy for social and economic development.

To stimulate, promote, finance, organize, coordinate, administer, etc., the research and development programmes in science and technology.

The legislative resolution laid down by *CONCYTEC*, also establishes the mechanisms for the development of plans and programmes of research and development as well as determining the relations that this body will maintain with other organizations – governmental, universities and private. The diagram shows the institutional organization of science and technology at government and public level and the relationship with *CONCYTEC*.

In Peru there are seven science academies; forty-one professional associations; eleven professional colleges and fifty-seven scientific and technological societies.

Of the thirty-three universities, twenty-two are state-run and eleven are private. There are 297 institutes and centres dedicated to research and development, distributed as follows:

Table 11.2

	Number	Per cent
Public firms and institutes	38	12.70
University system	151	50.70
Productive sector	87	29.20
Non-profit-making private institutions	8	2.70
State sector	13	4.70
Total	297	100.00

The public firms and institutes are fundamentally concerned with research activities in the areas of health (fourteen institutes); mining and petroleum (six institutes); fisheries (five institutes); and industry (four institutes). The university system is involved in the areas of agriculture and food (fifty-two institutes); health (thirty-three institutes); industry (twelve institutes); mining and petroleum (eleven institutes). The private sector mainly develops activities related to industry (seventy-eight institutes).

Policies and Financing in Science and Technology

In 1978 the Peruvian Government established a multi-sectorial commission responsible for studying the most adequate structure for the National System of Science and Technology; this commission completed its work in 1979. Establishing a national system of science and technology was considered a priority area in the *Plan Nacional 1977-80* (National Plan 1977–80) but the work instigated by *CNI* does not reflect positive results. The council administered and controlled training of staff and applied research in mathematics, biology, physiology, geophysics, medicine, oceanography and engineering in five national universities. The council also formed the following institutes: *Instituto de Tecnologías Avanzadas* (Institute of Advanced Technology); *Instituto de Tecnologías Intermedias* (Institute of Intermediate Technology); *Instituto Nacional de Física* (National Institute of Physics); *Instituto de Investigaciones Históricas, Económicas y Sociales* (Institute of Historical, Economic and Social Research); *Instituto Peruano de la Ciencia* (Peruvian Institute of Science) and *Instituto Nacional de Investigación Matemática* (National Institute of Mathematical Research).

Although the plan for 1982–85 is still being drawn up, some of the political objectives in science and technology are as follows:

To intensify teaching and training of human resources in science and technology to complement national incentives.

To aid scientific and technological activities with investments, incentives and exemptions; and to eliminate bureaucratic hindrances.

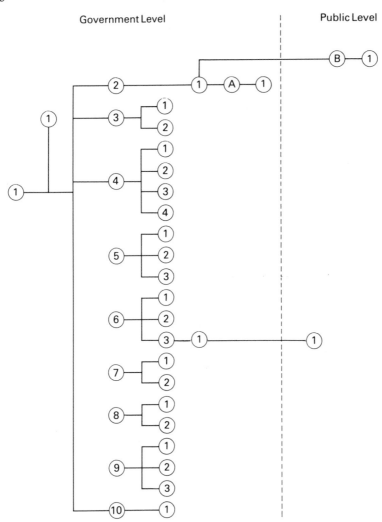

Fig. 13 Organization of science and technology

1	Presidencia (Presidency)
1.1	Consejo Nacional de Ciencia y Tecnología (CONCYTEC) (National Council of Science and Technology)
1.2	Ministerio de Educación (Ministry of Education)
1.2.1	Consejo Nacional de Universidades Peruanas (National Council of Peruvian Universities)
A	Universidades Nacionales (National Universities)
A.1	Institutos de Investigación (Research Institutes)
B	Universidades Privadas (Private Universities)
B.1	Institutos de Investigación (Research Institutes)
1.3	Ministerio de Industria, Comercio, Turismo e Integración (Ministry of Industry, Commerce, Tourism and Integration)
1.3.1	Instituto de Investigaciones Tecnológicas y Normas Técnicas (Institute of Technological Research and Technical Standards)
1.3.2	Servicio Nacional de Aprendizaje y Trabajo Industrial (National Service of Apprenticeship and Industrial Work)

Fig. 13 - *continued*

1.4	Oficina Primer Ministro (Office of the Prime Minister)
1.4.1	Oficina Nacional de Evaluación de Recursos Naturales (National Office for the Evaluation of Natural Resources)
1.4.2	Centro Nacional de Productividad (National Centre of Productivity)
1.4.3	Servicio Nacional de Meteorología e Hidrología (National Meteorological and Hydrology Service)
1.4.4	Instituto Nacional de Estadísticas (National Statistics Institute)
1.5	Ministerio de Salud (Ministry of Health)
1.5.1	Institutos Nacionales de Salud (National Health Institutes)
1.5.2	Instituto de Neonatología y Protección Materno-infantil (Institute of Neonatal Care and Maternal-Infant Protection)
1.5.3	Escuelas de Salud (Schools of Health)
1.6	Ministerio de Agricultura y Alimentación (Ministry of Agriculture and Food)
1.6.1	Instituto de Investigaciones Agroindustriales (Institute of Agro-industrial Research)
1.6.2	Centro Nacional de Capacitación e Investigación para la Reforma Agraria (National Centre of Education and Research for Agrarian Reform)
1.6.3	Dirección General de Investigaciones Agropecuarias (General Board of Agricultural and Livestock Research)
1.6.3.1	Centros de Investigación Agropecuaria (4) (Centres of Agricultural and Livestock Research)
1.6.3.1.1	Estaciones Experimentales (Experimental Stations)
1.7	Ministerio de Pesquería (Ministry of Fisheries)
1.7.1	Instituto del Mar del Peru (Marine Institute of Peru)
1.7.2	Dirección General de Investigaciones Científicas y Tecnológicas Pesqueras (General Board of Fisheries Scientific and Technological Research)
1.8	Ministerio de Transportes y Comunicaciones (Ministry of Transport and Communications)
1.8.1	Oficina de Investigación y Desarrollo (Office of Research and Development)
1.8.2	Instituto Nacional de Investigaciones y Capacitación en Telecomunicaciones (National Institute of Research and Education in Telecommunications)
1.9	Ministerio de Energía y Minas (Ministry of Energy and Mining)
1.9.1	Instituto de Investigaciones Científicas y Tecnológicas Mineras (Institute of Mining Research and Development)
1.9.2	Instituto Peruano de Energía Nuclear (Peruvian Institute of Nuclear Energy)
1.9.3	Instituto de Geología y Minería (Institute of Geology and Mining)
1.10	Ministerio de Vivienda y Construcción (Ministry of Housing and Construction)
1.10.1	Dirección General de Normalización e Investigación (General Board of Standards and Research)

To activate and facilitate the development of information services and systems in the field of science and technology.

To develop a national fund for scientific and technological development to function as an intersectorial body to finance the development of research as well as to survey the economic situation of individual researchers.

To coordinate the sectorial activities in science and technology and to promote intersectorial cooperation.

To channel a significant percentage of international technology and finance towards national development in science and technology.

To stimulate and strengthen research activities in science and technology in the universities, providing the appropriate mechanisms with the objective of obtaining national benefits from the results.

Although the priority areas have not been established, there is no doubt

Table 11.3 Human resources in science and technology

Year	Population	Scientists and engineers		Those employed in research and development†					Technicians	
		Total number of scientists and engineers (in thousands)	Total number (in units)	Breakdown in fields of specialization (in units)					Total (in thousands)	Those engaged in research and development (in units)
				Natural sciences	Engineering and technology	Agricultural sciences	Medicine	Social sciences		
1965	11.2	1238								
1970	13.4	1686	1596	496	13	507	330	180		
1975	15.4	3705	2625	647	406	662	495	415	2108	1028
1980	17.7	6167 (1976)	3932	745	832	923	756	675	2235	1382
1985*	20.4									
1990*	23.3									

* Based on a projection.
† Full-time basis.
Sources: UNESCO Statistics on Science and Technology 1980; Scientific and Technological Potential 1970.

that the development of the National Fund will be in this category, as the whole system of science and technology depends on this.

With respect to human resources, these are relatively scarce and could not cover the possible demand as a result of an intense development campaign in research activities in science and technology.

The country has only 3706 scientists and engineers in the field of research and development (1976) of whom 1326 work in the public sector, 320 in the private sector and 2060 in the area of higher education (see Table 11.2).

The number of graduates over the past six years has shown an average growth of 24 per cent. In 1975 7811 graduated, which was an increase of 59.96 per cent over the 1970 figures.

Of the total graduates in 1975 (doctors and bachelors) 5533 were men and 2278 women.

Research activities in the universities face grave problems of insufficient material and financial resources as well as an acute shortage of qualified staff.

An average of 2763 million soles per year is diverted to research and development activities and is distributed as shown in Table 11.4:

Table 11.4

Sector	Peruvian soles	Per cent
Public	2 277 000 000	82.4
University	348 000 000	12.6
Productive	140 000 000	5.0

Investment in the public sector is as shown in Table 11.5:

Table 11.5

	Percentage
Ministry of Energy and Mines	44
Ministry of Agriculture and Food	15
Ministry of Industry and Tourism	7
Ministry of Education	7
Ministry of Health	11
Ministry of Fisheries	8
Others	8

The sectors that utilize a major part of resources for research and development are: the government with 82.4 per cent of the universities with 12.6 per cent. The productive sector, on the other hand, finds itself short of resources for research and development.

Science and Technology at Government Level

The governing body in the policy of science and technology is the *Consejo*

Nacional de Investigación (National Research Council) which is a governmental organization at the highest level of decision.

The role that the governmental organizations play in the development of research and development can be seen in Figure 13. A great deal of research is carried out in institutes and centres under public administration, many of which are covered in the following sections.

Oficina Nacional de Estadística (National Office of Statistics)
Founded in 1975, it collects statistics on population, housing and the economic and agricultural census. *Indice de Precios al Consumidor* (Index of Consumer Prices) is published monthly.

Instituto Geofísico del Peru (Geophysics Institute of Peru)
This institute carries our research into the equatorial magnetic field; and publishes magnetic, ionospheric, seismological, meteorological data, etc. The observatories of Huancayo, Jicamarca, Ancon, Talara and Lima are affiliated to the institute.

Instituto Geológico, Minero y Metalúrgico (Institute of Geology, Mining and Metallurgy)
Coordinates and carries out research into mineral resources. *Boletín* (Bulletin) is published.

Research in Science and Technology at Academic Level

Consejo Nacional de la Universidad Peruana (National Peruvian University Council)
The council coordinates academic activities in the universities of Peru.

Instituto Experimental de Educación Primaria No. 1 (Experimental Institute of Primary Education No. 1)
The institute studies and analyses methods and systems of teaching and evaluates the results.

Universities

Universidad Nacional Agraria (National Agrarian University)
Founded in 1902, it was originally called the *Escuela Nacional de Agricultura* (National School of Agriculture)
There is at present a staff of 388 teachers and 3510 students in the fields of agriculture, sciences, agricultural engineering, zootechny, forestry and economic planning.

Universidad Nacional Agraria de la Selva (National Agrarian University of the Selva)
Established in 1964, it is run under state control with an academic year between April and December and with a staff of forty-six teachers in charge of 650 students. Courses are offered in agriculture, food industry and livestock production.

Universidad de Lima (Lima University)
Founded in 1963, it is a private university with 334 lecturers and 6500 students. Its areas of science and technology are represented by the faculties of communications, industrial engineering, computer engineering, and metallurgical engineering.

Universidad Nacional de la Amazonía Peruana (National University of the Peruvian Amazon)
Founded in 1962 as a state institute in Iquitos. The academic year is between April and February and the student population numbers 3274 with 164 teachers. Departments are sciences, agronomy, forestry sciences and chemistry.

Universidad Nacional del Centro del Perú (National University of Central Peru)
A state university founded in 1962, it has an academic year between April and December. Publications include *Anales Científicos* (Scientific Annals), which is produced regularly, and *Cuardernos Científicos* (Scientific Notebooks) irregularly. Students number 6161 with 279 teachers, developing activities in the areas of agronomy, forestry sciences, zootechny, nursing, electronics and mechanics, chemistry, metallurgy and mining, and architecture.

Universidad Nacional de Huánuco 'Hermilio Valdizán' (Hermilio Valdizán National University of Huánuco)
Established in 1964, it is state controlled, with an academic year between May and February. The teaching staff numbers 135 for a student population of 5545 in the departments of economics and commerce, agronomy, nursing and obstetrics, civil engineering and industrial engineering.

Universidad Nacional 'Federico Villarreal' (Federico Villarreal National University)
Founded in 1963 under state control, it has 1386 teachers and 22 000 students in the following scientific areas: odontology; oceanography and fish; human medicine; medical technology; geographical engineering; psychology; civil engineering; industrial engineering; social sciences and business studies.

Universidad Nacional de Ingeniería (National University of Engineering)
Although it has been recognized in its present status only since 1955, the university has been functioning since 1896 when it was known as the *Escuela Nacional de Ingenieros* (National School of Engineers). It is a state university with 10 431 students and 800 teachers in the areas of architecture, urban planning and fine arts, economics, civil engineering, metallurgical engineering, geology and mine studies, industrial engineering and computing systems, petroleum engineering and petrochemistry, chemical engineering, and sanitary engineering.

Universidad Nacional 'Jose Faustino Sánchez Carrión' (Jose Faustino Sánchez Carrión National University)
Founded in 1968, it has 2675 students with ninety-seven teachers in the

areas of fishery, industrial engineering, nutrition, sociology, and accounting and business studies.

Universidad Nacional 'Pedro Ruiz Gallo' (Pedro Ruiz Gallo National University)
Established in 1970, at present it has a total of 261 teachers for its 5460 students in the departments of agriculture, biology, nursing, agricultural engineering, civil engineering, veterinary medicine and zootechny.

Universidad Nacional Mayor de San Marcos de Lima (Greater National University of San Marcos de Lima)
Founded in 1551, it is one of the oldest universities in the Americas. Teachers number 2394, with 22 260 students in the faculties of physics and mathematics, chemistry and chemical engineering, geology and geography, biology, pharmacy and biochemistry, veterinary sciences, medicine, dentistry, social sciences, engineering, metallurgy, and psychology as well as the faculties of the arts and humanities.
Affiliated to the university are the following institutes:

Instituto de Biología Andina (Institute of Andean Biology)

Instituto de Medicina Tropical (Institute of Tropical Medicine)

Centro de Investigación de Recursos Naturales (Research Centre of Natural Resources)

Instituto de Patología (Institute of Pathology)

Instituto de Zootécnica (Institute of Zootechny)

Instituto de Bioquímica y Nutrición (Institute of Biochemistry and Nutrition)

Universidad Nacional de San Antonio de Abad (San Antonio de Abad National University)
Originally founded in 1962, it was reorganized in 1969. Student population numbers 15 000 with 425 teachers. Apart from the humanities, activities are developed in the areas of chemistry, civil engineering, architecture and agronomy.

Universidad Nacional de San Agustín (National University of San Augustín)
Situated in Arequipa since 1828, it is administered under state control. The library houses 127 000 volumes for its 10 900 students and 480 teachers. Apart from the humanities, there are faculties of biology, medicine, geology, chemistry and general studies.

Universidad Nacional de San Cristóbal de Huamanga (San Cristóbal de Huamanga National University)
Founded in 1677 in Ayacucho, it is a state-run university. It was closed for a period and reopened in 1959. The academic year extends for a continuous ten-month period with 7209 students and 416 teachers in the departments of agriculture and zootechny, medicine, physics and mathematics as well as the arts and humanities.

Universidad Nacional 'San Luis Gonzaga' (San Luis Gonzaga National University)
Founded in 1961 in Ica with a staff of 459 teachers and 6295 students in its departments of agronomy, pharmacy, biochemistry, dentistry, civil engineering, mechanical engineering, electrical engineering, veterinary medicine, medicine, biological sciences and fisheries, as well as the faculties of the arts and humanities.

Universidad Nacional de Trujillo (Trujillo National University)
Founded by Simon Bolívar in 1824, it is a state-run university with 553 teachers and 8721 students. Apart from the area of the humanities academic courses are offered in medicine, engineering, biological sciences, physics and mathematics, pharmacy and biochemistry.

Universidad Nacional Técnica de Cajamarca (Technical University of Cajamarca)
Founded in 1962, it is a state university. Its staff of ninety teachers supervises the education of 1561 students in the areas of education, agronomy, civil engineering, nursing and veterinary medicine.

Universidad Nacional de Piura (Piura National University)
A state university founded in 1962. The staff numbers 244 teachers with a student population of 5414 developing activities in agronomy and phytology, biological sciences, chemistry and mining, crop control, fisheries, engineering, physical engineering and general engineering, mathematics, computing and statistics.

Universidad Particular 'San Martín de Porres' (San Martín de Porres Private University)
This has almost 10 000 students and 256 teaching staff.

Universidad Particular Peruana 'Cayetano Heredia' (Cayetano Heredia Private University)
Founded in 1961, it is made up of the schools of pre-medical studies, and medicine, and the institute of High Altitude Studies (Space), and is particularly involved in the promotion of medical research.

Pontificia Universidad Católica del Perú (Pontifical Catholic University of Peru)
Established in 1917 with private funds. Its library houses 140 000 volumes to aid the 700 teachers and 7000 students. Departments include sciences and engineering as well as the faculties of humanities and arts.

Universidad Nacional Técnica del Callao (Technical University of Callao)
Founded in 1966, it develops its activities under state control with 6600 students and 252 teachers. Departments include chemical engineering, electrical engineering, fisheries studies, and mechanical engineering.

Instituto Superior de Administración y Tecnología (Higher Institute of Business Studies and Technology)
Founded in 1905, it has a staff of sixty-five teachers and 970 students and its library houses 4800 volumes.

Industrial Research and Technology

Instituto de Investigaciones Tecnológicas y Normas Técnicas (Institute of Technological Research and Technical Standards)
A dependant of the Ministry of Industry, it is dedicated to the study of industrial products and administers quality control.

Instituto Birchner-Benner (Birchner-Benner Institute)
Founded in 1979 it carries out research into substitute meat products and the composition of food of a high nutritive quality and low cost.

Agricultural and Marine Science and Technology

Instituto Nacional de Investigación Agraria (National Institute for Agrarian Research)
Founded in 1927, it produces various publications: *Avances en Investigación* (Research Progress); *Serie de Boletín Téchnico* (Technical Bulletin Series); *Informes Especiales* (Special Reports) among others. There is a technical staff of 250.

Estación Experimental Agropecuaria de Tulumayo (Tulumayo Agricultural and Livestock Research Station)
Founded in 1942, it is dependent on the General Board of Agricultural and Livestock Research and the Ministry of Agriculture and Food.

Estación Experimental Agrícola del Norte (Agricultural Experimental Station of the North)

Instituto de Biología Andina (Institute of Andean Biology)
The institute studies biological problems and the behaviour of organisms at high altitudes in the Andes.

Estación Altoandina de Biología y Reserva Zoo-Botánica de Checayani (High Andes Biology Station and Zoo-botanical Reserve of Checayani)

Instituto del Mar del Perú (Peruvian Marine Institute)

Instituto de Zoonosis e Investigación Pecuaria (Institute of Zoonosis and Livestock Research)
Carries out research into infectious diseases in man and animals.

Medical Science and Technology

A major part of research in this field is carried out in the specialized hospital units and medical centres and in the research institutes specializing in medical research.

Instituto Nacional de Salud (National Health Institute)

Instituto de Investigaciones Alérgicas (Institute of Allergy Research)

Nuclear Science and Technology

Instituto Peruano de Energía Nuclear (Peruvian Institute of Nuclear Energy)
Founded in 1955, it carries out research into the peaceful use of nuclear energy in medicine, biology, agriculture and industry.

Military Research in Science and Technology

Centro de Estudios Historico-Militares del Perú (Centre of Historical–Military Studies of Peru)

Instituto Geográfico Militar (Military Geographical Institute)
Publishes topographical, physical and political maps of the national territory.

Socio-Economic Research

Instituto de Ciencias de la Comunicación (Institute of Communication Sciences)
Carries out research and development into the methods of communication, studies on freedom of speech in Latin America and the interchange of journalists among different countries.

Meteorological and Astrophysical Research

Dirección General de Meteorología del Perú (General Board of Meteorology of Peru)
Founded in 1928, it has seventy-nine observation stations throughout the country and publishes meteorological and climatological reports.

Professional Societies and Associations in the Field of Science and Technology

There are forty-eight institutions dedicated to promotional activities in science and technology. Among them are:

Academia Nacional de Ciencias Exactas, Físicas y Naturales de Lima (National Academy of Exact, Physical and Natural Sciences of Lima)

Academia Nacional de Medicina (National Academy of Medicine)

Sociedad Nacional Agraria (National Agrarian Society)
Acts as an advisory service in different areas (cotton; coffee; grapes and wines; potato products, etc.).

Federación Médica Peruana (Peruvian Medical Federation)

Sociedad Nacional de Minería (National Society of Mining)

Sociedad de Ingenieros del Perú (Society of Engineers of Peru)

Information Services in Science and Technology

There are fifteen libraries, data archives and documentation centres and twenty-two museums with specialized information including:

Biblioteca Nacional (National Library)
Founded in 1821, it has 662 000 volumes, 11 770 maps and more than 1 500 000 reviews and periodicals.

Biblioteca Central de la Universidad Nacional Mayor de San Marcos (Central Library of the San Marcos National University)
Founded in the sixteenth century, it has more than 450 000 volumes.

Biblioteca Central de la Pontificia Universidad Católica del Perú (Central Library of the Pontifical Catholic University of Peru)
Houses 130 000 volumes.

Biblioteca de la Universidad Nacional de San Agustín (Library of the National University of San Agustín)
Has twelve specialized libraries with a total of 94 000 volumes, which when added to those in the Central Library make a total of more than 140 000 volumes.

International Cooperation in Science and Technology

Peru is one of the most active members of the Andean Group or Cartagena Agreement, and within this group of countries there is cooperation in matters of scientific and technological research in the area of industrial production, new technology, its applications, general problems, etc.

Mission ORSTOM au Perou – Cooperation auprès du Ministerio de Energía y Minas (ORSTOM Mission in Peru – Cooperation with the Ministry of Energy and Mining)

Puerto Rico

Geographical, Demographical, Political and Economic Features

The island of Puerto Rico is almost rectangular in shape and is the most easterly and smallest of the Greater Antilles. Situated at 129 kilometres to the east of Hispaniola and 74 kilometres to the west of Santo Tomas (Virgin Islands), it has 1126 kilometres of coastline.

Official estimates of 1979 show the population to be 3 410 000 with a density of 383 people per square kilometre. Urban population is 61.8 per cent and the people are mainly of Spanish and African descent. The indigenous population was annihilated at the beginning of colonization.

Government

The executive power is exercised by the Governor (elected by popular vote for four years) and a cabinet of fourteen secretaries or ministers. The legislative power is in the hands of the Senate (twenty-seven members) and a Chamber of Representatives (fifty-one members). The judicial power is administered by the Supreme Court, the High Court; a District Court and thirty-seven municipal judges. The federal government is represented by two district judges and an attorney appointed by the President of the United States of America. The autonomy of Puerto Rico is limited in matters of national defence, minting of currency, foreign relations and postal and customs administration.

AGRICULTURE
The main products are cotton, sugar cane, citrus fruits, coffee, bananas, yucca and tobacco.

MINERALS
Production includes clay, sand, lime, marble, copper, iron, manganese, gold and salt.

Education

According to figures of 1970, 89.3 per cent of the population is literate; in 1977 the government channelled US$662.15 million into education.

Scientific and Technological Research at an Academic Level

Universidad de Puerto Rico (University of Puerto Rico)
Researchers and teachers number 2722 for a student population of 50 492.
The university is divided into three campuses, a university college and six
regional colleges.

Rio Piedras Campus has faculties in business studies, humanities, law,
education, natural sciences, architecture, social sciences and schools of
planning, public communication, librarianship and public administration.

Mayaguez Campus has faculties in arts and sciences, agricultural sciences,
business studies, and engineering.

Medical Science Campus has schools of medicine, dentistry, public health
and preventive medicine, pharmacy and the college of biomedical sciences.

In addition there is Cayey University College and six regional seats in
Humacao, Arecibo, Ponce, Bayamon, Aquadilla, and Carolina.

Universidad Católica de Puerto Rico (Catholic University of Puerto Rico)
Has campuses in Arecibo, Guayama and Mayaguez. The student population
totals 11 698, with 583 teachers. Its colleges comprise arts and humanities,
science, law, education, and business administration.

Bayamón Central University
Has a staff of ninety lecturers and 2000 students and colleges of business
studies, natural sciences, education, and arts and humanities.

Inter-American University of Puerto Rico
With 1125 teachers and 28 725 students, it is divided into two campuses,
one regional college and one law school. *San Germán Campus* has
departments of languages, natural sciences, economics and business studies,
education, religion and philosophy, social sciences and fine arts. *San Juan
Campus* has departments in humanities, natural sciences, economics and
business studies, education, English, and social sciences.

There are regional colleges at Aquadillas, Arecibo, Barranquitas, Fajardo,
Guayama and Ponce.

University of the Sacred Heart
Has a staff of 100 full-time and 180 part-time teachers and 6425 students.
Departments include business studies, education, humanities, natural
sciences, and social sciences.

Agricultural Science and Technology

Institute of Tropical Forestry (USDA Forestry Service)
The institute carries out research into wood commerce and administration
and ecosystems; courses are offered to foreign students in cooperation with
the FAO and USAID. It also provides cooperative assistance to the State.
The library has 14 000 volumes.

Nuclear Science and Technology

Puerto Rican Nuclear Centre
Housed in the University of Puerto Rico, it is affiliated to the United States of America Commission for Atomic Energy and serves as a specialized research and teaching centre for Latin America in subjects such as nuclear engineering, nuclear science and technology, physics and radiological therapy, clinical uses of radioactive radio-isotopes, etc. The institute also possesses an oceanographical research ship and a nuclear reactor.

Meteorological and Astrophysical Research

Arecibo Observatory
Part of the National Astronomy and Ionosphere Center of Cornell University. The observatory has the largest radio/radar telescope in the world and receives transmissions from satellites. It is used for radar studies of plants, to examine the properties of the earth's upper atmosphere and to receive radio emissions from celestial bodies.

Professional Associations and Societies in the General Field

There are more than fourteen institutions involved in research and development, among which are:

Academia Puertoriqueña de la Lengua Española (Puerto Rican Academy of the Spanish Language)
The academy corresponds with the *Real Academia Española* and publishes *Boletín* (Bulletin), quarterly.

Instituto de Lexicografía Hispanoamericiano Augusto Malaret (Augusto Malaret Hispano-American Lexicographic Institute)
The institute studies and conserves indigenous languages, it also studies the African influence on the American language, and the appropriate use of technical terms.

Information Services in Science and Technology

There are twenty-four information centres among the museums and libraries, including:

Caribbean Regional Library
Has 116 000 volumes. It publishes *Current Caribbean Bibliography* annually.

Biblioteca General de la Universidad de Puerto Rico (University of Puerto Rico General Library)
Has more than 360 000 volumes, and the José M. Lázaro Library of the

same university has 1 500 000 volumes. There is also an Agricultural Experiment Station Library; a College of Social Science Reserve Collection; and a Law Library.

Biblioteca de la Universidad Católica de Puerto Rico (Library of the Catholic University of Puerto Rico)
Has 150 000 volumes. It publishes *Horizontes* (Horizons), twice a year.

Biblioteca Madre María Teresa Guevara (Madre María Teresa Guevara Library)
Has 150 000 volumes.

Surinam

Geographical, Demographical, Political and Economic Features

Surinam lies on the north-east coast of South America between Guyana, on the west, and French Guiana, on the east. Brazil lies to the south. The climate is tropical with quite heavy rainfall.

The area of the country is 163 265 square kilometres and the population in 1980 was 352 041.

Natural Resources

MINERALS

The economy of Surinam is based on bauxite, which together with its derivatives (alumina and aluminium) provides about 80 per cent of export earnings and 30 per cent of the gross domestic product. In 1979 production of bauxite was 4 760 000 metric tons. The industry is controlled by Suralco, a subsidiary of the United States company Alcoa, and by Billiton, part of the Royal Dutch/Shell group. Other minerals include iron ore, manganese, copper, nickel, platinum, gold and kaolin. In 1981 petroleum-bearing sand was discovered in the Saramacca district by the Gulf Oil Corporation and a pilot research project is planned in the area by the Surinam State Oil Corporation.

AGRICULTURE

Surinam is self-sufficient in sugar, rice, edible oil, citrus fruits, coffee and bananas. Plantains, pulses, maize, coconuts and groundnuts are also grown. Oil palm is a comparatively new crop and a large oil palm installation was set up at Victoria in the Borkopondo area in 1977. Production of rice in 1979 reached 236 000 metric tons and sugar cane 160 000. Fishing, particularly for shrimps, is growing in importance and its contribution to export earnings rose to 8 per cent in 1977 as compared to 2 per cent in 1973.

LIVESTOCK

The livestock count in 1980 was estimated as: cattle 40 000; goats 7000; sheep 2000; pigs 2000; poultry 1 050 000.

FORESTRY

Approximately 90 per cent of Surinam is covered by forest but only 10 per cent is commercially exploited. The forestry industry is dominated by Bruynzeel (formerly a Dutch company, but now the Surinam government own 50 per cent of the shares).

INDUSTRY

The country's industrial sector consists mainly of the foodstuff and consumer industries.

Development aid from the Netherlands will be used to make the country more self-sufficient in food and to increase hydroelectric output and export potential. A major project for the construction of an 800 MW hydroelectric power dam on the Kabalebo river was completed in 1981 and smaller projects undertaken, including a hydroelectric complex at Phedra-Jay Creek and the construction of a harbour in Apoera on the Corantijn river.

International Relations

Surinam is a member of the Organization of American States, the Inter-American Development Bank and the International Bauxite Association and in 1978 joined GATT and the IMF.

Education

Education is compulsory for children between six and twelve. It is free up to and including higher education, provided by the University of Surinam. There is a literacy rate of 80 per cent. The university has faculties of law and medicine. There are 550 students and a teaching staff of thirty-four.

Scientific activity is conducted in several establishments in the country. The Centre for Agricultural Research in Surinam is a branch of the University of Surinam and researches into tropical agriculture. Agricultural research is also carried out at the *Landbouweproefstation* (Agricultural Experiment Station) in Paramaribo. This is attached to the Department of Agriculture, Animal Husbandry and Fisheries. It publishes an annual report and *Surinam Agriculture* (two to three times a year). The *Dienst Lands Bosbeheer* (Surinam Forest Service) has a nature conservation department, which studies the ecological effects of agriculture, husbandry, mining, hydro-electric power, and management of nature reserves, and unspoilt areas and wildlife.

The *Geologisch Mijnbouwkundige Dienst* (Geological Mining Service) houses a library of 13 102 volumes and publishes a Yearbook and geological maps.

The *Stichting voor Wetenschappelijk Onderzoek van de Tropen* (Netherlands Foundation for the Advancement of Tropical Research) has its headquarters in the Hague. The institution's aim is the advancement of tropical research both pure and applied, by the awarding of grants.

The *Stichting Surinaams Museum* (Surinam Museum) has a library of 4500 volumes and concentrates on archaeology, and cultural and natural history.

Uruguay

Geographical, Demographical, Political and Economic Features

Uruguay is situated at the entrance of the River Plate waterway and is in a favourable maritime position with access to the Atlantic Ocean and to the heart of the South American continent through the River Plate and its tributaries. The total area of the country is 176 215 square kilometres.

The country shares its border to the south with Argentina, with the estuary of the River Plate forming a natural dividing line between the two countries. To the north lies Brazil with 773 kilometres of frontier. Because of its geographical position the country experiences a temperate climate. Due to its hilly nature and the dense network of rivers, Uruguay has rugged characteristics which favour agricultural and livestock development.

According to official estimates in 1979 the population is 2 878 000 with a density of 16.2 per square kilometre. Eighty-four per cent of the population is urban, and 90 per cent of the population is white of Spanish descent.

Government

The present form of government is republican. The legislative powers lie with the *Consejo de Estado* (Council of State) and the executive power is exercised by the President of the Republic, who is advised by the *Consejo de Seguridad Nacional* (Council of National Security) or his ministers of state.

Natural Resources

AGRICULTURE AND LIVESTOCK

Agriculture and livestock are two of the main sectors of activity, with beef the principal product. Out of a total area of 46 million acres, 41 million are dedicated to agriculture and livestock, 90 per cent of this acreage to livestock alone and only 10 per cent to arable farming. Cattle represent 70 per cent of Uruguay's exports.

The livestock census in 1977 showed 11 362 million head of cattle; 15 million head of sheep; 420 972 horses; 418 709 pigs and 10 461 goats.

FISHERIES

In 1978 the Swiss Bank authorized a loan of $28 million to develop the fishing industry and in the same year exports of fish totalled 30 000 tons valued at $22 million.

INDUSTRY

The industrial structure can be divided as shown in Table 12:

Table 12

	Percentage		*Percentage*
Food	20	Non-metallic minerals	4
Chemicals and oil		Paper and print	5
derivatives	12	Shoes and clothes	5
Drinks	12	Tobacco	4
Textiles	10		

Education

Primary education is obligatory and, like higher education, is free. Estimates in 1975 indicate that 94 per cent of the population is literate.

International Relations

Uruguay is a member of the United Nations Organization, the Organization of American States and the Latin American Free Trade Association.

Organization of Science and Technology

The central body in the national system of science and technology is the *Consejo Nacional de Investigaciones Científicas y Tecnológicas (CONICYT)* (National Council of Scientific and Technological Research), which is a dependant of the Ministry of Education and Culture. *CONICYT* acts as an advisory body in scientific and technological policy planning and in the execution of specific programmes in science and technology.

There are other planning organizations such as the *Departamento de Ciencia y Tecnología de la Dirección de Planeamiento de la Universidad de la República* (Department of Science and Technology of the Directorate of Planning of the University of the Republic), which plans and coordinates university activities in the field of science and technology, and the *Departamento de Ciencia y Tecnología de la División de Cooperación Internacional de la Secretaría de Planeamiento, Coordinación y Difusión* (Department of Science and Technology of the International Cooperation Division of the Secretariat of Planning, Coordination and Diffusion) which is responsible for structuring policies in science and technology.

Policies and Financing in Science and Technology

In 1972 *CONICYT* presented to the Executive the *Bases del Plan de*

Government Level | Professional Associations

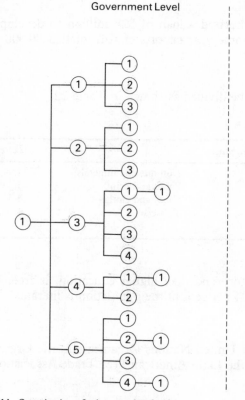

Fig. 14 Organization of science and technology

1	Presidencia (Presidency)
1.1	Ministerio de Educación (Ministry of Education)
1.1.1	Institutos y Comisiones Específicas (Specific Institutes and Commissions)
1.1.2	Consejo Nacional de Investigaciones Científicas y Técnicas (National Council of Scientific and Technological Research)
1.1.3	Universidad de la República (University of the Republic)
1.2	Ministerio de Industria y Energía (Ministry of Industry and Energy)
1.2.1	Instituto Geológico (Institute of Geology)
1.2.2	Comisión Nacional de Energía Atómica (National Commission for Nuclear Energy)
1.2.3	Centro Nacional de Tecnología y Productividad Industrial (National Centre of Industrial Technology and Productivity)
1.3	Ministerio de Agricultura y Pesca (Ministry of Agriculture and Fisheries)
1.3.1	Coordinación y Experimentación Agropecuaria (Agricultural and Livestock Coordination and Experimentation)
1.3.1.1	Centros y Estaciones Experimentales (Experimental Centres and Stations)
1.3.2	Departamento de Economía Agraria (Department of Agrarian Economy)
1.3.3	Oficina de Programación y Política Agropecuaria (Office of Agricultural and Livestock Programming and Policy)
1.3.4	Dirección Forestal, Parques y Fauna (Directorate of Forestry, Parks and Fauna)
1.4	Secretaria de Planeamiento, Coordinación y Difusión (Secretariat of Planning, Coordination and Diffusion)
1.4.1	División de Cooperación Internacional (Division of International Cooperation)
1.4.1.1	Departamento de Ciencia y Tecnología (Department of Science and Technology)
1.4.2	Dirección General de Estadísticas y Censos (General Directorate of Statistics and Census)
1.5	Ministerio de Salud Pública (Ministry of Public Health)

1.5.1 División de Asistencia (Assistance Division)
1.5.2 Departamento de Investigaciones Clínicas y de Laboratorio (Department of Clinical and Laboratory Research)
1.5.2.1 Instituto de Endocrinología (Institute of Endocrinology)
1.5.3 División de Higiene (Hygiene Division)
1.5.4 Hospitales (Hospitals)
1.5.4.1 Institutos y Centros Especializados (Institutes and Specialized Centres)

Desarrollo Científico (Basis of a Plan for Scientific Development) which established objectives for revitalizing the existing system. The objectives were listed in the *Plan General de Desarrollo* (General Development Plan) which determined the priority area to be the agriculture and livestock sector and guidelines were stated as follows:

To improve quality, quantity, situation and conditions of work, and to increase manpower in research and development.

To modernise existing facilities and to use available resources to meet present requirements.

To extend the information system in science and technology.

To create a balance between applied and basic research.

To achieve better utilization of national scientific and technological potential.

In order to reach these objectives, specific programmes have been instigated, such as: training of researchers, formulating a subsidy policy, acquiring equipment and other such elements to develop an adequate information service, etc.

All the proposed objectives were introduced into the *Plan Nacional de Desarrollo 1973-1977* (National Development Plan 1973-1977).

In 1977 the *Conclave Gubernamental de Solis* (Solis Governmental Committee) analysed the objectives afresh and determined the following recommendations:

(a) to integrate the productive sector into the national system, so that resources can be pooled, thus gaining the optimum possible results;
(b) the Ministry of Education and Culture together with the Secretariat of Planning, Coordination and Diffusion, to put forward to the Executive specific qualities that the national system of science and technology must possess;
(c) once the national system has been approved, to establish a *Plan Nacional de Desarrollo Científico y Tecnológico* (National Plan for Scientific and Technological Development) in collaboration with organizations within the system.

Human Resources

Latest estimates show a total of about 1500 scientists and engineers employed on a full-time or part-time basis in research and development. Of this total 36 per cent work in the production sector, 53 per cent in higher education and 21 per cent in the service sector. In 1980 a total of 2293 professionals and technicians graduated from university schools and faculties, of which 21 per cent studied subjects biased towards the productive sector; 10 per cent economics; 46 per cent health; and the rest followed courses in law, social sciences and the humanities.

Financial resources destined to the system of research and development are of the order of US$9 500 000 per year, which represents 0.2 per cent of the GNP. Of these funds, 67 per cent is provided by the public sector; 18 per cent by the private sector and 16 per cent from foreign sources. Of the total investment in research and development 12 per cent is channelled into basic research, 42 per cent into applied research and 46 per cent into experimental development.

Science and Technology at Government Level

Consejo Nacional de Investigaciones Científicas y Tecnológicas (CONICYT) (National Council of Scientific and Technological Research)
Founded in 1961, it is administratively a dependant of the Ministry of Education and Culture, but maintains a certain degree of autonomy. Its functions are '. . . to promote and develop scientific activity in all fields'. The official role of *CONICYT* is to advise the Executive, and, in particular, the Ministry of Foreign Relations and the *Oficina de Planeamiento y Presupuesto* (Office of Planning and Budgeting).

Dirección de los Programas de Investigación del Ministerio de Agricultura y Pesca (Directorate of Research Programmes of the Ministry of Agriculture and Fisheries)
The directorate programmes and coordinates research activities in this ministry (see section on Agricultural and Marine Science and Technology).

Dirección General de Estadísticas y Censos (General Directorate of Statistics and Census)

Instituto Artigas del Servicio Exterior (Artigas Institute of Foreign Affairs)

Instituto de Investigaciones Biológicas 'Clemente Estable' (Clemente Estable Institute of Biological Research)
Carries out research in the field of biological sciences, trains researchers and technicians and contributes to policy-making in this area. Twenty-four researchers are employed by the institute.

Instituto Geologico 'Ing Eduardo Terra Arocena' (Ing Eduardo Terra Arocena Geological Institute)
Affiliated to the Ministry of Industry and Energy, it has several specialized laboratories for research into mineral resources.

Research in Science and Technology at Academic Level

Sector Ciencias de la Oficina de Planeamiento de la Universidad de la República (Sector of Sciences of the Office of Planning of the University of the Republic)

This body plans internal university activities. Until 1973 there was a *Comisión de Investigaciones Científicas y Tecnológicas* (Commission of Scientific and Technological Research) located in the university itself for this purpose.

Universidad de la República (University of the Republic)
Founded in 1849 in Montevideo, it has a staff of 2822 researchers and lecturers and a student population of 39 000. Faculties include agronomy, architecture, economics and business studies, law and social sciences, humanities, sciences, engineering, medicine, dentistry, chemistry and veterinary medicine.

The university carries out experimental research and development in twenty-seven specialized institutes, thirty-six sections, forty-three departments and fourteen other units. Research results from the university are put to use in areas of public health, agriculture and livestock, etc. The faculty of agronomy has experimental stations in Paysandú, San Antonio, Salto, Banado de Medina and Cerro Largo.

Universidad del Trabajo del Uruguay (University of Labour of Uruguay)
Founded in 1878, it offers 220 different courses in its eighty-one schools of agriculture, industry and commerce, education and training. Teaching staff numbers 4200 with 50 000 students.

Instituto de Estudios Superiores (Institute of Higher Education)
Topics include geography and geomorphology, palaeontology, biological climatology and mathematics. Publications include *Boletines* (Bulletins) and *Revista* (Review).

Instituto de Enseñanza de Mecánica y Electrónica 'Profesor Dr José F. Arias' (Professor Dr José F. Arias Institute of Teaching in Mechanics and Electronics)

Industrial Research and Technology

Instituto Uruguayo de Normas Técnicas (Uruguayan Institute of Technical Standards)
Founded in 1939, it is responsible for supervising standards and control of weights and measures.

Instituto de Tecnología y Química (Institute of Technology and Chemistry)
Affiliated to the university, it carries out research and acts as an advisory body.

Laboratorio de Análisis Tecnológico del Uruguay (LATU) (Technological Analysis Laboratory of Uruguay)
The laboratory administers quality control in products for export and analyses raw materials and machinery for production processes.

Instituto Nacional de Carnes (National Meat Institute)
The institute studies and analyses meat products for quality control and conservation and oversees industrial processes.

Secretariado Uruguayo de la Lana (SUL) (Uruguayan Wool Secretariat)
The secretariat advises on and promotes all matters connected with wool, from its initial production to its use in the textile industry.

Centro Nacional de Tecnología y Productividad Industrial (CNTPI) (National Centre of Technology and Industrial Productivity)
The centre acts as the consultative entity between the public and private sectors to gain maximum efficiency in the field of technology, company management and industrial activity in general.

Centro de Investigaciones Tecnológicas (Centre of Technological Research)
Affiliated to the *Administración Nacional de Combustibles, Alcoholes y Portland (ANCAP)* (National Administration of Combustibles, Alcohol and Cement)
An autonomous organization that carries out research into cement, minerals, chemical products, and alcohol and derivatives. A sector of agronomy research is also attached to the centre.

Agricultural and Marine Science and Technology

Instituto Nacional de Pesca (National Fisheries Institute)

Centro de Investigaciones Agrícolas 'Alberto Boerger' (Alberto Boerger Centre of Agricultural Research)
Affiliated to the Ministry of Agriculture and Fisheries, its main objectives are to contribute to improving agricultural and livestock production at regional and national level. The staff numbers seventy researchers and there are six experimental stations *(La Estanzuela, Agropecuaria del Este, Agropecuaria del Norte, Granjera las Brujas, Citricola del Salto,* and *Animales de Granja).*

Centro de Investigaciones Veterinarias 'Miguel C. Rubino' (Miguel C. Rubino Centre of Veterinary Research)
Affiliated to the Ministry of Agriculture and Fisheries, it carries out research into animal health, with a staff of twenty-nine researchers.

Centro de Investigaciones Agrícolas de la Nación (National Centre of Agricultural Research)
Most work in this field is undertaken in the research stations affiliated to the Ministry of Agriculture.

Medical Science and Technology

There are at least ten important organizations dedicated to activities in this field, including:

Instituto de Endocrinología 'Profesor Dr Juan C. Mussio Fournier' (Professor Dr Juan C. Mussio Fournier Institute of Endocrinology)

Consejo Nacional de Higiene (National Hygiene Council)

Instituto Interamericano del Niño (Inter-American Child Institute)

Sociedad de Cirugía del Uruguay (Uruguayan Surgical Society)

Sociedad Uruguaya de Pediatría (Uruguayan Society of Paediatrics)

Sociedad de Radiología del Uruguay (Uruguayan Society of Radiology)

Liga Uruguaya contra la Tuberculosis (Uruguayan League against Tuberculosis)

Nuclear Science and Technology

Comisión Nacional de Energía Atómica (National Commission for Atomic Energy)
Founded in 1955, it has a staff of seven dedicated to research into the peaceful use of atomic technology.

Military Science and Technology

Instituto Geográfico Militar (Military Geographical Institute)
Founded in 1913, it undertakes cartographic and aero-photometric surveys.

Meteorological and Astrophysical Research

Dirección Nacional de Meteorología del Uruguay (National Meteorological Directorate of Uruguay)

Observatorio Astronómico (Astronomical Observatory)

Professional Societies and Associations

There are seventeen important organizations, among which are:

Academia Nacional de Ingeniería (National Academy of Engineering)

Asociación Odontológica Uruguaya (Odontology Association of Uruguay)

Information Services in Science and Technology

There are about 200 public and private institutions in Uruguay involved in documentation and information – 161 libraries, forty-nine centres of information; thirty-four archives and twenty-eight centres of documentation.

Uruguay makes use of various international information systems. In the field of agrarian sciences there is the *Subsistema Nacional de Información Agropecuaria del Uruguay (SNIAF)* (National Sub-system of Agricultural and Livestock Information of Uruguay), which is affiliated to the

AGRINTER System at Latin American level and to AGRIS at world level. In the area of health there is the *Centro Nacional de Documentación e Información de Medicina y Ciencias de la Salud (CEMDIN)* (National Centre of Medicine and Health Sciences Documentation and Information), affiliated to the *Biblioteca Regional de Medicina (BIREME)* (Regional Medical Library) in São Paulo, Brazil. In the field of industrial technology, the services of the United Nations Organization Industrial Development Programme are available.

Other information centres include:

Centro de Documentación Científica, Técnica y Económica (Centre of Scientific, Technical and Economic Documentation)
Founded in 1953, it forms part of the National Library. *Boletín Informativo* (Informative Bulletin) is published.

Biblioteca Nacional del Uruguay (National Library of Uruguay)
Founded in 1816, and it has a total of 700 000 volumes.

Centro de Estadísticas Nacionales y Comercio Internacional del Uruguay (CENCI) (Centre of National Statistics and International Trade of Uruguay)
Established in 1956, it is financed by the Uruguayan Government. Its main function is to provide statistical and economic information on Latin American countries. Several publications are issued: *Anuario Estadístico sobre el Intercambio Comercial* (Annual Statistical Report on Commercial Exchange); *Noticias Latinoamericanas* (Latin American News); *Estudios de Mercado* (Market Reports).

International Cooperation in Science and Technology

There are several bilateral agreements presently in operation notably in the fields of rural development, animal health, industrial service, agriculture, foreign trade, etc.

The government authorities have created a state unit to monitor international programmes of cooperation and to act as an advisory body to the executive in matters of administration of programmes with international institutions and the exchange of technology.

Among the various international agreements for cooperation is:

Oficina Regional de Ciencia y Tecnología de la UNESCO para America Latina y el Caribe (UNESCO Regional Office of Science and Technology for Latin America and the Caribbean)
General information on the 60 000 scientific institutes of the area is indexed, as is a collection of more than 1000 Latin American scientific publications. The office coordinates and carries out UNESCO programmes in engineering, marine sciences, science policy, earth sciences, natural resources, scientific information and science teaching. It organizes meetings and conferences and publishes *Boletín* (Bulletin) and *Boletín Internacional de Ciencias del Mar* (International Bulletin of Marine Sciences).

Venezuela

Geographical, Demographical, Political and Economic Features

Venezuela's borders to the north and north-west are formed by the Antilles Sea and the Atlantic Ocean, with 2813 kilometres of coastline, the border to the east is shared with Guyana, with 743 kilometres of frontier; to the south with Brazil (2000 kilometres of frontier); to the west and south-west with Colombia (2050 kilometres of frontier). The total area is 916 700 square kilometres.

Government

According to the Constitution of 1961, Venezuela is a federal republic comprising 20 states, 1 federal district, 2 territories and 72 islands (federal dependencies). The executive power is exercised by the President of the Republic, who is elected by universal suffrage for a period of 5 years. Adults over 18 years of age have the right to vote. The President is advised by a council of ministers. The legislative power is in the hands of the Congress, through a Senate of 43 members and a Chamber of Deupties of 182 members of parliament. Each state has a legislative assembly. The judicial power is exercised by the High Court of Justice and the Courts of Justice. There are 5 High Courts in the Federal District (3 Penal and 2 Civil Courts).

The states (20), divided into 166 districts and 656 municipalities, and the federal district with 2 departments and 24 parishes, as well as the territories, have governors nominated by the President of the Republic. Each district has a municipal council and each municipality has a *Junta Comunal* (Communal Council).

Natural Resources

AGRICULTURE

The country is divided into three natural zones – agricultural; pastoral and forestry. The agricultural zone produces (in thousand tons): coffee (66); cacao (17); sugar cane (5.4); maize (900); rice (534); tobacco (15); cotton (20) (these figures are according to estimates made in 1980).

The pastoral zone has more than 6 million head of livestock of all types and numerous horses. Finally, the third zone has extensive forest land, mainly of cedar, mahogany and pine and also tropical species such as rubber, balata and vanilla as well as others which are mainly exploited by the natives in the region.

MINERALS

The country is rich in oil deposits and is the fifth world producer. Exploitation started in 1917 with an annual production of 18 000 cubic metres; 30 000 square miles is at present being exploited. Almost one-third of oil is produced from the region of Maracaibo. Venezuela has good prospects of expanding its potentialities in oil production in the twenty-first century. Deposits in the Orinoco region are extensive and could be one of the major world sources. Although production was formerly dominated by the powerful foreign oil companies, these were nationalized in January 1976. PETROVEN, the most important State company, produces about 2.1 million barrels per day, of which 1.9 are destined for export.

Venezuela also has important gold mines in the south-east region of the state of Bolivar, and also in Callao, Sosa Mendez and the region of Guyana. In the region of Amazonas there are significant deposits of diamonds. Deposits of manganese are estimated at several million tons, as well as reserves of tri-calcic phosphate and sulphur. The nickel mines are estimated to have 600 000 tons of pure nickel. Finally, there are deposits of copper and asbestos.

Education

Primary education is compulsory and is provided by the state. According to estimates in 1977, 82 per cent of the population is literate.

International Relations

The Republic of Venezuela is a member of the United Nations Organization and the Latin American Free Trade Association (LAFTA).

Organization of Science and Technology

The *Consejo Nacional de Investigaciones Científicas y Tecnológicas (CONICIT)* (National Council for Scientific and Technological Research) and the *Comisión de Ciencia y Tecnología* (Commission for Science and Technology) are the organizations that are the driving force behind the network of research and development in science and technology. In 1976 a national strategy was drawn up in which it was proposed to form an institutional network composed of *CONICIT*; the *Superintendencia de Inversiones Extranjeras (SIEX)* (the Superintendence of Foreign Investment); the *Oficina General de Evaluación* (General Evaluation Office); *Registro Nacional de Proyectos Industriales* (National Register of Industrial Projects); *Banco de la Tecnología* (Technology Bank); *Consejo Venezolano de Normas Industriales (COVENIN)* (Venezuelan Council of Industrial Standards) and the various research institutions. Also proposed was the establishment of organizations such as: an *Oficina Coordinadora de Negociaciones del Estado Venezolano* (Office to Coordinate State Negotiations), *Oficina Nacional de Ingenieria* (National Office of Engineering) and a *Centro Nacional de Evaluación de Tecnologías* (National Centre of Technological Evaluation).

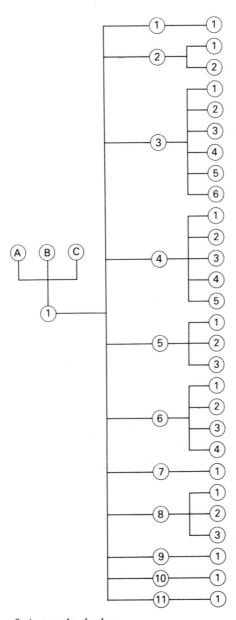

Fig. 15 Organization of science and technology

Fig. 15 – *continued*

1	Ministerio Secretaría de la Presidencia de la República (Secretariat of the President of the Republic)
A	Comisión de Ciencia y Tecnología (Commission of Science and Technology)
B	CORDIPLAN
C	CONICIT
1.1	Ministerio de la Juventud (Ministry of Youth)
1.1.1	Fundacion G. M. de Ayac (G. M. de Ayac Foundation)
1.2	Ministerio de Educación (Ministry of Education)
1.2.1	EDUPLAN
1.2.2	Consejo Nacional de Universidades (National Council of the Universities)
1.3	Ministerio de Agricultura y Ganadería (Ministry of Agriculture and Livestock)
1.3.1	Consejo Nacional de Investigaciones Agrícolas (National Council of Agricultural Research)
1.3.2	Fondo Nacional de Investigaciones Agropecuarias (National Fund of Agricultural and Livestock Research)
1.3.3	CIARA
1.3.4	Centro de Investigaciones Agrícolas (Centre for Agricultural Research)
1.3.5	Centro de Investigaciones Veterinarias (Centre for Veterinary Research)
1.3.6	Centro de Investigaciones Pesqueras (Centre for Fish Research)
1.4	Ministerio de Fomento (Ministry of Development)
1.4.1	Corporación Venezolana de Fomento (Venezuelan Development Corporation)
1.4.2	Registro de la propiedad IND (Property Register IND)
1.4.3	Compania Anónima de Administración y Fomento Eléctrico (CADAFE) (Stock company for the administration and development of electricity)
1.4.4	Consejo Venezolano de Normas Industriales (COVENIN) (Venezuelan Council for Industrial Standardization)
1.4.5	IMPRO
1.5	Ministerio de Sanidad y Asistencia Social (Ministry of Health and Social Security)
1.5.1	Instituto Venezolano de Investigaciones Científicas (IVIC) (Venezuelan Institute of Scientific Investigation)
1.5.2	Comision de Fomento y Becas de S.A.S. (Commission for Support and Grants S.A.S.)
1.5.3	Instituto Nacional de Nutricion (National Institute of Nutrition)
1.6	Ministerio de Minas e Hidrocarburos (Ministry of Mines and Hydrocarbons)
1.6.1	FORINVES
1.6.2	Centro de Evaluaciones Mineras (Centre for Mineral Evaluation)
1.6.3	C.A. Petroleos de Venezuela (Oil Companies of Venezuela)
1.6.4	Centro de Investigaciones de SIDOR (SIDOR Research Centre)
1.7	Ministerio de Defensa (Ministry of Defence)
1.7.1	Hidrologia y Navegacion (Hydrology and Navigation)
1.8	Ministerio de Obras Publicas (Ministry of Public Works)
1.8.1	Corporacion Venezolana de Guayana (Venezuelan Guyana Corporation)
1.8.2	Fundacion Instituto Nacional de Historia (Institute of National History)
1.8.3	COPLAMAR CONAVIAL
1.9	Ministerio de Comunicaciones (Ministry of Communication)
1.9.1	Centro de Investigaciones de la CAREV (CAREV Research Centre)
1.10	Ministerio de Hacienda (Chancellor of the Exchequer)
1.10.1	Superintendencia de Investigaciones Extranjeras (Superintendence of Foreign Research)
1.11	Ministerio de Relaciones Exteriores (Ministry of Foreign Relations)
1.11.1	Instituto de Comercio Exterior (Institute of Foreign Trade)

Financing and Policies of Science and Technology

For the period 1976–80 the first plan for science and technology was drawn up.

The document analysed and proposed various policies, among which were:

the development of national technology which incorporates and assimilates world technological progress;

the use of technology emphasizing the optimum use of the scarce specialist human resources available.

With these objectives in mind, the following methods were proposed:

to establish a *Sistema de Planificación de Ciencia y Tecnología* (System of Planning in Science and Technology) which is incorporated into the *Sistema Nacional de Planificación* (National Planning System);

to guarantee a flow of funds to basic scientific research and to maintain a balance between these activities and the area of applied research;

to determine policies in the educational sector, etc.

Priority sectors are defined as: hydrocarbons; petrochemicals; energy; metallurgy; capital goods; agriculture; food technology; health; nutrition; housing; construction; telecommunications and electronics.

Human resources in science and technology in 1975 were as shown in Table 13.1.

Table 13.1 Scientists and technologists in research and development

Exact and natural sciences	930
Engineering and technology	441
Medicine	699
Agriculture	573
Social science and law	406
Humanities, education and the arts	173
Architecture	77
Others	33
Total	3332

Table 13.2 Financial resources for 1976–80

Social development	1339
Development of agriculture and food industry	2557
Industrial development	1849
Protection and improvement of environment and natural resources	832
*Total (millions of Bolivars)	6637

*Of this total, 18 per cent was from the industrial sector and 82 per cent from the public sector.

Science and Technology at Government Level

Government organizations and the universities carry out most of the research activities in the country. Below are listed a few specialist organizations involved in research, documentation, etc.

Consejo Nacional de Investigaciones Científicas y Tecnológicas (CONICIT) (National Council for Scientific and Technological Research)
Founded in 1967, it is involved in the promotion of research into all fields. It is also the governmental body that outlines the general policy for the development of research in science and technology.

Dirección General de Estadísticas y Censos (National Statistical and Census Office)
Established in 1938, it is affiliated to the Ministry of Development. Statistics are published on population, housing, industry and foreign trade.

Dirección de Cartografía Nacional del Ministerio de Obras Públicas (National Office of Cartography of the Ministry of Public Works)
Its main function is to produce geographical maps and charts of the national territory.

Instituto Venezolano de Petroquímica (Venezuelan Petrochemical Institute)
Founded in 1956, it acts as 'overseer' in the business aspects of the state petrochemical industry and by-products.

Dirección de Geología del Ministerio de Energía y Minas (Geology Office of the Ministry of Energy and Mines)
Founded in 1936, it carries out research in the fields of geotechnology, marine geology and mineralogy in general, and the drawing of geological surveys, maps and charts.

Dirección de Minas del Ministerio de Minas e Hidrocarburos (Mining Office of the Ministry of Mines and Hydrocarbons)

Scientific and Technological Research at Academic Level

Consejo de Desarrollo Científico y Humanístico (Council of Scientific and Humanistic Research)
Founded in 1958, it promotes research, finances publications, invites foreign scientists, sends representatives to international congresses and authorizes grants.

Instituto Latinoamericana de Investigaciones Científicas en Educación a Distancia (Latin American Institute of Scientific Research in Correspondence Education)
Established in 1980, it offers courses and carries out research on a correspondence basis.

Instituto Venezolano de Investigaciones Científicas (Venezuelan Institute of Scientific Research)
Founded in 1959, it conducts post-graduate research in biology, medicine, chemistry, physics, mathematics and technology.

Centro de Estudios del Desarrollo (CENDES) (Centre of Development Studies)
Established in 1960, it is a centre for research, teaching and planning in

matters relating to the economic, social and administrative development of Venezuela.

Instituto Venezolano de Análisis Económico y Social (Venezuelan Institute of Social and Economic Analysis)
Founded in 1960, it is a private institute dedicated to the study of and reporting on economic and financial matters.

Universidad de Carabobo (Carabobo University)
A state-run university founded in 1852 which has a student population of 24 000 with 500 teachers. Several publications are produced, among which are *Revista de la Facultad de Ciencia* (Faculty of Science Review) and *Boletín Universitario* (University Review). Its faculties are economics, engineering, law and medicine.

Universidad Católica Andrés Bello (Andrés Bello Catholic University)
Founded in 1953, it is a private university with an academic year between October and July. Students number 8060 with a staff of 580 teachers. Apart from the arts and humanities faculties there are schools of industrial engineering, basic engineering, and civil engineering.

Universidad Central de Venezuela (Central University of Venezuela)
A state university founded in 1725, it has 51 169 students and 6477 teachers for its faculties of agriculture, science, engineering, medicine, dentistry, veterinary sciences, pharmacy and faculties in humanities. Also affiliated to the university are the faculty of agriculture and veterinary sciences in Maracay and the faculty of engineering in Cagua.

Universidad Centro-Occidental (Central-Western University)
Founded in 1936 as a Centre of Advanced Education, it attained university status in 1968. Run under state control with 9712 students and 506 teachers in its faculties of agronomy, sciences, medicine, engineering, and veterinary medicine. Publications include *El Veterinario* (The Veterinary Surgeon) six times yearly.

Universidad de los Andes (University of the Andes)
Founded in 1785 as the *Real Colegio Seminario de San Buenaventura de Merida* (San Buenaventura de Merida Royal College), it became a university in 1810. There is a campus in Trujillo and one in San Cristobal. The university is state-controlled, with an academic year from January to June and August to December. Among its publications are: *Revista Forestal* (Forestry Review); *Revista de la Farmacia* (Pharmacy Review); and annual reports from the faculties. Students number 24 185, with 2301 teachers for the faculties of humanities, pharmacy, odontology, medicine, engineering, forestry, and sciences.

Universidad Metropolitana (Metropolitan University)
A private university founded in 1970, it has an academic year from October to March and March to July. Its departments are: physics, humanities, mathematics, chemistry, chemical engineering, mechanical engineering, electronics, and computer systems.

Universidad Nacional Experimental Francisco de Miranda (Francisco de Miranda Experimental University)
Founded in 1977, it has a small student population of only 300 and 74 lecturers for its faculties of civil and industrial engineering, computing, and medicine.

Universidad Nacional Experimental de los Llanos Centrales 'Romulo Gallegos' (Romulo Gallegos Central Plains Experimental University)
Established in 1977, it has 340 students and 65 teaching staff with faculties in agriculture and health sciences.

Universidad Nacional Experimental de Táchira (Táchira Experimental University)
Functions under state control since its founding in 1974 and has an academic year from February to June and August to December. Students number 1514 for the faculties of mechanical, agricultural and industrial engineering, and zoology.

Universidad de Oriente (University of the East)
Established in 1958, it is run under state control with an academic year from March to December. The staff numbers 1181 teachers for 16 463 students. Publications include *Boletín del Instituto Oceanográfico* (Bulletin of the Oceanography Institute); *Oriente Agropecuario* ('Oriente' Agriculture and Livestock) and *UDO Investiga* (UDO Investigates). Its schools are in the humanities, agricultural science and general sciences. Affiliated to the university is the Institute of Oceanography; the *Monagas Campus* which has schools of agriculture and livestock; the Institute of Agriculture and Livestock; the *Bolívar Campus*, with schools of Medicine, geology and mining; the *Nueva Esparta Campus*, with a school of biology; and a Research Centre.

Universidad Rafael Urdaneta (Rafael Urdaneta University)
Founded in 1973, it has faculties in engineering, animal feeding and general sciences.

Universidad de Santa María (Santa María University)
A private university founded in 1953, it has 4500 students and 300 teachers for its faculties of humanities, engineering and pharmacy.

Universidad Simón Bolívar (Simón Bolívar University)
Established in 1970, it functions under state control with 7800 students and 860 lecturers, between the months of September and July. There are several Institutes which collaborate together in research projects: *Instituto de Petróleo* (Petroleum Institute); *Instituto de Tecnología y Ciencias Marinas* (Institute of Marine Science and Technology); *Instituto de Recursos Naturales Renovables* (Institute of Renewable Natural Resources); *Instituto de Investigaciones Metalúrgicas* (Institute of Metallurgical Research); *Instituto de Energía Eléctrica* (Institute of Electrical Energy); *Instituto de Investigación y Desarrollo Industrial* (Institute of Industrial Research and Development).

Universidad del Zulia (Zulia University)
Originally founded in 1891, it was shut down in 1904 and reopened in 1946. A state-run university with 26 882 students and 1415 lecturers, its academic year runs from February to December. Apart from the faculties of humanities, there are faculties of engineering, agriculture, medicine, dentistry and veterinary sciences, and schools of medicine, dentistry; dietetics and nutrition, nursing, agricultural engineering, petrochemical, civil, geodetic, chemical, and mechanical engineering, and schools of veterinary sciences, bio-analysis, and basic engineering. The following institutes are affiliated to the university: Institute of Clinical Research; Institute of Applied Calculus; Institute of Odontological Research; Institute of Mathematical Studies; Institute of Biological Research; Institute of Genetic Medicine; Centre of Experimental Surgery; 'Kasmera' Biological Centre; Institute of Petroleum Research; Centre of Biological Studies; Institute of Agronomy Research, and the Institute of Animal Production.

Instituto Universitario Politécnico (Polytechnic College)
Founded in 1962, it has a staff of 160 teachers for its 2600 students. The polytechnic offers courses in mechanical, electrical, chemical, and metallurgical engineering, electronics and electrical technology, and mechanics. *Catálogo* (Catalogue) is published annually.

Agricultural and Marine Science and Technology

Centro Nacional de Investigaciones Agropecuarias (National Centre for Agricultural and Livestock Research)
Founded in 1937, it is a dependant of the *Fundación Nacional para la Investigación Agrícola* (National Foundation for Agricultural Research).

Consejo Nacional de Investigaciones Agrícolas (National Council for Agricultural Research)
Established in 1959, it promotes research in agriculture and runs conventions for specific research projects. Among its functions the Council also manages the National Foundation for Agricultural Research.

Estación Experimental de Café (Experimental Coffee Station)
Publishes an annual report.

Instituto Agrario Nacional (National Agrarian Institute).
The institute deals principally with matters related to agrarian reform.

Fundación La Salle de Ciencias Naturales (La Salle Foundation of Natural Sciences)
Founded in 1957, it carries out oceanographic research. The foundation has marine and hydrobiological research stations on the island of Margarita. Affiliated to the foundation are: the *Instituto Universitario de Tecnología del Mar* (University Institute of Marine Technology); the *Estación de Investigaciones Marinas de Margarita* (Margarita Marine Research Station); and *Liceo Naútico Pesquero* (Institute of Sea Fishing).

Medical Science and Technology

A major part of research in this field is carried out in specialized hospital units and university departments.

Instituto de Medicina Experimental (Institute of Experimental Medicine)
Founded in 1949, it carries out research in biochemistry, pharmacology, physiology, neurology and general and applied pathology. *Boletín Informativo Sistema Nacional de Documentación e Información Biomédica* (Informative Bulletin on the National System of Bio-medical Documentation and Information) is published by the institute.

Instituto Nacional de Nutrición (National Institute of Nutrition)
It has a library of over 10 000 volumes and publishes *Archivos Latinoamericanos de Nutrición* (Latin American Archives on Nutrition).

Nuclear Science and Technology

Consejo Nacional para el Desarrollo de la Industria Nuclear (National Council for the Development of the Nuclear Industry)

Military Research in Science and Technology

Observatorio Naval 'Juan Manuel Cagigal' (Juan Manuel Cagigal Naval Observatory)
Founded in 1890, it carries out research in astronomy, geophysics, oceanography, hydrography, seismology and meteorology and publishes monthly bulletins.

Economic and Social Research

Centro de Estudios Venezolanos Indígenas (Venezuelan Centre of Indigenous Studies)

Fundación 'Lisandro Alvarado' (Lisandro Alvarado Foundation)

Instituto Caribe de Antropología y Sociología (Caribbean Institute of Anthropology and Sociology)

Meteorological and Astrophysical Research

Estación Meteorológica (Meteorological Station)
Founded in 1940, it carries out meteorological and hydrographical research programmes, mainly in the Orinoco Basin.

Professional Associations and Societies

There are more than seventy organizations in Venezuela that carry out activities related to research and development. Among them are:

Academia de Ciencias Exactas, Físicas y Matemáticas (Academy of Exact Sciences, Physics and Mathematics)
Established in 1917, it promotes study and research in various fields of science. The academy also approves school science textbooks.

Academia Nacional de Medicina (National Academy of Medicine)
Publishes *Gaceta Médica de Caracas* (Medical Gazette of Caracas).

Asociación Venezolana para el Avance de la Ciencia (Venezuelan Association for the Advancement of Science)

Sociedad Venezolano de Ciencias Naturales (Venezuelan Society of Natural Sciences)
Founded in 1931, it organizes conferences on scientific topics. There is a department for natural caves exploration and study, and a biological station for the study of Venezuelan flora and fauna. The society also carries out research into environmental pollution. *Boletin de la SVCN* (SVCN Bulletin) and *Ciencia del Día* (Science Today) are published.

Asociación Venezolana de Ingenieria Eléctrica y Mecanica (Venezuelan Association of Electrical and Mechanical Engineering)
Publishes *Acta Científica Venezolana* (Venezuelan Scientific Acts) six times a year.

Information Services in Science and Technology

There are more than thirty-five libraries and eleven museums in Venezuela housing specialist documentation. Among them are:

Biblioteca Nacional (National Library)
Founded in 1883, it has 400 000 volumes and has recently become the central organization of the *Instituto Autónomo de la Biblioteca Nacional y Servicios de Bibliotecas* (Autonomous Institute of the National Library and Library Services), which follows the NATIS system designed by UNESCO.

There are also the following university libraries:

Universidad Central de Venezuela with 150 000 volumes.

Universidad Catolica 'Andrés Bello' with 112 000 volumes.

Universidad de Oriente in Cumana with 43 000 volumes.

Universidad de Zulia in Maracaibo with 20 000 volumes.

Universidad de los Andes in Merida with 114 000 volumes.

Universidad de Carabobo in Valencia with 11 000 volumes.

Biblioteca Técnica Científica Centralizada (Central Scientific and Technical Library)
Founded in 1966 in Barquisimeto.

International Cooperation in Science and Technology

Office de la Recherche Scientifique et Technique Outre-Mer (ORSTOM) (Office of Overseas Scientific and Technological Research) – Venezuela branch.

Venezuela is an active member in most of the United Nations and Organization of American States programmes and organizations. *CONICIT* maintains relations with scientific organizations in Latin America and with the universities of North America and Europe.

Appendix 1

Regional Statistics

Source
Sixth Meeting of the Permanent Conference of National Scientific and Technological Organizations of Latin America and the Caribbean.

Table 1 Socio-economic indicators

Country	Area (square kilometres)	Population estimate (1978) (thousands)	Density (1978) (per square kilometre)	Average annual population growth 1975–78	GNP 1978 (US$) (thousands)	GNP per inhabitant (US$)
Argentina	2 766 889	26 393	10	1.3	67 193 880	2 546
Bahamas	13 935	225	16	3.4	524 700	2 332
Barbados	431	250	580	0.8	508 337	2 033
Bolivia	1 098 581	5 137	5	—	4 066 450	792
Brazil	8 511 965	115 396	14	2.8	187 260 136	1 623
Chile	756 945	10 857	14	1.9	15 127 131	1 393
Colombia	1 138 914	25 645	23	2.7	22 498 246	877
Costa Rica	50 700	2 111	41	2.4	3 419 927	1 620
Cuba	114 524	9 728	84	1.4	—	—
Dominican Republic	48 734	5 124	105	3.0	4 581 900	894
Ecuador	283 561	7 814	28	3.4	7 280 600	932
El Salvador	21 041	4 354	207	2.8	3 042 800	699
Granada	344	97	282	-2.8	61 296	632
Guatemala	108 889	6 621	61	2.9	6 207 600	938
French Guiana	91 000	66	1	3.2	—	—
Guyana	214 969	820	4	1.7	473 729	578
Haiti	27 750	4 833	174	1.8	1 094 000	226
Honduras	112 088	3 439	31	3.6	1 740 500	506
Jamaica	10 991	2 133	195	1.5	2 540 258	1 191
Mexico	1 972 547	66 944	34	3.6	89 918 997	1 343
Nicaragua	130 000	2 395	18	3.6	2 087 834	872
Panama	75 650	1 826	24	3.1	2 270 000	1 243
Paraguay	406 752	2 888	7	3.0	2 536 973	878
Peru	1 285 216	16 819	13	2.8	10 337 013	615
Surinam	163 265	374	2	0.8	838 089	2 241
Uruguay	176 215	2 864	16	0.6	4 925 102	1 720
Venezuela	912 050	13 122	14	3.0	39 105 830	2 980

US$ calculated on rate of exchange in October 1981.

Table 2 Retrospective data on human and financial resources in research destined to higher education – third grade

Country	Year	National currency	Personnel in research and development		Scientists and engineers per million population	Expenditure in research and development				Expenditure in higher education third grade	
			Total	Scientists and engineers		National currency (thousands)	US$ (thousands)	US$ per inhabitant	% GNP	Total % GNP	Current % GNP
Argentina	1970	Peso	21 250[a]	6 500[a]	274	207 500	55 197	2.0	0.2	0.4	0.4
	1972		24 200[a]	7 350[a]	301	246 100	30 516	1.0	0.1	0.4	0.4
	1975		25 000[a]	7 500[a]	295	—	—	—	—	0.8	0.7
	1976		26 500[a]	8 000[a]	311	23 171 700	176 105	7.0	0.3	0.4	0.4
	1977		27 000[a]	8 100[a]	312	—	—	—	—	0.5	0.4
	1978		27 500[a]	8 250[a]	313	195 278 000	253 861	10.0	0.4	0.6	0.5
	1980		—	9 500[a]	351	1 258 700 000	629 350	—	—	—	—
Brazil	1975	Cruzeiro	—	—	—	8 020 500	986 522	9.0	0.8	—	0.7
	1976		—	—	—	10 346 500	972 571	9.0	0.7	—	0.7
	1977		—	17 187	153	10 599 000	752 529	7.0	0.5	—	0.8
	1978		—	24 015	208	20 781 000[b]	1 163 736[b]	10.0	0.6	—	0.9
Chile	1978	Peso	—	—	—	2 416 606	76 340	7.0	0.5	1.1	1.1
Colombia	1971	Peso	—	1 140[c, d]	54	210 614[d]	10 497[d]	0.5	0.1	0.3[e]	0.3[e]
	1978		—	3 404[c]	183	735 783	18 718	0.7	0.1	—	—
Costa Rica	1974	Colon	—	—	—	35 201[f]	4 503[f]	2.0	0.3	—	—
	1975		—	—	—	39 113[f]	4 564[f]	2.0	0.2	—	1.6
	1976		—	—	—	43 459[f]	5 071[f]	3.0	0.2	—	1.9
	1977		—	—	—	48 288[f]	5 635[f]	3.0	0.2	—	—
	1978		—	—	—	53 653[f]	6 261[f]	3.0	0.2	—	—

Table 2 – *continued*

Country	Year	National currency	Personnel in research and development			Expenditure in research and development				Expenditure in higher education third grade	
			Total	*Scientists and engineers*	*Scientists and engineers per million population*	*National currency (thousands)*	*US$ (thousands)*	*US$ per inhabitant*	*% GNP*	*Total % GNP*	*Current % GNP*
Cuba	1969	Peso	12 361	1 850	220	91 735	91 735	11.0[g]	1.3[g]	—	—
	1977		19 659	4 959	517	74 258	89 852	9.0[g]	0.5[g]	—	—
	1978		19 868	4 972	511	83 163	112 270	12.0[g]	—	—	—
	1979		21 786	5 680	581	91 049				—	0.3
Ecuador	1970	Sucre	1 103[h]	595	100	90 515	4 442	1.0	0.3	0.7	0.2
	1973		761[h]	544	82	142 310	5 692	1.0	0.2	—	—
Guatemala	1970	Quetzal	364[h]	230	44	3 008	3 008	1.0	0.2	—	0.2
	1972		522[h]	267	48	3 932	3 932	1.0	0.2	—	—
	1974		747[h]	310	52	5 139	5 139	1.0	0.2	—	—
	1978		981[h]	549	83	13 504	13 504	2.0	0.2	—	0.2
Jamaica	1969	Dollar	—	—	—	833[i]	1 000[i]	1.0	0.1	—	0.2
	1970		—	—	—	1 060[i]	1 272[i]	1.0	0.1	—	0.3
	1971		—	—	—	1 095[i]	1 333[i]	1.0	0.1	—	0.4
Mexico	1969	Peso	—	3 665[c]	78	519 134[i]	41 531[i]	1.0	0.1	0.2	0.2
	1970		12 456[c]	3 743[c]	74	761 611	60 929	1.0	0.2		0.4
	1971		13 525[c]	4 064[c]	77	1 034 124	82 730	2.0	0.2	0.5	0.4
	1973		—			1 277 618	102 209	2.0	0.2	0.4	0.3
	1974		—	8 446[c]	145	—	—			—	0.2
Nicaragua	1971	Cordoba	—	—		7 847[i]	1 121	1.0	0.1		

Country	Currency	Year									
Panama	Balboa	1974	—	—	—	2 908	2 908	2.0	0.2	0.7	0.6
		1975	982	204	122	3 296	3 296	2.0	0.2	0.7	0.6
Paraguay	Guaraní	1971	—	134	57	167 265	1 328	1.0	0.2	0.4	0.3
Peru	Sol	1970	5 295[k]	1 686[k]	125	984 636[k]	25 443[k]	2.0	0.4	—	0.1
		1976	8 984	3 932	247	2 314 754	41 804	3.0	0.3	—	0.1
Uruguay	Peso	1971	3 033	1 150	394	1 673	6 508	2.0	0.2	—	—
		1972	—	—	—	1 858	3 437	1.0	0.2	—	—
Venezuela	Bolívar	1970	4 608	1 779	173	102 270	22 726	2.0	0.2	1.3	1.2
		1973	5 109	2 720	241	289 697	67 314	6.0	0.4	—	1.6

(a) Data for scientists and engineers is presented in net year/man.
(b) Not including private production firms.
(c) Full-time and part-time basis.
(d) Excluding data on law, human sciences and education.
(e) Data refers to 1970.
(f) Data only refers to public expenditure.
(g) Global social product.
(h) Not including auxiliary staff.
(i) Data refers to general expenditure only.
(j) Data refers to research centres only.
(k) Data does not include human sciences and education.

Table 3 Total staff employed in research and development activities by category. Last available year on full-time basis

Country	Year	Scientists and engineers	Technicians	Auxiliary staff	No. of technicians per scientist and engineer
Argentina[a]	1980	9 500	13 300	—	1.40
Bahamas[b]	1970	19	1	38	0.05
Bolivia[c]	1967	400	800[d]	—[d]	2.00[d]
Brazil	1978	24 015[e]	34 559[f]	—[d]	—
Chile	1975	5 948	—	—	—
Colombia	1978	1 449	704	463	0.50
Cuba	1979	5 680	6 593	9 513	1.20
Ecuador[g]	1976	469	411	158	0.90
El Salvador[h]	1974	802	519	—	0.60
Guatemala	1978	549	432	—	0.80
Honduras[i]	1974	5[j]	1	1	0.20
Mexico	1974	5 896	—	—	—
Panama	1975	204	301	477	1.50
Paraguay	1971	134	—	—	—
Peru	1976	3 932	2 235	2 817	0.60
Uruguay	1971	1 150	1 087	796	0.90
Venezuela[k]	1977	1 627	2 500	3 391	1.50

(a) Data in net year/man.
(b) Data only refers to the Central Government.
(c) Data excludes law; human sciences and the arts.
(d) Data for auxiliary staff is included with the technicians.
(e) Not including data relative to the general service sector.
(f) The figure refers to technicians and auxiliary staff of the productive sector only.
(g) Data refers to research and development in the agricultural sciences only.
(h) Data refers to twenty-eight centres out of a total of forty-one that carry out activities in research and development.
(i) Data refers to a research centre only.
(j) Data refers to scientists and engineers on full-time basis only.
(k) Data refers to 167 centres out of a total of 406 involved in research and development activities.

Table 4 Number of scientists and engineers employed in research and development activities by field of study. Last available year on full-time basis

Country	Year	All sectors	Exact and natural sciences	Engineering and technology	Medical sciences	Agricultural sciences	Social and human sciences
Argentina[a]	1976	8 000	2 960	1 150	1 750	1 110	1 030
Bahamas[b]	1970	19	2	—	—	10	7
Brazil[c]	1974	7 725	3 660	1 088	818	785	1 374
Chile	1975	5 948	1 885	1 367	1 562	411	723
Colombia	1971	1 140[d,e]	188	154	127	348	323[e]
Cuba	1978	4 972	668	1 441	1 112	1 257	494
Ecuador[f]	1976	469	100	45	—	304	20
El Salvador[g]	1974	802	190	120	183	85	224
Guatemala	1974	310	43	98	20	61	88
Honduras[h]	1974	5[i]	—	5	—	—	—
Mexico	1974	5 896[j]	1 523	1 170	648	765	1 535
Panama	1975	204[k]	39	41	40	33	28
Peru	1970	1 686[l,m]	496	83	330	507	180[l]
Uruguay	1971	1 150[n]	142	285	239	253	109
Venezuela[o]	1977	4 060[d,p]	1 033	730	944	774	560

General note – scientists and engineers are classified according to sector of study (that is, equivalent qualification or course of study). The six main scientific and technological sectors incorporate the main sectors of study of *CINE*.

(a) Data presented in net year/man.
(b) Data refers to Central Government only.
(c) Data refers to basic research at post-graduate level only and to teachers in post-graduate sectors of the universities.
(d) Data refers to the number of scientists and engineers on full-time and part-time basis.
(e) Excluding data referring to law, human sciences and education.
(f) Data refers to research and development in agricultural sciences only.
(g) Data refers to twenty-eight centres out of a total of forty-one that carry out research and development.
(h) Data refers to research centres only.
(i) Data refers to scientists and engineers on full-time basis only.
(j) The total includes 255 scientists and engineers for whom no breakdown into fields of specialization is available.
(k) The total includes twenty-three scientists and engineers for whom no breakdown into fields of specialization is available.
(l) Excluding data relating to human sciences and education.
(m) The total includes ninety scientists and engineers for whom no breakdown into fields of specialization is available.
(n) The total includes 122 scientists and engineers for whom no breakdown into fields of specialization is available.
(o) Data refers only to 167 centres out of a total of 406 that develop activities in research and development.
(p) The total includes nineteen scientists and engineers for whom no breakdown into fields of specialization is available.

Table 5 Scientific and technological personnel employed in research and development activities per sector of execution. Last available year on full-time basis

Country	Year	All sectors		Productive sectors		Higher education		General service sector	
		Scientists and engineers	Technicians	Scientists and engineers	Technicians	Scientists and engineers	Technicians	Scientists and engineers	Technician
Argentina	1980	9 500[a]	13 300	1 900[a]	3 000	4 750	6 300	2 850[a]	4 000
Brazil	1978	—	—	8 497	5 392	15 518	—	—	—
Chile	1975	5 948	—	627	—	4 975	—	346	—
Colombia	1978	3 404[b]	704	50[b]	3	2 004[b]	462	1 350[b]	239
Cuba	1979	5 680	6 593	2 938	3 700	624	248	2 118	2 645
Ecuador[c]	1976	469	411	378	329	91	82	—	—
Guatemala	1974	310	439	166	124	43	70	101	245
Honduras[d]	1974	5[e]	1	—	—	—	—	5[e]	1
Mexico	1974	5 896	—	1 973	—	1 968	—	1 955	—
Panama	1975	204	301	116	194	62	80	26	27
Paraguay	1971	134	—	38	—	49	—	47	—
Peru	1976	3 932	2 235	1 295	863	1 544	6	1 093	1 366
Uruguay	1971	1 150	1 087	394	379	537	336	219	372
Venezuela	1977	4 060[b]	2 500	966[b]	1 230	2 881[b]	1 033	213[b]	237

(a) Data presented in net year/man.
(b) Data refers to number of scientists and engineers on full-time and part-time basis.
(c) Data refers to research and development in the agricultural sector only.
(d) Data refers to a research centre only.
(e) Data refers to scientists and engineers on full-time basis.

Table 6 Total expenditure destined to research and development

Country	Year	National currency	All sectors (thousands)	Productive sector (thousands)	Higher education (thousands)	General service sector (thousands)
Argentina	1980	Peso	1 258 700 000	433 000 000	419 100 000	406 600 000
Brazil	1978	Cruzeiro	20 781 000[a]	6 442 000[a]	5 548 000	8 791 000
Chile	1978	Peso	–	1 120 300[b]	1 296 300	1 296 300
Colombia	1978	Peso	735 783	3 121	215 280	517 382
Cuba	1979	Peso	91 049	56 002	4 782	30 265
Ecuador[c]	1976	Sucre	258 366	218 508	39 858	–
Guatemala	1974	Quetzal	5 139	1 175	392	3 572
Jamaica	1971	Dollar	1 095[d,e]	156[e,f]	325[e]	539[e,f]
Mexico	1973	Peso	1 277 618	611 021	332 194	334 493
Panama	1975	Balboa	3 296	1 429	503	3 304
Paraguay	1971	Guaraní	123 959[e]	18 560[e]	19 984[e]	84 415[e]
Peru	1976	Sol	2 314 754	1 515 281	169 366	630 107
Uruguay	1972	Peso	1 858	1 177	362	319
Venezuela[g]	1977	Bolívar	883 545	618 236	248 829	16 480

(a) Not including private production firms.
(b) Data refer to the productive sector (activities in research and development).
(c) Data only refer to research and development in agricultural sector.
(d) The total includes $74,000, whose breakdown has not been provided.
(e) Data refer to general expenditure only.
(f) Data relative to the productive sector refer to private firms and to the general service sector.
(g) Data refer to 167 centres out of a total of 406 that develop activities in research and development.

Table 7 Higher education third grade: students graduating from national institutions by sector of study

| Country | Year | Total | Sector of study | | | | | | No. of graduates per million inhabitants |
			Natural and exact sciences	Engineering and technology	Medical sciences	Agricultural sciences	Social and human sciences	Others	
Argentina	1970	23 991	617	2 486	5 884	722	14 282	—	1 010
Bolivia	1976	1 542[a]	329	199	505	46	463	—	300
Brazil	1978	198 716	3 711	21 341	19 380	5 931	144 594	3 759	1 722
Chile	1979	17 640[b]	322	2 917	2 798	787	10 816	—	1 616
Colombia	1977	18 780[b]	656	2 544	2 022	822	12 736	—	750
Costa Rica	1978	4 146[b]	239	395	699	177	2 636	—	1 964
Cuba	1979	20 615	766	2 434	1 332	1 660	14 423	—	2 109
Dominican Republic	1978	2 452[a]	145	473	823	76	935	—	479
Ecuador	1973	3 660	85	440	394	295	2 446	—	555
El Salvador	1978	1 487	22	806	164	205	293	—	342
French Guiana	1976	437	22	—	—	—	402	13	7 048
Guatemala	1979	1 340	32	156	354	71	702	25	190
Guyana	1978	965	93	56	—	—	816	—	1 177
Haiti	1978	344	31	83	151	38	41	—	71
Honduras	1978	747	13	206	128	104	278	18	217
Jamaica	1977	1 281[c]	282	113	105	33	748	—	609
Mexico	1978	59 254[b]	1 573	15 663	15 269	3 646	23 103	—	885
Panama	1978	2 298	228	267	385	169	1 172	77	1 258
Paraguay	1973	552[d]	67	30	76	74	260	45	221
Peru	1978	6 108[b]	161	883	1 378	609	2 791	286	363
Uruguay	1979	1 987	73	170	920	162	615	47	690
Venezuela	1978	13 668	448	2 625	1 693	840	8 062	—	1 042

(a) Data refer to universities and equivalent establishments only.
(b) Data refer to universities only.
(c) University of the West Indies only.
(d) Universidad Nacional de Asunción (Asunción University) only.

Table 8 Higher education third grade: students graduating from national institutions from 1970

Country	1970	1975	1976	1977	1978	1979
Argentina	23 991	—	—	—	—	—
Bolivia	1 313[a]	1 093[a]	1 973[a]	—	1 542[a]	—
Brazil	64 049	153 065[b]	185 015	199 907	198 716	—
Chile	8 255	11 900	—	14 474	—	17 640[c]
Colombia	8 209	13 616	—	18 780[c]	—	—
Costa Rica	—	—	2 307	4 471[a]	4 146[c]	—
Cuba	3 003[d]	—	9 233	—	15 343	20 615
Dominican Republic	665[d]	1 173[e]	—	1 692[a]	2 452[a]	—
El Salvador	428[f]	1 102	—	1 190[f]	1 487	—
Guatemala	514[g]	1 021[g]	1 020[g]	1 443[g]	1 099[g]	1 340
Guyana	160	834	—	—	965	—
Haiti	—	—	265[c]	331	344	—
Honduras	105[d, h]	580	—	242[h]	747	—
Jamaica[i]	663	1 160	1 275	1 281	—	—
Mexico	9 478[d]	24 674[i]	—	—	59 254[c]	—
Panama	589	1 130[b]	1 668	2 062	2 298	—
Peru[c]	6 914[d]	6 335	4 700	5 450	6 108	—
Uruguay	1 065	2 049	—	—	—	1 987
Venezuela	4 927	7 986	12 940	14 030	13 668	—

(a) Data refers to universities and equivalent establishments only.
(b) For 1974.
(c) Data refers to universities only.
(d) For 1969.
(e) Universidad Autónoma de Santo Domingo (Autonomous University of Santo Domingo) only.
(f) Universidad de Salvador (Salvador University) only.
(g) Universidad de San Carlos (San Carlos University) only.
(h) Universidad de Honduras (Honduras University) only.
(i) For 1973.

Appendix 2

Directory of Selected Establishments

In the section for Universities and Colleges the order follows that adopted by the country concerned.

Argentina

Academia Nacional de Ciencias de Buenos Aires
(National Academy of Sciences of Buenos Aires)
Address: Junin 1278, Buenos Aires

Academia Nacional de Medicina
(National Academy of Medicine)
Address: Las Heras 3092, 1425 Buenos Aires

Biblioteca del Servicio Geológico Nacional
(National Geological Service Library)
Address: Santa Fé 1548/3 piso, Buenos Aires

Biblioteca Nacional
(National Library)
Address: Mexico 564, Buenos Aires

Centro Argentino de Información Científica y Tecnológica – UNESCO
(Argentine Centre for Scientific and Technological Information)
Address: Moreno 431/33, Buenos Aires

Centro de Altos Estudios de Ciencias Exactas
(Centre for Higher Studies in Exact Sciences)
Address: Solis 550, Buenos Aires

Centro de Estudios Urbanos y Regionales (CEUR)
(Centre of Urban and Regional Studies)
Address: Bartolomé Mitre 2212, 1039 Buenos Aires

Centro de Información de las Naciones Unidas
(Information Centre of the United Nations)
Address: Marcelo T. de Alvear 684, Buenos Aires

Centro de Investigación de Biología Marina
(Research Centre of Marine Biology)
Address: C.C. 157, 1650 San Martín, Buenos Aires

Centro de Investigación Documentaria del Instituto Nacional de Tecnología Industrial – UNIDO
(Centre for Documentary Research of the National Institute of Industrial Technology)
Address: Libertad 1235, Casilla 1359, Buenos Aires

Centro de Investigaciones Bella Vista
(Bella Vista Research Centre)
Address: Jose Manuel Estrada 66, Bella Vista, Corrientes

Centro de Investigaciones de Recursos Naturales
(Natural Resources Research Centre)
Address: Castelar, Provincia de Buenos Aires

Centro Espacial San Miguel – Observatorio Nacional de Física Cósmica
(San Miguel Space Centre - National Observatory of Cosmic Physics)
Address: Avenida Mitre 3100, 1663 San Miguel

Centro Espacial Vicente Lopez
(Vincente Lopez Space Centre)
Address: Avenida del Libertador 1513, Vicente Lopez

Centro Nacional de Información Educativa – UNESCO
(National Centre of Educational Information)
Address: Avenida Eduardo Madero 235, Buenos Aires

Chacra Experimental de Barrow
(Barrow Experimental Farm)
Address: C.C. 216, 7500 Tres Arroyos

Comisión Nacional de Energía Atómica (CNEA)
(National Commission for Atomic Energy)
Address: Avenida del Libertador 8250, 1429 Buenos Aires

Comisión Nacional de Investigaciones Espaciales (CNIE)
(National Commission of Space Research)
Address: Avenida Pedro Zanni 250, 1104 Buenos Aires

Consejo Nacional de Investigación Científica y Técnica (CONICET)
(National Council for Scientific and Technical Research)
Address: Rivadavia 1917, 1033 Buenos Aires

Departamento de Estudios Históricos Navales
(Department of Historical Naval Studies)
Address: Comando en Jefe de la Armada, Avenida Com. Py y Corbeta Uruguay, Buenos Aires

Departamento Efectos del Medio Contaminado, Dirección Nacional de Estudios y Proyectos – UNEP
(Department for Measuring the Effects of Pollution, National Direction of Studies and Projects)
Address: Calle Alsina 301, 187 Buenos Aires

Escuela Nacional de Educación Técnica No 1
(National School of Technical Education No 1)
Address: Ruta Nacional No 22, Plaza Huincul, Neuquen

Escuela Nacional de Educación Técnica No 4
(National School of Technical Education No 4)
Address: Lacarra 535, Buenos Aires

Estación Experimental Agro-Industrial
(Agro-Industrial Experimental Station)
Address: Obispo Colombres C.C. 71, San Miguel de Tucumán

Estación Experimental Agropecuaria de Salta - INTA
(The Salta Experimental Station of Arable and Livestock Farming)
Address: C.P. 228, Salta

Estación Experimental Agropecuaria Mendoza - INTA
(Mendoza Experimental Station of Arable and Livestock Farming)
Address: C.C. 3, Lujan de Cuyo, Mendoza

Estación Experimental Regional Agropecuaria - INTA
(Regional Experimental Station of Arable and Livestock Farming)
Address: C.C. 31, 2700 Pergamino

Estación Hidrobiológica
(Hydro-biological Station)
Address: 7631 Quequen, Provincia de Buenos Aires

Facultad de Ciencias Aplicadas a la Industría
(Faculty of Applied Science for Industry)
Address: Cnte. Salas 227, San Rafael, Mendoza

Fundación Cosio
(The Cosio Foundation)
Address: Las Heras 2603, Buenos Aires

Instituto Agrario Argentino de Cultura Rural
(Argentine Agrarian Institute of Rural Culture)
Address: Peru 277, Buenos Aires

Instituto Argentine de Racionalización de Materiales (IARM)
(Argentine Institute of Standards)
Address: Chile 1192, Buenos Aires

Instituto de Biología y Medicina Experimental
(Institute of Biology and Experimental Medicine)
Address: Obligado 2490, 1428 Buenos Aires

Instituto de Desarrollo Económico y Social
(Institute of Economic and Social Development)
Address: Guemes 3950, Buenos Aires

Instituto Geográfico Militar
(Military Geographical Institute)
Address: Avenida Cabildo 301, 1426 Buenos Aires

Instituto de Hematología, Instituto Nacional de la Salud
(Haematology Institute, National Health Institute)
Address: Martínez de Hoz y Marconi, Haedo, Provincia de Buenos Aires

Instituto de Investigación Aeronáutica y Espacial (IIAE)
(Institute of Aeronautic and Space Research)
Address: Guarnición Aérea, Avenida Fuerza Aérea Km 5½, Córdoba

Instituto de Investigaciones Médicas
(Institute of Medical Research)
Address: Boulevar Wilde 761, Rosario

Instituto de Mecánica Aplicada y Estructuras
(Institute of Applied and Structural Mechanics)
Address: Avenida Pellegrini 250, Rosario

Instituto de Microbiología e Industrias Agropecuarias – INTA
(Institute of Microbiology and Arable and Livestock Farming)
Address: Villa Udaondo, Castelar, Provincia de Buenos Aires

Instituto de Planeamiento Regional y Urbano (IPRU)
(Institute of Regional and Urban Planning)
Address: Calle Mexico 625/5 piso, Buenos Aires

Instituto de Suelos y Agrotécnica
(Institute of Soil and Agro-technology)
Address: Cerviño 3101, Buenos Aires

Instituto Nacional de Desarrollo Pesquero
(National Institute of the Development of Fisheries)
Address: Casilla 175, 7600 Mar del Plata

Instituto Nacional de Estadística y Censos
(National Institute of Statistics and Census)
Address: Hipólito Irigoyen 250/12 piso, Buenos Aires

Instituto Nacional de Limnología
(National Institute of Limnology)
Address: José Macia 1933/43, Santo Tomé, Provincia de Santa Fé

Instituto Nacional de Microbiología
(National Institute of Microbiology)
Address: Avenida Vélez Sarsfield 563, Buenos Aires

Instituto Nacional de Tecnología Agropecuaria (INTA)
(National Institute of Arable and Livestock Technology)
Address: Rivadavia 1439, 1033 Buenos Aires

Instituto Nacional de Tecnología Industrial (INTI)
(National Institute of Industrial Technology)
Address: Avenida Leandro N. Alem 1067, 1001 Buenos Aires

Instituto Nacional de Vitivinicultura
(National Institute of Viticulture and Viniculture)
Address: San Martín 430, 5500 Mendoza

Instituto Superior de Ciencias
(Higher Institute of Sciences)
Address: General Paz 1010, Rio Cuarto, Córdoba

Instituto Tecnológico de Buenos Aires
(Buenos Aires Technological Institute)
Address: Avenida Emilio Madero 351/99, Buenos Aires

Instituto Torcuato di Tella
(Torcuato di Tella Institute)
Address: 11 de Septiembre 2139, 1428 Buenos Aires

Observatorio Astronómico
(Astronomical Observatory)
Address: Laprida 854, Córdoba

Servicio de Endocrinología y Metabolismo
(Endocrinology and Metabolism Centre)
Address: Martinez de Hoz y Marconi, Villa Sarmiento, Buenos Aires

Servicio Meteorológico Nacional
(National Meteorological Service)
Address: 25 de Mayo 658, Buenos Aires

Subsecretaria de Planeamiento Ambiental – UNEP
(Sub-secretariat of Environmental Planning)
Address: Santa Fé 1548, 1060 Buenos Aires

National and Provincial Universities

Escuela de Ingenieria Aeronáutica
(School of Aeronautical Engineering)
Address: Guarnición Aérea, 5103 Córdoba

Universidad de Buenos Aires
(University of Buenos Aires)
Address: Viamonte 444/430, Buenos Aires

  Escuela Superior de Comercio 'Carlos Pellegrini'
  (Carlos Pellegrini Higher School of Commerce)
  Address: Marcelo T. de Alvear 1851, Buenos Aires

  Colegio Nacional de Buenos Aires
  (National College of Buenos Aires)
  Address: Bolivar 263, Buenos Aires

  Centro de Investigaciones Médicas
  (Centre of Medical Research)
  Address: General Donato Alvarez 3000, Buenos Aires

  Instituto Bibliotecnológico
  (Institute of Librarianship)
  Address: Azcuenaga 280, Buenos Aires

Universidad Nacional de Catamarca
(National University of Catamarca)
Address: Republica 350, 5000 Catamarca

Universidad Nacional del Centro de la Provincia de Buenos Aires
(Central National University of the Province of Buenos Aires)
Address: General Pinto 399, 7000 Tandil

Universidad Nacional del Comahué
(National University of Comahué)
Address: Buenos Aires 1400, Neuquen

Universidad Nacional de Córdoba
(National University of Córdoba)
Address: Calle Obispo Trejo y Sanabria 242, Córdoba

Universidad Nacional de Cuyo
(National University of Cuyo)
Address: Centro Universitario, Parque General San Martín, Mendoza

Universidad Nacional de Jujuy
(National University of Jujuy)
Address: Gorriti 237, 4600 San Salvador de Jujuy

Universidad de la Pampa
(University of La Pampa)
Address: 9 de Julio 149, 6300 Santa Rosa, La Pampa

Universidad Nacional de La Plata
(National University of La Plata)
Address: Calle 7, 776, La Plata

Universidad Nacional del Litoral
(National University of the Litoral)
Address: Boulevar Pellegrini 2750, 3000 Sante Fé

Universidad Nacional de Mar del Plata
(National University of Mar del Plata)
Address: Juan Bautista Alberdi 2695, Mar del Plata, Provincia de Buenos Aires

Universidad Nacional del Nordeste
(National University of the North-East)
Address: 25 de Mayo 868, Corrientes

Universidad Nacional de la Patagonia San Juán Bosco
(San Juán Bosco National University of Patagonia)
Address: C.C. 786 Correo Central, 9000 Comodoro Rivadavia, Chubut

Universidad Nacional de Río Cuarto
(National University of Río Cuarto)
Address: Campus Universitario, Enlace 8 Y Km.603, Rio Cuarto

Universidad Nacional de Rosario
(National University of Rosario)
Address: Córdoba 1814, Rosario

Universidad Nacional de Salta
(National University of Salta)
Address: Buenos Aires 177, 4400 Salta

Universidad Nacional de Santiago del Estero
(National University of Santiago del Estero)
Address: Avenida Belgrano (S) 1912, 4000 Santiago del Estero

Universidad Nacional del Sur
(National University of the South)
Address: Avenida Colon 80, 8000 Bahía Blanca

Universidad Nacional de Tucumán
(National University of Tucumán)
Address: Ayacucho 491, 4000 San Miguel de Tucumán

Universidad Tecnológica Nacional
(National University of Technology)
Address: 25 de Mayo 564, Buenos Aires

Private Universities

Universidad del Aconcagua
(University of Aconcagua)
Address: Catamarca 147, 5500 Mendoza

Universidad Argentina de la Empresa
(The Argentine University of 'La Empresa')
Address: Libertad 1340, Buenos Aires

Universidad Argentina 'John F. Kennedy'
(Argentine University of John F. Kennedy)
Address: Calle Bartolomé Mitre 1407, Buenos Aires

Universidad de Belgrano
(University of Belgrano)
Address: Federico Lacroze 1959, Buenos Aires

Universidad Católica Argentina 'Santa María de los Buenos Aires'
(Santa María of Buenos Aires Argentinian Catholic University)
Address: Juncal 1912, Buenos Aires

Universidad del Museo Social Argentino
(University of the Social Museum of Argentina)
Address: Corrientes 1723, Buenos Aires

Universidad del Salvador
(University of Salvador)
Address: Alberti 158, Buenos Aires

Universidad Católica de Córdoba
(Catholic University of Córdoba)
Address: Obispo Trejo 323, Córdoba

Universidad Católica de Cuyo
(Catholic University of Cuyo)
Address: Avenida Ignacio de la Roza 1516 Oeste, 5400 Rivadavia,
San Juan

Universidad Católica de La Plata
(Catholic University of La Plata)
Address: Calle 13, 1227, La Plata

Universidad Notarial Argentino
(Notarial University of Argentina)
Address: Calle 51, 435, 1900 La Plata

Universidad de Mendoza
(University of Mendoza),
Address: Diagonal Dag Hammarskjold 750, 5500 Mendoza

Universidad 'Juan Agustín Maza'
(Juan Agustín Maza University)
Address: Salta 1690 – Urquiza 350, 5500 Mendoza

Universidad de Morón
(University of Morón)
Address: Cabildo 134, Morón, Provincia de Buenos Aires

Universidad del Norte Santo Tomás de Aquino
(Santo Tomás de Aquino University of the North)
Address: C.P. 32, San Miguel de Tucumán

Universidad de la Patagonia 'San Juán Bosco'
(San Juán Bosco University of Patagonia)
Address: General Mosconi, Comodoro Rivadavia, Provincia del Chubut

Universidad Católica de Salta
(Catholic University of Salta)
Address: Ciudad Universitaria, Castañares, Salta

Universidad Católica de Santa Fé
(Catholic University of Santa Fé)
Address: Echague 7151, 3000 Santa Fé

Universidad Católica de Santiago del Estero
(Catholic University of Santiago del Estero)
Address: Libertad 321, 4200 Santiago del Estero

Belize

Baron Bliss Institute
Address: POB 990, Belize City

National Library Service
Address: POB 990, Belize City

Universities and Colleges

Belize College of Arts, Science and Technology
Address: POB 990, Belize City

Fletcher College
Address: Corozal, Belize

University of the West Indies Extra-Mural Department
Address: University Centre, Belize City

Wesley College
Address: Belize City

Bolivia

Academia Nacional de Ciencias de Bolivia
(Bolivian National Academy of Sciences)
Address: POB 5829, La Paz

Biblioteca Central de la Universidad Mayor de 'San Andrés'
(Central Library of San Andrés University)
Address: Casilla 6548, La Paz

Biblioteca Central de la Universidad Mayor de 'San Simón'
(Central Library of San Simón University)
Address: Avenida Oquendo Sucre, Casilla 992, La Paz

Centro Nacional de Documentación Científica y Tecnológica
(National Centre for Scientific and Technological Documentation)
Address: Casilla 3383, La Paz

Centro Nacional de Documentación e Información Educativa
(National Centre for Educational Documentation and Information)
Address: c/o Ministerio de Educación y Cultura, La Paz

Comisión Boliviana de Energía Nuclear
(Bolivian Commission for Nuclear Energy)
Address: Avenida 6 de Agosto 2905, La Paz

Instituto Boliviano de Tecnología Agropecuaria (IBTA)
(Bolivian Institute of Agriculture and Livestock Technology)
Address: Avenida Camacho 1471, POB 5783, La Paz

Instituto Boliviano del Petróleo (IBP)
(Bolivian Oil Institute)
Address: Casilla 4722, La Paz

Instituto Geográfico Militar y de Catastro Nacional
(Military Institute of Geography and National Property Register)
Address: Cuartel General Miraflores

Instituto Nacional de Estadística
(National Statistics Institute)
Address: Casilla 6129, La Paz

Observatorio 'San Calixto'
(San Calixto Observatory)
Address: Casilla 5939, La Paz

Servicio Geológico de Bolivia
(Bolivian Geological Service)
Address: Federico Suazo 1673, Casilla 2729, La Paz

Universities

Universidad Boliviana Mayor Real y Pontificia de San Francisco Javier
(San Francisco Javier Royal and Pontifical University of Bolivia)
Address: Apartado 212, Sucre

Universidad Boliviana Mayor de 'San Andrés'
(San Andrés University)
Address: Casilla 6548, La Paz

Universidad Mayor de 'San Simón'
(San Simón University)
Address: Casilla 992, Cochabamba

Universidad Técnica de Oruro
(Oruro Technical University)
Address: Casilla de Correos 49, Oruro

Universidad Boliviana 'Tomas Frías'
(Tomas Frías University)
Address: Edificio Central, Avenida del Maestro, Casilla 36, Potosí

Universidad Boliviana Mayor 'Gabriel René Moreno'
(Gabriel René Moreno University)
Address: Casilla Postal 702, Santa Cruz de la Sierra

Universidad Boliviana 'Juan Misael Saracho'
(Juan Misael Saracho University)
Address: POB 51, Tarija

Universidad Boliviana 'Mariscal Jose Ballivián'
(Mariscal Jose Ballivián University)
Address: Casilla 38, Trinidad, Beni

Universidad Católica Boliviana
(Bolivian Catholic University)
Address: Casillo Postal 4805, La Paz

Brazil

Academia Brasileira de Ciências
(Brazilian Academy of Sciences)
Address: Caixa Postal 229, 20000

Academia Nacional de Medicina
(National Academy of Medicine)
Address: Caixa Postal 459, ZC-00

Asociação Brasileira de Odontologia
(Brazilian Odontology Association)
Address: Avenida 13 de Maio, 13-10 Rio de Janeiro RJ

Asociação Brasileira de Química
(Brazilian Association of Chemistry)
Address: Avenida Rio Branco 156, Caixa Postal 550, Rio de Janeiro RJ

Asociação Brasileira de Psiquiatria
(Brazilian Psychiatry Association)
Address: Alvaro Ramos 405, Rio de Janeiro RJ

Asociação Internacional de Lunologia
(International Association of Lunology)
Address: Caixa Postal 322 Francia, São Paulo

Asociação Médica Brasileira
(Brazilian Medical Association)
Address: Rua São Carlos do Pinhal 324, Caixa Postal 8094, São Paulo

Biblioteca Central Universidade de Brasília
(Central Library of the University of Brasília)
Address: Apartado 15, 70910 Brasília DF

Biblioteca Nacional
(National Library)
Address: Avenida Rio Branco 219-30, Rio de Janeiro

Centro de Energia Nuclear na Agricultura (CENA)
(Centre of Nuclear Energy in Agriculture)
Address: Caixa Postal 96, 13400 Piracicaba SP

Centro de Pesquisa Agropecuária do Tropico Umido – EMBRAPA
(Centre for Research in Agriculture and Livestock Farming in Tropical Zones)
Address: Caixa Postal 48, 66000 Belem, Pará

Centro de Pesquisas e Desenvolvimento (CEPED)
(Research and Development Centre)
Address: Caixa Postal 09, 42800 Camaçari, Bahia

**Centro de Pesquisas e Desenvolvimento 'Leopoldo A. Miguez de Mello'
– PETROBRAS**
(Leopoldo A. Miguez de Mello Research and Development Centre)
Address: Ilha do Fundao, Quadra 7, Caixa Postal 809, Rio de Janeiro

Centro de Tecnología Agrícola e Alimentar – EMBRAPA
(Agricultural and Food Technology Centre)
Address: Rua Jardim Botãnico 1024, 22460 Rio de Janeiro

Centro Nacional de Pesquisa de Mandioca e Fruticultura – EMBRAPA
(National Centre for Cassava and Fruit Research)
Address: Ministerio da Agricultura 44380, Cruz Das Almas, Bahia

Comissão Nacional de Energia Nuclear (CNEN)
(National Commission for Nuclear Energy)
Address: Rua General Severiano 90, Botafogo ZC82 20000, Rio de Janeiro RJ

Departamento Nacional de Meteorología
(National Department of Meteorology)
Address: Praça 15 de Novembro 2, 50A Rio de Janeiro

Emprêsa de Pesquisa Agropecuária da Bahia
(Enterprise for Agricultural and Livestock Research for the State of Bahia)
Address: Avenida Ademar de Barros 967, Caixa Postal 1222 Ondina,
4000 Salvador, BA

Emprêsa de Pesquisa Agropecuária de Minas Gerais (EPAMIG)
(Enterprise for Agricultural and Livestock Research for the State of Minas Gerais)
Address: Avenida Amazonas 115, Caixa Postal 515, 30000 Belo Horizonte, MG

Instituto 'Adolfo Lutz'
(Adolfo Lutz Institute)
Address: Caixa Postal 7027, 01000 São Paulo

Instituto Agronómico
(Institute of Agronomy)
Address: Caixa Postal 28, 13100 Campinas, SP

Instituto Brasileiro de Desenvolvimento Forestal (IBDF)
(Brazilian Institute of Forestry Development)
Address: Palacio do Desenvolvimento 12/13 Andares, Brasilia DF

Instituto Brasileiro de Estudos e Pesquisas de Gastroenterologia
(Brazilian Institute for Gastroenterological Study and Research)
Address: Rua Dr Seng 320, Bairro da Bela Vista, Caixa Postal 6209,
01331 São Paulo

Instituto Brasileiro de Petróleo
(Brazilian Petroleum Institute)
Address: Avenida Rio Branco 156-10-S, 1035, Rio de Janeiro

Instituto Brasileiro do Café
(Brazilian Coffee Institute)
Address: Avenida Rodrigues Alves 129, 20081, Rio de Janeiro RJ

Instituto Butantan
(Butantan Institute)
Address: Caixa Postal 65, Butantan SP

Instituto de Engenharia de São Paulo
(Engineering Institute of São Paulo)
Address: Palacio Maua, 8 e 9 andares, São Paulo

Instituto de Energía Atómica
(Institute of Atomic Energy)
Address: Caixa Postal 11049, Rio de Janeiro RJ

Instituto de Engenharia Nuclear
(Nuclear Engineering Institute)
Address: Caixa Postal 2180, Rio de Janeiro

Instituto de Pesquisas do Experimentação Agropecuario do Nordeste (IPEANE)
(Experimental Agricultural and Livestock Research Institute of the Northeast)
Address: Caixa Postal 205 Curado, Pernambuco

Instituto de Pesquisas Espaciais (INPE)
(Institute of Space Research)
Address: Caixa Postal 515, 12200 São Jose Dos Campos SP

Instituto de Pesquisas Radioativas
(Radioactive Research Institute)
Address: Caixa Postal 1941 Belo Horizonte

Instituto de Pesquisas Technológicas
(Institute of Technological Research)
Address: Ciudade Universitaria Armando de Salles Oliveira, Caixa Postal 7141, 01000 São Paulo

Instituto Evandro Chagas
(Evandro Chagas Institute)
Address: Avenida Almirante Barroso 492, Belém, Pará

Instituto Florestal-Estado do São Paulo
(São Paulo State Forestry Institute)
Address: Rua do Orto 1197, São Paulo

Instituto Geológico
(Institute of Geology)
Address: Caixa Postal 329, São Paulo

Instituto 'Nami Jafet' para o Progresso da Ciência e Cultura
(Nami Jafet Institute for the Advancement of Science and Culture)
Address: Rua Agostinho Gomes 1455 São Paulo SP

Instituto Nacional de Tecnologia
(National Institute of Technology)
Address: Avenida Venezuela 82, 70 Andar, Rio de Janeiro

Instituto 'Oscar Freire'
(Oscar Freire Institute)
Address: Caixa Postal 4350, São Paulo

Instituto 'Oswaldo Cruz'
(Oswaldo Cruz Institute)
Address: Avenida Brasil 4365, Rio de Janeiro

Instituto Regional de Meteorologia 'Coussirat Araújo'
(Coussirat Araújo Regional Institute of Meteorology)
Address: Ministerio de Agricultura, Porto Alegre

Laboratorio Central Gonçalo Moniz
(Gonçalo Moniz Central Laboratory)
Address: Rua Pedro Less, Canela 40000, Salvador, Bahia

Observatorio Nacional do Brasil
(National Observatory of Brazil)
Address: Rua General Bruce 586, Rio de Janeiro RJ

Serviço de Documentação Geral da Marinha
(General Documentation Centre of the Navy)
Address: Rua D. Manuel 15, Centro 20010 Rio de Janeiro

Serviço de Pesquisa e Experimentação de Cancer
(Cancer Research Centre)
Address: Praça Cruz Vermelha 23, Rio de Janeiro RJ

Serviço Nacional de Levantamento e Conservação de Solos – EMBRAPA
(National Service for the Surveying and Conservation of Soil)
Address: Rua Jardim Botânico 1024, 22460, Rio de Janeiro RJ

Sociedade Brasileira para o Progresso da Ciência
(Brazilian Society for Scientific Progress)
Address: Caixa Postal 11008, 01008 São Paulo

Sociedade Científica de São Paulo
(Scientific Society of São Paulo)
Address: Caixa Postal 1904, São Paulo

Colleges and Universities

Universidade Federal do Acre
(Federal University of Acre)
Address: Avenida Getulio Vargas 654, Centro Rio Branco, Acre

Universidade Federal de Alagoas
(Federal University of Alagoas)
Address: Praça Visconde de Sinimbu 206, 5700 Maceio, Alagoas

Universidade do Amazonas
(Amazonas University)
Address: Rua Jose Paranagua 200, Caixa Postal 348, 6900 Manaus

Universidade Federal da Bahia
(Federal University of Bahia)
Address: Rua Augusto Viana Sin Numero, Canela, Salvador, Bahia

Universidade Regional de Blumenau
(Regional University of Blumenau)
Address: Rua Antonio Da Veiga 140, Caixa Postal 7E, 89100 Blumenau, Santa Caterina

Universidade de Brasília
(University of Brasília)
Address: Agencia Postal 15, 70910 Brasilia DF

Pontifícia Universidade Católica de Campinas
(Pontifical Catholic University of Campinas)
Address: Rua Marechal Deodoro 1099, 13100, Campinas-São Paulo

Universidade Estadual de Campinas
(State University of Campinas)
Address: Cidade Universitaria, Barao Geraldo, Caixa Postal 1170, Campinas-São Paulo

Universidade de Caxias do Sul
(University of Caxias do Sul)
Address: Rua Francisco Getulio Vargas Sin Numero, 95100 Caxias do Sul, RS

Universidade Federal do Ceará
(Federal University of Ceará)
Address: Avenida Da Universidade 2853, Caixa Postal 1000, 60000 Fortaleza, CE

Universidade Federal do Espírito Santo
(Federal University of Espírito Santo)
Address: Avenida Fernando Ferrari-Goiabeiras 29000, Vitoria, ES

Universidade Federal Fluminense
(Federal University of Fluminense)
Address: Rua Miguel De Frias 9, 24220 Niteroi, RJ

Universidade de Fortaleza
(University of Fortaleza)
Address: Caixa Postal 1258, Fortaleza, Ceara

Universidade Gama Filho
(Gama Filho University)
Address: Rua Manoel Vitorino 625, Piedade, CEP 20740, Rio de Janeiro, RJ

Universidade Federal de Goiás
(Federal University of Goiás)
Address: Campus II Km 13, Rodovia Goiania, Necropolis, Caixa Postal 130, 74000 Giania, Goiás

Universidade de Itaúna
(Itaúna University)
Address: Rua Capitão Vicente 10, Caixa Postal 100, 35680 Itaúna MG

Universidade Federal de Juiz de Fora
(Juiz de Fora Federal University)
Address: Rua Benjamin Constant 790, 36100 Juiz de Fora, Minas Gerais

Universidade Estadual de Londrina
(State University of Londrina)
Address: Caixa Postal 2111, 86600 Londrina, Paraná

Universidade Mackenzie
(Mackenzie University)
Address: Rua Maria Antonia 403, 01222 São Paulo

Universidade Federal do Maranhão
(Federal University of Maranhão)
Address: Largo Dos Amores 351, 65000 São Luis, Maranhão

Universidade Estadual de Maringá
(State University of Maringá)
Address: Avenida Colombo 3960, Caixa Postal 331, 87100 Maringá, PR

Universidade Estadual de Mato Grosso do Sul
(State University of the Southern Mato Grosso)
Address: Caixo Postal 649, Cidade Universitaria, 79100 Campo Grande MTS

Universidade Federal de Mato Grosso
(Federal University of the Mato Grosso)
Address: Avenida Fernando Correa Da Costa, 78000 Cuiaba

Universidade Católica de Minas Gerais
(Catholic University of Minas Gerais)
Address: Avenida Dom Jose Gaspar 500, Caixa Postal 2686, 30000 Belo
Horizonte, Minas Gerais

Universidade Federal de Minas Gerais
(Federal University of Minas Gerais)
Address: Cidade Universitaria, Pampulha, Caixa Postal 1621–1622,
30000 Belo Horizonte, Minas Gerais

Universidade de Mogi das Cruzes
(Mogi das Cruzes University)
Address: Caixa Postal 411, Avenida Candido Xavier De Almeida Souza 200,
08700 Mogi Das Cruzes, SP

Universidade Regional do Nordeste
(Regional University of the North-East)
Address: Avenida Mar Floriano Peixoto 178, 58100 Campina Grande, Paraíba

Fundação Norte Mineira de Ensino Superior
(North Mineira Foundation for Higher Education)
Address: Vila Mauriceia s/n, Caixa Postal 126, 39400 Montes Claros, Minas
Gerais

Universidade Federal de Ouro Prêto
(Federal University of Ouro Prêto)
Address: Praça Tiradentes 20, 35400 Ouro Prêto MG

Universidade Federal do Pará
(Federal University of Pará)
Address: Avenida Governador Jose Malcher 1192, Belém, Pará

Universidade Federal da Paraíba
(Federal University of Paraíba)
Address: Campus Universitario 58000, Joao Pessoa, Paraíba

Universidade Católica do Paraná
(Catholic University of Paraná)
Address: Rua Imaculada Conceiçao 1115, Caixa Postal 2293, 80000 Curitiba

Universidade Federal do Paraná
(Federal University of Paraná)
Address: Rua 15 de Novembro 1299, Curitiba, Paraná

Universidade de Passo Fundo
(Passo Fundo University)
Address: Bairro San Jose, Caixa Postal 566, 9100 Passo Fundo,
Rio Grande do Sul

Universidade Católica de Pelotas
(Catholic University of Pelotas)
Address: Rua Feliz Da Cunha 412, 96100 Pelotas, Rio Grande do Sul

Universidade Federal de Pelotas
(Federal University of Pelotas)
Address: Campus Universitario, 96100 Pelotas, Rio Grande do Sul

Universidade Católica de Pernambuco
(Catholic University of Pernambuco)
Address: Rua Do Principe 526, Boa Vista, 50000 Recife, Pernambuco

Universidade Federal de Pernambuco
(Federal University of Pernambuco)
Address: Avenida Prof. Moares Rego, Cidade Universitaria, Recife, Pernambuco

Universidade Federal Rural de Pernambuco
(Rural Federal University of Pernambuco)
Address: Rua D. Manuel Madeiros, Dois Irmaos, Caixo Postal 2071,
50000 Recife PE

Universidade Católica de Petrópolis
(Catholic University of Petrópolis)
Address: Rua Benjamin Constant 213, Centro Caixa Postal 944,
25600 Petropolis, RJ

Universidade Federal do Piauí
(Federal University of Piauí)
Address: Campus Universitario, Barrio Ininga 64000 Teresina, Piauí

Universidade Estadual de Ponta Grossa
(State University of Ponta Grossa)
Address: Praca Santos Andrade, 84100 Ponta Grossa, Paraná

Fundação Universidade do Rio Grande
(University Foundation of Rio Grande)
Address: Rua Luis Lorea 261, 96200 Rio Grande, Rio Grande do Sul

Universidade Federal do Rio Grande do Norte
(Federal University of Rio Grande do Norte)
Address: Campus Universitario s/n, 59000 Natal, RN

Pontifícia Universidade do Rio Grande do Norte
(Pontifical University of Rio Grande do Norte)
Address: Avenida Ipiranga 6681, Caixa Postal 1429, 90000 Porto Alegre

Universidade Federale do Rio Grande do Sul
(Federal University of Rio Grande do Sul)
Address: Avenide Paulo Gama, 90000 Porto Alegre, Rio Grande do Sul

Pontificia Universidade do Rio de Janeiro
(Pontifical University of Rio de Janeiro)
Address: Rua Marques de São Vicente 225, 22453 Rio de Janeiro, RJ

Universidade do Estado do Rio de Janeiro
(State University of Rio de Janeiro)
Address: Rua San Francisco Xavier 524, Maracana ZC11 CEP 20000,
Rio de Janeiro

Universidade Federal do Rio de Janeiro
(Federal University of Rio de Janeiro)
Address: Ilha Da Cidade Universitaria, Rio de Janeiro

Universidade Federal Rural do Rio de Janeiro
(Rural Federal University of Rio de Janeiro)
Address: Km 47 da Antigua Rodovia, Rio/São Paulo, 23460 Seropedica,
Itaguay RJ

Universidade Católica do Salvador
(Catholic University of Salvador)
Address: Praça 2 de Julho 7, 40000 Campo Grande, Salvador, Bahia

Universidade Federal de Santa Catarina
(Federal University of Santa Catarina)
Address: Campus Universitario, Trindade Caixa Postal 476,
88000 Florianopolis SC

Universidade para o Desenvolvimiento do Estado de Santa Catarina
(Development University for the State of Santa Catarina)
Address: Campus Universitario, Rua Madre Benvenuta s/n, Itacorobi 88000,
Florianopolis SC

Universidade Federal de Santa Maria
(Federal University of Santa Maria)
Address: Paixa de Camobi Km 9, Edificio da Admin. Central,
97100 Santa Maria, RS

Universidade Federal de São Carlos
(Federal University of São Carlos)
Address: Via Washington Luiz Km 235, Caixa Postal 676, 13560 São Carlos,
São Paulo

Universidade de São Paulo
(São Paulo University)
Address: Cidade Universitaria Caixa Postal 8191, São Paulo SP

Universidade Estadual Paulista 'Julio de Mesquita Filho'
(Julio de Mesquita Filho State University)
Address: Praça de SE 108, CEP 01001-Centro, Caixa Postal 30919 São Paulo SP

Pontificia Universidade Católica de São Paulo
(Pontifical University of São Paulo)
Address: Rua Monte Alegre 984, Bairro Perdizes, 05014 São Paulo

Universidade Federal de Sergipe
(Federal University of Sergipe)
Address: Rua Lagarto 952, Aracaju, Sergipe

Universidade Federal de Uberlândia
(Federal University of Uberlândia)
Address: Avenida Para 1720, Jardin Umuarama, Caixa Postal 593, 38400
Uberlândia MG

Universidade do Vale do Rio dos Sinos
(University of the Rio dos Sinos Valley)
Address: Praça Tiradentes 35, Caixa Postal 275, 93000 São Leopoldo RS

Universidade Federal de Viçosa
(Federal University of Viçosa)
Address: Avenida PH Rolfs s/n, Viçosa, Minas Gerais

Chile

Asociación Chilena de Sismología e Ingenieria Antisísmica
(Chilean Association of Seismology and Anti-Seismological Engineering)
Address: Beauchef 851, Casilla 2777, Santiago

Biblioteca del Congreso Nacional
(Library of the National Congress)
Address: Compania 1175/2 piso, Cassilla 1199, Santiago

Biblioteca Nacional
(National Library)
Address: Avenida Bernardo O'Higgins 651, Santiago

Biblioteca Central de la Universidad de Chile
(Central Library of the University of Chile)
Address: Arturo Prat 23, Casilla 10-D, Santiago

Centro de Documentación Pedagógica – UNESCO
(Centre of Pedagogic Documentation)
Address: Alarife Gamboa 071, Santiago

Centro Nacional de Información y Documentación (CENID) – UNESCO
(National Centre of Information and Documentation)
Address: Casilla 297-V, Correo 15, Santiago

Comisión Chilena de Energía Nuclear
(Chilean Commission for Nuclear Energy)
Address: Avenida Salvador 943, Santiago

Comisión Nacional de Investigaciones Científicas y Tecnológicas (CONICYT)
(National Commission of Scientific and Technological Research)
Address: Calle Canada 308, Santiago

Comité de Investigaciones Tecnológicas – INTEC-UNIDO
(Committee of Technological Research)
Address: Avenida Santa María 06500, Casilla 667, Santiago

Consejo de Rectores de las Universidades Chilenas
(Council of Rectors of Chilean Universities)
Address: Moneda 673/8 piso, Casilla 14798, Santiago

Escuela Agrícola 'El Vergel'
(El Vergel School of Agriculture)
Address: Casilla 2-D, Angol

Escuela Militar 'General Bernardo O'Higgins'
(Bernardo O'Higgins Military School)
Address: Casilla 174, Las Condes, Satiago

Estacion Experimental 'Las Vegas' de la Sociedad Nacional de Agricultura
(Las Vegas Experimental Station of the National Society of Agriculture)
Address: Huelquen-Paine, Santiago

Fundación Gildemeister
(Gildemeister Foundation)
Address: Amunategui 178/5 piso, Casilla 99-D, Santiago

Instituto Agrario de Estudios Económicos – INTAGRO
(Agrarian Institute of Economic Studies)
Address: Tenderini 187, Casilla 13907, Santiago

Instituto Antártico Chileno
(Chilean Antarctic Institute)
Address: Luis Thayer Ojeda 814, Casilla 16521, Correo 9, Santiago

Instituto Bacteriológico de Chile
(Bacteriological Institute of Chile)
Address: Casilla 48, Santiago

Instituto Chile (Composed of 6 national academies)
(Chile Institute)
Address: Almirante Montt 453, Clasificador 1349, Correo Central, Santiago

Instituto Científico de Lebu
(Lebu Scientific Institute)
Address: Apartado 123, Lebu

Instituto de Fomento Pesquero
(Institut of Fisheries Development)
Address: Jose Domingo Canas 2277, Casilla 1287, Santiago

Instituto Forestal
(Forestry Institute)
Address: Huerfanos 554, Casilla 3085, Santiago

Instituto Geográfico Militar
(Military Geographical Institute)
Address: Nueva Santa Isabel 1640, Santiago

Instituto de Higiene del Trabajo y Contaminación Atmosférica
(Institute of Occupational Hygiene and Environmental Contamination)
Address: Casilla 3979, Santiago

Instituto de Investigaciones Agropecuarias (INIA)
(Institute of Agricultural and Livestock Research)
Address: Casilla 5427, Santiago

Instituto de Investigaciones y Ensayo de Materiales (IDIEM)
(Institute of Research and Examination of Materials)
Address: Plaza Ercilla 883, Casilla 1420, Santiago

Instituto de Investigaciones Geológicas (IIG)
(Institute of Geological Research)
Address: Augustinas 785/6 piso, Casilla 10465, Santiago

Instituto de Investigación de Recursos Naturales
(Natural Resources Research Institute)
Address: Casilla 14995, Correo Central, Santiago

Instituto de Medicina Experimental del SNS
(Institute of Experimental Medicine of the National Health Service)
Address: Avenida Irarrazabal 849, Casilla 3401, Santiago

Instituto Nacional de Investigaciones Tecnológicas y Normalización (INDITECNOR)
(The National Institute of Technological Research and Standards)
Address: Casilla 995, Correo 1, Santiago

Observatorio Astrofísico 'Manuel Foster'
(Manuel Foster Astrophysics Observatory)
Address: Casilla 6014, Santiago

Observatorio Astronómico Nacional
(National Astronomic Observatory)
Address: Casilla 36–D, Santiago

Observatorio Europeo Austral
(European Observatory for the Southern Hemisphere)
Address: Casilla 16317, Correo 9, Santiago

Observatorio Interamericano de Cerro Tololo
(Cerro Tololo Interamerican Observatory)
Address: Casilla 603, La Serena

Oficina Meteorológica de Chile
(Meteorological Office of Chile)
Address: Casilla 717, Santiago

Sociedad Chilena de Química
(Chilean Chemistry Society)
Address: Casilla de Correo 2613, Concepción

Sociedad Científica de Chile
(Chilean Scientific Society)
Address: Rosa Eguiguren 813, Casilla 696, Santiago

Sociedad Científica Chilena 'Claudio Gay'
(Claudio Gay Chilean Scientific Society)
Address: Casilla 2974, Santiago

Sociedad de Biología de Chile
(Chilean Biology Society)
Address: Casilla 16164, Santiago 9

Universities

Universidad Austral de Chile
(Southern University of Chile)
Address: Casilla 567, Valdivia

Universidad Católica de Chile
(Catholic University of Chile)
Address: Avenida Bernardo O'Higgins 340, Casilla 114-D, Santiago

Universidad Católica de Valparaiso
(Valparaiso Catholic University)
Address: Brasil 2950, Casilla 4059, Valparaiso

Universidad de Chile
(University of Chile)
Address: Bernardo O'Higgins 1058, Casilla 10-D, Santiago

Universidad de Concepción
(University of Concepción)
Address: Casilla 20-C, Concepción

Universidad del Norte
(University of the North)
Address: Casilla 1280, Antofagasta

Universidad Técnica del Estado
(State Technical University)
Address: Avenida Ecuador 3469, Casilla 4637, Santiago

Universidad Técnica 'Federico Santa María'
(Federico Santa María Technical University)
Address: Casilla 110-V, Valparaiso

Colombia

Academia Colombiana de Ciencias Exactas, Físicas y Naturales
(Colombian Academy of Exact, Physical and Natural Sciences)
Address: Carrera 8a, Calle 8a, Apartado Aéreo 2584, Bogotá

Academia Nacional de Medicina
(National Academy of Medicine)
Address: Carrera 9, No 20-13, Apartado Aéreo 23224, Bogotá

Archivo Nacional de Colombia
(National Archives of Colombia)
Address: Calle 24, No 5-60, Cuarto Piso, Bogotá

Biblioteca Central de la Pontificia Universidad Javeriana
(Central Library of the Javeriana Pontifical University)
Address: Carrera 7a, No 40-62, Apartado Aéreo 5315, Bogotá

Biblioteca Central de la Universidad Nacional de Colombia
(Central Library of the National University of Colombia)
Address: Apartado Aéreo 14490, Bogotá

Biblioteca Nacional de Colombia
(National Library of Colombia)
Address: Calle 24, No 5-60, Apartado 27600, Bogotá

Departamento Administrativo Nacional de Estadísticas
(Administrative Department of National Statistics)
Address: Apartado Aéreo 80043, Bogotá

Instituto Colombiano Agropecuario
(Colombian Agricultural and Livestock Institute)
Address: Apartado Aéreo 7984, Calle 37, No 8-43, Bogotá

Instituto Colombiano de Normas Técnicas (ICONTEC)
(Colombian Institute of Technical Standards)
Address: Carrera 19, No 39B-16, Bogotá

Instituto de Asuntos Nucleares
(Institute of Nuclear Affairs)
Address: Avenida Eldorado Carrera 50, Apartado Aéreo 8595, Bogotá

Instituto de Investigaciones Marinas de Punta Betín
(Punta Betín Institute of Marine Research)
Address: Apartado Aéreo 1016, Bogotá

Instituto de Investigaciones Tecnológicas
(Institute of Technological Research)
Address: Apartado Aéreo 7031, Bogotá

Instituto Geográfico 'Agustin Codazzi'
(Agustin Codazzi Geographical Institute)
Address: Avenida Ciudad de Quito 48-51, Bogotá

Instituto Nacional de Cancerología
(National Institute of Cancer)
Address: Calle 1, No 9-85, Bogotá

Instituto Nacional de Investigaciones Geológico Mineras
(National Institute of Geological and Mineral Research)
Address: Apartado Aéreo 48-65, Bogotá

Instituto Nacional de Medicina Legal
(National Institute of Forensic Medicine)
Address: Carrera 13, No 7-30, Bogotá

Instituto Nacional de Salud (INPES)
(National Health Institute)
Address: Avenida Eldorado Carrera 50, Apartado Aéreo 80334, Bogotá

Observatorio Astronómico Nacional
(National Astronomical Observatory)
Address: Apartado Aéreo 2584, Bogotá

Sociedad Colombiana de Cancerología
(Colombian Cancer Society)
Address: Hospital Militar, Piso 13, Bogotá

Sociedad Colombiana de Ingenieros
(Colombian Society of Engineers)
Address: Carrera 4, No 10-41, Bogotá

Sociedad Colombiana de Patología
(Colombian Pathology Society)
Address: Hospital San José, Calle 10, No 18-75, Bogotá

Sociedad Colombiana de Psiquiatría
(Colombian Psychiatry Society)
Address: Carrera 18, No 84-87, Oficina 203, Bogotá

Sociedad Colombiana de Químicos e Ingenieros Químicos
(Colombian Society of Chemists and Chemical Engineers)
Address: Avenida Jimenez 8-74, Oficina 501-513, Bogotá

National Universities

Colegio Mayor de Nuestra Señora del Rosario
(College of Our Lady of the Rosary)
Address: Velle 14, No 6-25, Bogotá

Escuela de Administración y Finanzas y Tecnologías
(School of Administration, Finance and Technology)
Address: Apartado Aéreo 3300, Medellín

Universidad de Antioquia
(University of Antioquia)
Address: Apartado Aéreo 1226, Ciudad Universitaria, Medellín

Universidad del Atlántico
(Atlantic University)
Address: Carrera 43, No 50-53, Apartado Aéreo 1890, Barranquilla

Universidad de Caldas
(Caldas University)
Address: Apartado Aéreo 275, Manizales, Caldas

Universidad de Cartagena
(University of Cartagena)
Address: Apartado Aéreo 1382, Cartagena

Universidad del Cauca
(Cauca University)
Address: Apartado Nacional 113, Calle 5a, 4-70, Popayan

Universidad Francisco de Paula Santander
(Francisco de Paula Santander University)
Address: Avenida Gran Colombia 12E-96, Barrio Colsag, Apartado Aéreo 1055

Universidad Nacional de Colombia
(National University of Colombia)
Address: Ciudad Universitaria, Apartado Aéreo 14-490, Bogotá

Universidad de Córdoba
(Cordoba University)
Address: Apartado Aéreo 354, Carretera A. Cerete, Km 5, Montería

Universidad Distrital 'Francisco José de Caldas'
(Francisco José de Caldas University)
Address: Apartado Aéreo 8668, Carrera 8, No 40-78, Bogotá

Universidad de Nariño
(Nariño University)
Address: Carrera 22, No 18–109, Pasto, Nariño

Universidad de Pamplona
(Pamplona University)
Address: Apartado Aéreo 1046, Carrera 4, No 4–38, Pamplona

Universidad Pedagógica Nacional
(National Pedagogic University)
Address: Apartado Aéreo 53040, Calle 72, No 11–86, Bogotá

Universidad de Quindío
(Quindío University)
Address: Avenida Bolívar Calle 12, Apartado Aéreo 460, Armenia

Universidad Industrial de Santander
(Santander Industrial University)
Address: Apartado Aéreo 678, Bucaramanga, Santander

Universidad de Tolima
(Tolima University)
Address: Apartado Aéreo No 546, Ibagué

Universidad del Valle
(Valle University)
Address: Ciudad Universitaria Meléndez, Apartado Aéreo 2188, Cali

Universidad Pedagógica y Tecnológica de Colombia
(Pedagogic and Technological University of Colombia)
Address: Apartado Aéreo 1094, Ciudad Universitaria, Tunja, Boyaca

Universidad Tecnológica de Magdalena
(Magdalena Technological University)
Address: Carrera 2a, No 16–44, Santa Marta

Universidad Tecnológica de Pereira
(Pereira Technological University)
Address: Apartado Aéreo 97, Pereira

Private Universities

Fundación Universidad de Bogotá 'Jorge Tadeo Lozano'
(Jorge Tadeo Lozano University Foundation of Bogotá)
Address: Apartado Aéreo 3485, Calle 23, No 4–47, Bogotá

Universidad Autónoma Latinoamericana
(Autonomous University of Latin America)
Address: Carrera 55, No 49–51, Apartado Aéreo 3455, Medellín

Universidad Pontificia Bolivariana
(Bolivian Pontificial University)
Address: Apartado Postal 109, La Playa 40–102, Medellín

Universidad la Gran Colombia
(Gran Colombia University)
Address: Carrera 6, No 13–40, Apartado Aéreo 7909, Bogotá

Pontificia Universidad Javeriana
(Javeriana Pontificial University)
Address: Carrera 7a, No 40-62, Apartado Aéreo 5315, Bogotá

Universidad Libre de Colombia
(Free University of Colombia)
Address: Carrera 6, No 8-69, Bogotá

Universidad de los Andes
(University of the Andes)
Address: Carrera 1, No 18-A-82, Apartado Aéreo 4976, Bogotá

Universidad de Medellín
(Medellín University)
Address: Apartado Aéreo 1983, Calle 31, No 83-B-150, Medellín

Universidad del Norte
(University of the North)
Address: Apartado Aéreo 1569, Barranquilla

Universidad de San Buenaventura
(San Buenaventura University)
Address: Apartado Aéreo 053746, Bogotá

Universidad de Santo Tomás
(Santo Tomás University)
Address: Carrera 9a, No 51-23, Apartado Aéreo 21019, Bogotá

Universidad Social Católica de La Salle
(La Salle Catholic University)
Address: Calle 11, No 1-47, Bogotá

Costa Rica

Archivo Nacional de Costa Rica
(National Archives of Costa Rica)
Address: Calle 7, Avenida 4, Apartado 5028, San José

Asociación Costarricense de Cirugía
(Costa Rican Surgery Association)
Address: Apartado 2724, San José

Asociación Costarricense de Pediatría
(Costa Rican Paediatrics Association)
Address: Apartado 1654, San José

Biblioteca 'Carlos Monge Alfaro' de la Universidad de Costa Rica
(Carlos Monge Alfaro Library of the Costa Rica University)
Address: Ciudad Universitaria, San José

Biblioteca Nacional
(National Library)
Address: Apartado 10008, Calle 5, Avenidas 1/3, San José

Centro Agronómico Tropical de Investigación y Enseñanza (CATIE)
(Centre for Research and Teaching of Tropical Agronomy)
Address: Turrialba

Centro de Ciencia Tropical
(Tropical Science Centre)
Address: Apartado 8-3870, San José

Centro de Estudios Médicos 'Ricardo Moreno Canas'
(Ricardo Moreno Canas Centre for Medical Studies)
Address: Apartado 10151, San José

Comisión de Energía Atómica de Costa Rica
(Atomic Energy Commission of Costa Rica)
Address: Apartado 6681, San José

Dirección General de Geología, Minas y Petróleo
(Directorate-General of Geology, Mining and Petroleum)
Address: Apartado 2549, San José

Instituto Geográfico Nacional
(National Geographical Institute)
Address: Apartado 2272, San José

Instituto Meteorológico Nacional
(National Meteorological Office)
Address: Apartado 7-3350, San José

Organización de Estudios Tropicales
(Organization of Tropical Studies)
Address: Universidad de Costa Rica, Ciudad Universitaria, San José

Universities

Universidad de Costa Rica
(Costa Rica University)
Address: Ciudad Universitaria Rodrigo Facio, San Pedro de Montes de Oca, San José

Universidad Estatal a Distancia
(Open University)
Address: Apartado 2, Plaza Gonzales Viquez, San José

Universidad Nacional Autónoma de Heredia
(National Autonomous University of Heredia)
Address: Apartado 86, Heredia

Cuba

Biblioteca Central
(Central Library)
Address: Academia de Ciencias de Cuba, Capitolio Nacional, Habana

Biblioteca Nacional 'José Martí'
(José Martí National Library)
Address: Apartado Oficial No 3, Plaza de la Revolución José Martí, Habana

Biblioteca 'Rubén Martínez Villena'
(Rubén Martínez Villena Library)
Address: Obispo 160, Habana

Casa de las Américas
(Americas House)
Address: G y Tercera Vedado, Habana

Centro de Investigaciones para la Industria Minero Metalúrgica
(Centre of Research for the Mining and Metallurgical Industry)
Address: Finca La Luisa, Km 1, Carretera Varona, Arroyo Naranjo, Habana

Centro de Investigaciones Pesqueras
(Fisheries Research Centre)
Address: Avenida I y 26, Miramar, Marianao, Habana

Centro Nacional de Información de Ciencias Médicas
(National Centre of Medical Science Information)
Address: 23 No 177 entre N y O, Vedado, Habana

Grupo Nacional de Radiología
(National Radiology Group)
Address: Ministerio de Salud Pública, 23 y N, 3er Piso, Vedado, Habana

Instituto Cubano de Investigaciones Mineras y Metalúrgicas
(Cuban Institute of Mineral and Metallurgical Research)
Address: Aguiar 207 e/Empedrado y Tejadillo, Habana

Instituto de Botánica
(Botany Institute)
Address: Calzada del Cerro No 1257, Esquina de Buenos Aires, Habana 6

Instituto de Documentación e Información Científica y Técnica
(Institute of Scientific and Technical Documentation and Information)
Address: Calle C, No 351, Habana 4

Instituto de Geofísica y Astronomía
(Geophysics and Astronomy Institute)
Address: Calle 212, No 2906 entre 29 y 31, Reparto Cubanacan, Habana 16

Instituto de Investigaciones de la Caña de Azúcar
(Sugar Cane Research Institute)
Address: Avenida Van Troi No 17203, Rancho Boyeros

Instituto de Investigaciones Fundamentales del Cerebro
(Institute of Fundamental Research on the Brain)
Address: Calle Loma y 37, Nuevo Vedado, Habana 4

Instituto de Investigaciones Nucleares
(Nuclear Research Institute)
Address: Managua, Habana

Instituto de Matemática, Cibernética y Computación
(Mathematics, Cybernetics and Computing Institute)
Address: Capitolio Nacional, Habana 2

Instituto de Meteorología
(Meteorology Institute)
Address: Casablanca, Habana 2

Instituto de Oncología y Radiobiología de La Habana
(Oncology and Radiobiology Institute of Havana)
Address: 29 y F Vedado, Habana

Instituto de Química y Biología Experimental
(Experimental Chemistry and Biology Institute)
Address: Avenida 26, No 1605 entre calzada de Puentes Grandes y Boyeros,
Habana 6

Instituto de Zoología
(Zoology Institute)
Address: Calle 214, Esquina Ave 19, No 17A 09, Reparto Atabey, Habana 16

Instituto Nacional de Desarrollo y Aprovechamiento Forestales (INDAF)
(National Institute of Forest Development and Exploitation)
Address: Virtudes 680, Habana

Instituto Nacional de Higiene
(National Institute of Hygiene)
Address: Infanta y Crucero de Ferrocarril, Habana

Instituto Nacional de la Reforma Agraria (INRA)
(National Institute of Agrarian Reform)
Address: Calzada de Bejucal y Calle 100, Habana

Instituto Tecnológico de Electrónica 'Fernando Aguado Rico'
(Fernando Aguado Rico Electronic Technology Institute)
Address: Belascoain y Maloja, Habana

Instituto Tecnológico 'Mártires de Girón'
(Mártires de Girón Technology Institute)
Address: 5a Avenida 16607 Esquina 170, Marianao, Habana

Universities

Universidad de la Habana
(Havana University)
Address: Calle 1, No 302 Esquina A5, Vedado, Habana

Universidad de Oriente
(University of Oriente)
Address: Avenida Patricio Lumumba s/n, Santiago de Cuba, Oriente

Universidad de las Villas
(Las Villas University)
Address: Carretera de Camajuani, Km 10, Santa Clara, Las Villas

Universidad de Camaguey
(Camaguey University)
Address: Carretera de Circunvalación, Camaguey

Dominican Republic

Asociación Médica de Santiago
(Santiago Medical Association)
Address: Apartado 445, Santo Domingo

Asociación Médica Dominicana
(Dominican Medical Association)
Address: Apartado de Correos No 1237, Santo Domingo

Biblioteca de la Sociedad Amantes de la Luz
(Amantes de la Luz Society Library)
Address: España Esquina Avenida Central, Santo Domingo

Biblioteca de la Universidad Autónoma de Santo Domingo
(Library of the Autonomous University of Santo Domingo)
Address: Ciudad Universitaria, Apartado 1355, Santo Domingo

Instituto Azucarero Dominicano
(Dominican Sugar Institute)
Address: Centro de los Heroes, Apartado 667, Santo Domingo

Instituto de Cultura Dominicano
(Dominican Cultural Institute)
Address: Biblioteca Nacional Cesar Nicolas Penson, Santo Domingo

Instituto Cartográfico Militar de las Fuerzas Armadas
(Institute of Military Cartography of the Armed Forces)
Address: Santo Domingo

Instituto Superior de Agricultura (ISA)
(Higher Institute of Agriculture)
Address: Apartado 166, Santiago

Instituto Tecnológico de Santo Domingo
(Technological Institute of Santo Domingo)
Address: Avenida de los Proceres, Gala, Apartado 249, Zona 2, Santo Domingo

Universities

Universidad Autónoma de Santo Domingo
(Autonomous University of Santo Domingo)
Address: Ciudad Universitaria, Apartado 1355, Santo Domingo

Universidad Católica Madre y Maestra
(Madre y Maestra Catholic University)
Address: Autopista Duarte, Santiago de los Caballeros

Universidad Central del Este
(Central University of the East)
Address: Calle Duarte 36, San Pedro de Macoris

Universidad Nacional 'Pedro Henríquez Ureña'
(Pedro Henríquez Ureña National University)
Address: Apartado 1423, Santo Domingo

Universidad Tecnológica de Santiago
(Technological University of Santiago)
Address: 685 Santiago

Ecuador

Academia Ecuatoriana de Medicina
(Ecuadorean Academy of Medicine)
Address: Casa de la Cultura Ecuatoriana, Apartado 67, Quito

Comisión Ecuatoriana de Energía Atómica
(Ecuadorean Commission for Atomic Energy)
Address: Calle Cordero 779 y Avenida 6 de Diciembre, Casilla 2517, Quito

Dirección General de Geología y Minas
(General Directorate of Geology and Mines)
Address: c/o Ministerio de Recursos Naturales y Energia, Quito

Dirección General de Hidrocarburos
(General Directorate of Hydrocarbons)
Address: Avenida 10 de Agosto 321, Quito

Estación de Investigación Charles Darwin
(Charles Darwin Research Station)
Address: Pto Ayora, Santa Cruz

Instituto de Ciencias Nucleares
(Nuclear Sciences Institute)
Address: POB 2759, Quito

Instituto Ecuatoriano de Ciencias Naturales
(Ecuadorean Institute of Natural Sciences)
Address: POB 408, Quito

Instituto Interamericano de Agricultura Experimental
(Inter-American Experimental Agriculture Institute)
Address: Conocoto, Linea 63

Instituto Nacional de Estadística y Censos
(National Institute of Statistics and Census)
Address: Avenida 10 de Agosto 229, Quito

Instituto Nacional de Higiene 'Izquieta Pérez'
(Izquieta Pérez National Institute of Hygiene)
Address: Apartado 3961, Quito

Instituto Nacional de Investigaciones Agropecuarias
(National Institute for Arable and Livestock Research)
Address: San Javier 295 y Orellana, Apartado 2600, Quito

Instituto Nacional de Meteorología e Hidrografía
(National Institute of Meteorology and Hydrography)
Address: Calle Daniel Hidalgo 132, Quito

Instituto Nacional de Nutrición
(National Institute of Nutrition)
Address: Apartado 3806, Quito

Instituto Nacional de Pesca
(National Fisheries Institute)
Address: Casilla 5918, Quito

Instituto Oceanográfico de la Armada
(Naval Oceanographic Institute)
Address: Avenida 25 de Julio, POB 5940, Quito

Observatorio Astronómico de Quito
(Astronomical Observatory of Quito)
Address: Apartado 165, Parque Alameda, Quito

Universities and Colleges

Colegio Nacional de Agricultura 'Luis A. Martínez'
(Luis A. Martínez National College of Agriculture)
Address: Casilla 286, Ambato

Escuela Politécnica Nacional
(National Polytechnic)
Address: Isabel la Católica y Veintemilla, Apartado 2759, Quito

Escuela Superior Politécnica de Chimborazo
(Higher Polytechnic of Chimborazo)
Address: Casilla 4703, Riobamba

Escuela Superior Politécnica del Litoral
(Higher Polytechnic of the Coast)
Address: Rocafuerte 101 y Julian Coronel, Casilla 5863, Guayaquil

Universidad Central del Ecuador
(Central University of Ecuador)
Address: Ciudadela Universitaria, Apartado 166, Quito

Pontificia Universidad Católica del Ecuador
(Pontifical Catholic University)
Address: Avenida 12 de Octubre 1976, Apartado 21-84, Quito

Universidad Católica de Cuenca
(Catholic University of Cuenca)
Address: POB 19A, Cuenca

Universidad de Cuenca
(Cuenca University)
Address: Apartado 168, Cuenca

Universidad Estatal de Guayaquil
(State University of Guayaquil)
Address: Casilla 421, Guayaquil

Universidad Nacional de Loja
(National University of Loja)
Address: Casilla Letra S, Loja

Universidad Católica de Santiago de Guayaquil
(Catholic University of Santiago de Guayaquil)
Address: Casilla 4671, Guayaquil

Universidad Técnica de Babahoyo
(Technical University of Babahoyo)
Address: Via Flores, Babahoyo, Los Ríos

Universidad Técnica Particular de Loja
(Private Technical University of Loja)
Address: Apartado 608, Loja

Universidad Técnica de Machala
(Technical University of Machala)
Address: Casilla 466, Machala

Universidad Técnica de Manabí
(Technical University of Manabí)
Address: Casilla 82, Portoviejo, Manabí

Universidad Técnica 'Luis Vargas Torres'
(Luis Vargas Torres Technical University)
Address: Avenida 179, Esmeraldas

Universidad Laica 'Vicente Rocafuerte' de Guayaquil
(Vicente Rocafuerte Lay University of Guayaquil)
Address: Avenida de las Americas, Apartado 11-33, Guayaquil

El Salvador

Biblioteca Nacional
(National Library)
Address: 8a Avenida Norte y Calle Delgado, San Salvador

Centro de Investigaciones Geotécnicas
(Centre for Geotechnical Research)
Address: Apartado 109, San Salvador

Centro Nacional de Tecnología Agropecuaria (CENTA)
(National Centre for Agricultural and Livestock Technology)
Address: Final 1a Avenida Norte, San Salvador

Instituto Geográfico Nacional
(National Geographical Institute)
Address: Apartado Postal 247, Delgado, San Salvador

Instituto Salvadoreño de Investigaciones del Café
(Salvadorean Institute of Coffee Research)
Address: Ministro de Agricultura, 23 Avenida Norte No 114, San Salvador

Universities and Colleges

Escuela Nacional Agrícola
(National School of Agriculture)
Address: San Andres, Ciudad Arce, La Libertad, Apartado 2139, San Salvador

Universidad Nacional de El Salvador
(National University of El Salvador)
Address: Ciudad Universitaria, Final 25 Avenida Norte, San Salvador

French Guiana

Centre Spatial Guyanais
(Guiana Space Centre)
Address: Kourou

Centre Technique Forestier Tropical
(Tropical Forestry Technical Centre)
Address: Centre de Kourou, Boîte Postale 116, 97310 Kourou

Institut de Recherches Agronomiques Tropicales et des Cultures Vivrières
(Research Institute for Tropical Agronomy and Cultivation of Foodstuffs)
Address: Boîte Postale 60, 97301 Cayenne Cedex

Institut de Recherches sur les Fruits et Agrumes
(Fruit and Citrus Research Institute)
Address: Exploitations-Pilotes Agro-Industrielles de Guyane, Boîte Postale 1125
à Cayenne

Institut Pasteur
(Pasteur Institute)
Address: Boîte Postale 304, 97300 Cayenne

**Office de la Recherche Scientifique et Technique Outre-Mer, Centre
ORSTOM de Cayenne**
(Office for Overseas Scientific and Technical Research, Cayenne Centre)
Address: Boîte Postale 165, 97301 Cayenne Cedex

Colleges

Centre Universitaire Antilles-Guyane
(Antilles-Guiana University Centre)
Address: 85 rue Leopold Heder, Boîte Postale 718, 97300 Cayenne

Guatemala

Academia de Ciencias Médicas, Físicas y Naturales de Guatemala
(Guatemala Academy of Medicine, Physics and Natural Sciences)
Address: 13 Calle 1-25 Zona I, Apartado Postal 569, Ciudad de Guatemala

**Instituto Centroamericano de Investigación y Tecnología Industrial
(ICAITI)**
(Central American Institute of Industrial Research and Development)
Address: POB 1552, Avenida La Reforma 4-47, Zona 1Q, Ciudad de Guatemala

Instituto de Nutrición de Centroamerica y Panama
(Central America and Panamanian Institute of Nutrition)
Address: Carretera Roosevelt, Zona II, Ciudad de Guatemala

Instituto Geográfico Nacional
(National Geographic Institute)
Address: Avenida Las Americas 5-76, Zona 13, Ciudad de Guatemala

Instituto Nacional de Energía Nuclear
(National Institute of Nuclear Energy)
Address: 3 Avenida 'A' 2-68, Zona 1, Apartado Postal 1421, Ciudad de
Guatemala

Instituto Nacional de Sismología, Vulcanología, Meteorología, e Hidrología
(National Institute of Seismology, Vulcanology, Meteorology, and Hydrology)
Address: Ciudad de Guatemala

Universities

Universidad de San Carlos de Guatemala
(University of San Carlos de Guatemala)
Address: Ciudad Universitaria, Guatemala 12

Universidad del Valle de Guatemala
(University of Valle de Guatemala)
Address: Apartado Postal 82, Guatemala

Universidad Francisco Marroquin
(Francisco Marroquin University)
Address: 6 Avenida 0-28 Zona 10, Ciudad de Guatemala

Universidad Mariano Gálvez de Guatemala
(Mariano Gálvez de Guatemala University)
Address: Apartado Postal 1811, Ciudad de Guatemala

Universidad Rafael Landívar
(Rafael Landívar University)
Address: Vista Hermosa III, Zona 16, Apartado Postal 39, Guatemala CA

Guyana

Guyana Medical Science Library
Address: Georgetown Hospital Compound, Georgetown

Guyana Museum and Zoo
Address: Company Path, North Street, Georgetown

Guyana Society
Address: Company Path, Georgetown

Guyana Sugar Corporation Limited
Address: 22 Church Street, Georgetown

Guyanan Institute of International Affairs
Address: POB 812, 189 Charlotte Street, St Lacytown, Georgetown

National Library
Address: POB 110, Georgetown

National Science Research Council (CNIC)
Address: POB 689, Georgetown

Universities and Colleges

Guybau Technical Training Complex
Address: Mackenzie, Linden

Guyana School of Agriculture Corporation
Address: Mon Repos, East Coast Demerara

New Amsterdam Technical Institute
Address: New Amsterdam, Berbice

University of Guyana
Address: POB 10 1110, Georgetown

Institute of Development Studies
Address: POB 841, Georgetown

Haiti

Bibliotèque Nationale d'Haiti
(National Library of Haiti)
Address: Rue Hammerton Killick, Port au Prince

Universities and Colleges

Université d'Etat d'Haiti
(Haiti State University)
Address: 25 Rue Bonne Foi, Box 2279, Port au Prince

Ecole Polytechnique d'Haiti
(Polytechnic of Haiti)
Address: Port au Prince

Institut Supérieur Technique d'Haiti
(Higher Technical Institute of Haiti)
Address: 22 Avenue du Chili, POB 992, Port au Prince

Honduras

Biblioteca Nacional de Honduras
(Honduras National Library)
Address: 6a Avenida 'Salvador Mendieta', Tegucigalpa

Instituto de Ingenieros y Arquitectos de Honduras
(Institute of Engineers and Architects of Honduras)
Address: Tegucigalpa

Instituto Geográfico Nacional (IGN)
(National Geographic Institute)
Address: Barrio La Bolsa, Tegucigalpa DC

Universities and Colleges

Escuela Agrícola Panamericana
(Pan American School of Agriculture)
Address: Apartado 93, Tegucigalpa

Universidad Nacional Autónoma de Honduras
(Autonomous National University of Honduras)
Address: Ciudad Universitaria, Blvd Suyapa, Tegucigalpa DC

Jamaica

Caribbean Food and Nutrition Institute (CFNI)
(Instituto de Nutrición y Alimentos del Caribe)
Address: Jamaica Centre, POB 140, Mona, Kingston

Institute of Jamaica
Address: 12–16 East Street, Kingston

Jamaican Association of Sugar Technologists
Address: Sugar Industry Research Institute, Mandeville

Medical Research Council Laboratories
Address: University of West Indies, Mona, Kingston 7

Scientific Research Council
Address: POB 350, Kingston

Universities and Colleges

College of Arts, Science and Technology
Address: 237 Old Hope Road, Kingston 6

School of Agriculture
Address: Twickenham Park, Spanish Town

University of the West Indies
Address: Mona, Kingston 7

Mexico

Academia de la Investigación Científica
(Academy of Scientific Research)
Address: Avenida Revolución 1909, Octavo Piso, México, DF

Academia Nacional de Ciencias
(National Academy of Sciences)
Address: Apartado M-77-98, México 1, DF

Archivos Históricos y Bibliotecas
(Historical Archives and Libraries)
Address: Calzada M. Gandhi y Paseo de la Reforma, México 5, DF

Asociación de Médicas Mexicanas AC
(Mexican Association of Women Doctors)
Address: Oklahoma 151, México 18, DF

Asociación Mexicana de Geólogos Petroleros
(Mexican Association of Petroleum Geologists)
Address: Cipres 176, México 4, DF

Asociación Nacional de Universidades e Institutos de Enseñanza Superior (ANUIES)
(National Association of Universities and Institutes of Higher Education)
Address: Avenida Insurgentes Sur 2133, 30, México 20, DF

Biblioteca Central de la Universidad Nacional Autónoma de México
(Central Library of the National Autonomous University of Mexico)
Address: Ciudad Universitaria, México, DF

Biblioteca Nacional de México
(National Library of Mexico)
Address: República del Salvador 70, México, DF

Centro de Estudios Educativos
(Centre of Educative Studies)
Address: Avenida Revolución 1291, México 20, DF

Centro de Información Científica y Humanística – UNAM
(Scientific and Humanistic Information Centre)
Address: Ciudad Universitaria, Apartado 70-392, México 20, DF

Centro de Investigación y de Estudios Avanzados del Instituto Politécnico Nacional
(Research and Advanced Studies Centre of the National Polytechnic Institute)
Address: Apartado 14-740, México 14, DF

Centro Nacional de Ciencias y Tecnologías Marinas
(National Centre of Marine Science and Technology)
Address: Apartado 512, Veracruz

Centro Internacional del Mejoramiento de Maíz y Trigo
(International Centre for the Improvement of Maize and Wheat)
Address: Apartado Postal 6-641, Londres 40, México 6, DF

Consejo Nacional de Ciencia y Tecnología (CONACYT)
(National Council of Science and Technology)
Address: Insurgentes Sur 1677, México 20, DF

Dirección General de Relaciones Educativas, Científicas y Culturales
(General Board of Educational, Scientific and Cultural Relations)
Address: Brasil 31, Piso 2, México 1, DF

Dirección General del Servicio Meteorológico Nacional
(General Board of the National Meteorological Service)
Address: Avenida Observatorio 192, Tacubaya

Instituto de Astronomía
(Institute of Astronomy)
Address: Apartado Postal 70-264, México 20, DF

Instituto de Salubridad y Enfermedades Tropicales
(Institute of Public Health and Tropical Diseases)
Address: Calle de Carpio 470, México 17, DF

Instituto Mexicano de Investigaciones Tecnológicas, AC
(Mexican Institute of Technological Research)
Address: Calzada Legaria 694, México 10, DF

Instituto Mexicano de Recursos Renovables
(Mexican Institute for the Conservation of Natural Resources)
Address: Dr Veryiz 724, México 12, DF

Instituto Mexicano del Petróleo
(Mexican Petroleum Institute)
Address: Avenida 100 Metros 152, POB 14-805, México 14, DF

Instituto Miles de Terapéutica Experimental
(Miles Institute of Experimental Therapy)
Address: Calzada Xochimilco 77, Apartado 22026, México 22, DF

Instituto Nacional de Astrofísica, Optica y Electrónica
(National Institute of Astrophysics, Optics and Electronics)
Address: Apartados Postales 216 y 51 (Tonantzintla) Puebla PUE

Instituto Nacional de Cardiología
(National Cardiology Institute)
Address: Juan Badiano I, Tlalpan, México 22, DF

Instituto Nacional de Higiene
(National Institute of Hygiene)
Address: Calzada Mariano Escobedo 20, México 17, DF

Instituto Nacional de Investigaciones Agrícolas
(National Institute of Agricultural Research)
Address: Apartado 6-882, México 6, DF

Instituto Nacional de Investigaciones Forestales
(National Institute of Forestry Research)
Address: Avenida Progreso No 5, México 21, DF

Instituto Nacional de Investigaciones Nucleares
(National Institute of Nuclear Research)
Address: Apartado Postal 27-190, México 18, DF

Instituto Nacional de Investigaciones Pecuarias
(National Institute of Livestock Research)
Address: Km 15.5 Cerretera México-Toluca, Palo Alto, DF

Instituto Nacional de Neumología
(National Institute of Pneumonology)
Address: Tlalpan, México 22, DF

Instituto Nacional de Pesca
(National Fisheries Institute)
Address: Avenida Cuauhtemoc 80, México 7, DF

Sociedad Matemática Mexicana
(Mexican Society of Mathematics)
Address: Facultad de Ciencias UNAM, Apartado 70-450, México 20, DF

Sociedad Mexicana de Parasitología, AC
(Mexican Society of Parasitology)
Address: Nicolas de San Juan 1015, Apartado 12813, México 12, DF

Sociedad Mexicana de Pediatría
(Mexican Society of Paediatrics)
Address: Calzada de Madereros 240, México, DF

Sociedad Mexicana de Salud Pública
(Mexican Society of Public Health)
Address: Leibnitz 32, Primer Piso, México 5, DF

Sociedad Química de México
(Chemistry Society of Mexico)
Address: Cipres 176, Apartado Postal 4-875, México 4, DF

Universities and Colleges

Instituto Politécnico Nacional – Universidad Técnica
(National Polytechnic Institute – Technical University)
Address: Unidad Profesional Zacatenco, México 14, DF

Instituto Tecnológico y de Estudios Superiores de Monterrey
(Technological and Higher Studies Institute of Monterrey)
Address: Sucursal de Correos J, Monterrey, Nuevo León

Instituto Tecnológico y de Estudios Superiores de Occidente
(Technological and Higher Studies Institute of the West)
Address: Avenida Ninos Heroes 1342-8, Guadalajara JAL

Instituto Tecnológico Regional de Chihuahua
(Chihuahua Regional Institute of Technology)
Address: Apartado Postal 119, Chihuahua CHIH

Instituto Tecnológico Regional de Durango
(Durango Regional Institute of Technology)
Address: Blvd F. Pescados 1830 OTE, Apartado Postal 465, Durango

Instituto Tecnológico Regional de Mérida
(Mérida Regional Institute of Technology)
Address: Apartado Postal 561, Mérida, Yucatán

Instituto Tecnológico Regional de Morelia
(Morelia Regional Institute of Technology)
Address: Carretera Morelia Salvatierra, Apartado Postal 750, Morelia, Michoacan

Instituto Tecnológico Regional de Oaxaca
(Oaxaca Regional Institute of Technology)
Address: Calzado Tecnológico y Wilfredo-Massieu s/n, Oaxaco, OAX

Instituto Tecnológico Regional de Querétaro
(Querétaro Regional Institute of Technology)
Address: Apartado Postal 124, Avenida Tecnológico y General Mariano Escobedo, Queretaro QRO

Instituto Tecnológico Regional de Saltillo
(Saltillo Regional Institute of Technology)
Address: Apartado Postal 600, Entronque Carreteras Monterrey y Piedras Negras, Saltillo, Coahuila

Instituto Tecnológico de Sonora
(Sonora Institute of Technology)
Address: Avenida Rodolfo Elias Calles y Chihuahua, Ciudad de Obregon, SON

Universidad Nacional Autónoma de México (UNAM)
(National Autonomous University of Mexico)
Address: Ciudad Universitaria, Villa Obregon, DF

Universidad Autónoma del Estado de México
(Autonomous University of the State of Mexico)
Address: Constituyentes 100 Oriente, Toluca, Estado de México

Universidad Femenina de México
(Women's University of Mexico)
Address: Avenida Constituyentes 151, México 18, DF

Universidad Autónoma de Aguascalientes
(Autonomous University of Aguascalientes)
Address: Jardin del Estudiante 1, Aguascalientes, Ags

Universidad de las Américas
(University of the Americas)
Address: Apartado Postal 100, Santa Catarina Martir, Puebla

Universidad Anáhuac
(Anáhuac University)
Address: Apartado Postal 10-844, Lomas Anáhuac, México 10, DF

Universidad Autónoma de Baja California
(Autonomous University of Baja California)
Address: Apartado Postal 459, Rio Conchos y Paseo Del Valle, Mexicali, Baja California

Universidad Autónoma Chapingo
(Autonomous University of Chapingo)
Address: Chapingo, Estado de México

Universidad Autónoma de Chiapas
(Autonomous University of Chiapas)
Address: 2A Poniente Sur 118, 5to Piso, Apartado Postal 343, Tuxtla Gutierrez, Chiapas

Universidad Autónoma de Chihuahua
(Autonomous University of Chihuahua)
Address: Ciudad Universitaria, Chihuahua

Universidad Autónoma de Coahuila
(Autonomous University of Coahuila)
Address: Apartado Postal 308, Saltillo, Coahuila

Universidad de Colima
(Colima University)
Address: Avenida Universidad 333, Colima

Universidad Juárez del Estado de Durango
(Juárez University of the State of Durango)
Address: Constitucion 404 Sur, Durango

Universidad de Guadalajara
(Guadalajara University)
Address: Avenida Juarez 975, Guadalajara

Universidad Autónoma de Guadalajara
(Autonomous University of Guadalajara)
Address: Apartado Postal I-440, Guadalajara

Universidad de Guanajuato
(Guanajuato University)
Address: Lascurain de Retana 5, Guanajuato

Universidad Autónoma de Guerrero
(Autonomous University of Guerrero)
Address: Avenida Juarez 13, Chilpancingo, Guerrero

Universidad Autónoma de Hidalgo
(Autonomous University of Hidalgo)
Address: Abasolo 6000, Pachuca, Hidalgo

Universidad Iberoamericana
(Ibero-American University)
Address: Avenida Cerro de las Torres 395, México 21, DF

Universidad Intercontinental
(Intercontinental University)
Address: Insurgentes Sur 4135, México 22, DF

Universidad Autónoma de Ciudad Juárez
(Autonomous University of Juarez)
Address: Apartado Postal 1594-D, Ciudad Juárez, Chihuahua

Universidad del Valle de México
(University of Valle de Mexico)
Address: Sadi Carnot 57, Colonia San Rafael, México 4, DF

Universidad La Salle de México
(La Salle de México University)
Address: Benjamin Franklin 47, México 18, DF

Universidad Autónoma Metropolitana
(Autonomous Metropolitan University)
Address: Apartado Postal 325, México, DF

Universidad Michoacana de San Nicolás de Hidalgo
(University of Michoacana de San Nicolás de Hidalgo)
Address: Santiago Tapia 403, Morelia, Michoacan

Universidad de Montemorelos
(Montemorelos University)
Address: Apartado 16, Montemorelos, Nueva León

Universidad de Monterrey
(Monterrey University)
Address: Avenida Gonzalitos 300 Sur, Monterrey, Nueva León

Universidad Autónoma del Estado de Morelos
(State of Morelos Autonomous University)
Address: Cuernavaca, Morelos

Universidad Autónoma de Nayarit
(Nayarit Autonomous University)
Address: Ciudad de la Cultura Amado Nervo, Tepic, Nayarit

Universidad Autónoma del Noroeste
(Autonomous University of the North-West)
Address: Monclova 1430, Segundo Piso, Coronel República, Saltillo, Coahuila

Universidad Autónoma de Nueva León
(Nueva León Autonomous University)
Address: Torre de la Rectoría, Piso 8, Ciudad Universitaria, Monterrey,
Nueva León

Universidad del Norte
(University of the North)
Address: Cuahtémoc Sur 985, Monterrey, Nueva León

Universidad Autónoma 'Benito Juárez' de Oaxaca
(Benito Juarez University of Oaxaca)
Address: Apartado 76, Esquina Avenida Independencia y Macedonio Alcala, Oaxaca

Universidad Autónoma de Puebla
(Puebla Autonomous University)
Address: 4 Sur 104, Puebla

Universidad Popular Autónoma del Estado de Puebla
(Autonomous University of the State of Puebla)
Address: 9 Poniente 1508, Puebla

Universidad Autónoma de Querétaro
(Autonomous University of Queretaro)
Address: Apartado Postal 184, Centro Universitario, Cerro Las Campanas,
Querétaro

Universidad Regiomontana
(Regiomontana University)
Address: Villagran 238 Sur, Apartado Postal 243, Monterrey

Universidad Autónoma de San Luis de Potosí
(Autonomous University of San Luis de Potosí)
Address: Alvaro Obregon 64, San Luis, Potosí

Universidad Autónoma de Sinaloa
(Autonomous University of Sinaloa)
Address: Apartado Postal 1919, Culiacan, Sinaloa

Universidad de Sonora
(Sonora University)
Address: Apartado Postal 336, Hermosillo, Sonora

Universidad del Sudeste
(University of the South-East)
Address: Apartado Postal 204, Campeche

Universidad Autónoma Juárez de Tabasco
(Juárez Autonomous University of Tabasco)
Address: Zona de la Cultura, Villa Hermosa, Tabasco

Universidad Autónoma de Tamaulipas
(Tamaulipas Autonomous University)
Address: Apartado Postal 186, Ciudad Victoria, Tamaulipas

Universidad Veracruzana
(Veracruz University)
Address: Zona Universitaria, Lomas Del Estadio, Xalapa, Veracruz

Universidad de Yucatán
(Yucatán University)
Address: Apartado 415, Calle 57 por 60, Merida, Yucatán

Universidad Autónoma de Zacatecas
(Zacatecas Autonomous University)
Address: Galeana Numero 1, Zacatecas ZAC

Nicaragua

Instituto Agrario de Nicaragua
(Agrarian Institute of Nicaragua)
Address: Managua

Instituto Centroamericano de Administración de Empresas (INCAE)
(Central American Institute of Commercial Administration)
Address: Apartado 2485, Managua

Instituto Nicaraguense de Cine
(Nicaraguan Institute of the Cinema)
Address: Apartado Postal 4660, Managua

Servicio Geológico Nacional
(National Geological Service)
Address: Apartado Postal 13–47, Managua

Universities and Colleges

Escuela Nacional de Agricultura y Ganadería
(National School of Agriculture and Livestock)
Address: Apartado Postal 453, Kilometro 12, Carretera Norte, Managua

Universidad Nacional Autónoma de Nicaragua
(National Autonomous University of Nicaragua)
Address: León

Universidad Centroamericana (Sección de Nicaragua)
(Central American University)
Address: Apartado 69, Managua

Universidad Politécnica de Nicaragua
(Polytechnic of Nicaragua)
Address: Apartado 3595, Managua

Panama

Academia Nacional de Ciencias de Panamá
(National Academy of Sciences of Panama)
Address: Apartado 4570, Panamá City

Centro para el Desarrollo de la Capacidad Nacional de Investigación
(Centre for the Development of National Research Capacity)
Address: Estafeta Universitaria, Universidad de Panamá, Panamá City

Gorgas Memorial Laboratory of Tropical and Preventive Medicine
Address: Avenida Justo Arosemena 35-30, Apartado 6991, Panamá City

Smithsonian Tropical Research Institute
Address: POB 2072, Balboa

Universities

Universidad de Panamá
(University of Panama)
Address: Estafeta Universitaria, Panamá City

Universidad Santa María la Antigua
(Santa María la Antigua University)
Address: Apartado 6-1696, Panamá 6

Paraguay

Biblioteca de la Sociedad Científica del Paraguay
(Library of the Scientific Society of Paraguay)
Address: Avenida España 505, Asunción

Centro Paraguayo de Estudios de Desarrollo Economico y Social
(Paraguayan Centre of Economic and Social Development Studies)
Address: Casilla 1189, Asunción

Centro Paraguayo de Ingenieros
(Paraguayan Centre for Engineers)
Address: Avenida España 959, Casilla 336, Asunción

Instituto Geográfico Militar
(Military Geographical Institute)
Address: Avenida Peru y Artigas, Asunción

Instituto Nacional de Investigaciones Científicas
(National Institute of Scientific Research)
Address: Casilla de Correos 1141, Asunción

Instituto Nacional de Parasitología
(National Institute of Parasitology)
Address: Casilla de Correos 1102, Asunción

Servicio Tecnico Interamericano de Cooperacion Agrícola
(Inter-American Technical Service for Agricultural Cooperation)
Address: Casilla de Correos 819, Asunción

Sociedad Científica del Paraguay
(Scientific Society of Paraguay)
Address: Avenida España 505, Asunción

Sociedad de Pediatría y Puericultura del Paraguay
(Paediatrics and Child Welfare Society of Paraguay)
Address: 25 de Mayo y Tacuai, Asunción

Universities

Universidad Católica de Nuestra Señora de la Asunción
(Catholic University of Our Lady of the Assumption)
Address: Independencia Nacional y Comuneros, Asunción

Centro de Investigaciones
(Research Centre)
Address: Casilla de Correos 1718, Asunción

Universidad Nacional de Asunción
(National University of Asunción)
Address: Colon 73, Asunción

Peru

Academia de Estomatología del Perú
(Peruvian Academy of Stomatology)
Address: Apartado 2467, Lima

Academia Nacional de Ciencias Exactas, Físicas y Naturales de Lima
(National Academy of Exact, Physical and Natural Sciences of Lima)
Address: Casilla 1979, Lima

Academia Nacional de Medicina
(National Academy of Medicine)
Address: Apartado 987, Lima

Academia Peruana de Cirugía
(Peruvian Academy of Surgery)
Address: Camana 773, Lima

Biblioteca Central de la Universidad Nacional Mayor de San Marcos de Lima
(Central Library of the Greater National University of San Marcos de Lima)
See entry under university

Biblioteca Central de la Pontificia Universidad Católica del Perú
(Central Library of the Pontificial Catholic University of Peru)
Address: Ciudad Universitaria – Final de la Avenida Bolívar s/n Fundo Pando, Apartados 1761 y 5729, Lima

Biblioteca de la Universidad Nacional de San Agustín
(Library of the National University of San Agustín)
Address: Apartado 23, Lima

Biblioteca Nacional
(National Library)
Address: Avenida Abancay, Apartado 2335, Lima

Consejo Nacional de la Universidad Peruana
(National Peruvian University Council)
Address: Apartado 4664, Calle Aldabas 3ra Cuadra s/n Surco, Lima

Dirección General de Meteorología del Perú
(General Board of Meteorology of Peru)
Address: Avenida Arequipa 5200, Apartado 1308, Miraflores, Lima

Estación Altoandina de Biología y Reserva Zoo-Botánica de Checayani
(High Andes Biology Station and Zoo-botanical Reserve of Checayani)
Address: Checayani, Azangaro (Puno)

Estación Experimental Agrícola del Norte
(Agricultural Experimental Station of the North)
Address: Atahualpa 211, Lambayeque

Estación Experimental Agropecuaria de Tulumayo
(Tulumayo Agricultural and Livestock Research Station)
Address: Apartado 78, Tingo María, Huanuco

Federación Médica Peruana
(Peruvian Medical Federation)
Address: Apartado 4439, Lima

Instituto Birchner-Benner
(Birchner-Benner Institute)
Address: Schell 598, Miraflores, Lima

Instituto de Biología Andina
(Institute of Andean Biology)
Address: Apartado 5073, Lima

Instituto de Ciencias de la Comunicación
(Institute of Communication Sciences)
Address: Las Moreras 220, Urbanización Camacho, Ate, Lima 3

Instituto de Investigaciones Alérgicas
(Institute of Allergy Research)
Address: Avenida La Marina 2501, Maranga, San Miguel

Instituto de Zoonosis e Investigación Pecuaria
(Institute of Zoonosis and Livestock Research)
Address: Apartado 1128, Lima

Instituto del Mar del Perú
(Peruvian Marine Institute)
Address: Esquina General Valle y Gamarra, Callao, Apartado 22, Lima

Instituto Experimental de Educación Primaria No 1
(Experimental Institute of Primary Education No 1)
Address: Barranco Avenida Miraflores No 200, Lima

Instituto Geofísico del Peru
(Geophysics Institute of Peru)
Address: Apartado 3747, Lima

Instituto Geográfico Militar
(Military Geographical Institute)
Address: Apartado 2038, Lima

Instituto Geológico, Minero y Metalúrgico
(Institute of Geology, Mining and Metallurgy)
Address: Apartado 889, Pablo Bermudez 211, Lima

Instituto Nacional de Investigación Agraria
(National Institute for Agrarian Research)
Address: Sinchi Roca 2728, Oficina 802, Lima

Instituto Nacional de Salud
(National Health Institute)
Address: Capac Yupanqui 1400, Apartado 451, Lima

Instituto Peruano de Energía Nuclear
(Peruvian Institute of Nuclear Energy)
Address: Avenida Canada 1470, Urbanización Santa Catalina, Apartado 1687, Lima

Mission ORSTOM au Perou – Cooperation Auprès du Ministerio de Energía y Minas
(ORSTOM Mission of Peru, cooperating with the Ministry of Energy and Mines)
Address: La Mariscala 115, San Isidro, Lima

Oficina Nacional de Estadística
(National Office of Statistics)
Address: Avenida 28 de Julio 10 56, Lima

Sociedad de Ingenieros del Perú
(Society of Engineers of Peru)
Address: Avenida N. de Pierola 788, Casilla 1314, Lima

Sociedad Nacional Agraria
(National Agrarian Society)
Address: A. Miro Quesada 327–341, Lima

Sociedad Nacional de Minería
(National Society of Mining)
Address: Plaza San Martin 917, Lima

Universities

Universidad Nacional Agraria
(National Agrarian University)
Address: Apartado 456, La Molina, Lima

Universidad Nacional Agraria de la Selva
(National Agrarian University of the Selva)
Address: Apartado 156, Tingo Maria, Huanuco

Universidad de Lima
(Lima University)
Address: Prolongacion Javier, Prado s/n, Monterrico

Universidad Nacional de la Amazonía Peruana
(National University of the Peruvian Amazon)
Address: Apartado 496, Iquitos

Universidad Nacional del Centro del Perú
(National University of Central Peru)
Address: Calle Real 160, Apartado 138, Huancayo

Universidad Nacional de Huánuco 'Hermilio Valdizán'
(Hermilio Valdizán National University of Huánuco)
Address: JR Dos de Mayo 680, Apartado 278, Huanuco

Universidad Nacional 'Federico Villarreal'
(Federico Villarreal National University)
Address: Avenida Nicolas Pierola 412, Apartado 1518 6049, Lima

Universidad Nacional de Ingeniería
(National University of Engineering)
Address: Casilla 1301, Lima

Universidad Nacional 'Jose Faustino Sánchez Carrión'
(Jose Faustino Sánchez Carrión National University)
Address: Avenida Grau 592, Oficina 301, Apartado 81, Huacho

Universidad Nacional 'Pedro Ruiz Gallo'
(Pedro Ruiz Gallo National University)
Address: 8 de Octubre 637, Apartado 48, Lambayeque

Universidad Nacional Mayor de San Marcos de Lima
(Greater National University of San Marcos de Lima)
Address: Ciudad Universitaria, La Punta, Callao

Universidad Nacional de San Antonio de Abad
(San Antonio de Abad National University)
Address: Avenida de la Cultura s/n, Apartado 367, Cuzco

Universidad Nacional de San Agustín
(National University of San Agustín)
Address: Siglo XX 227, Apartado 23, Avequipa

Universidad Nacional de San Cristóbal de Huamanga
(San Cristóbal de Huamanga National University)
Address: Apartado 220, Ayacucho

Universidad Nacional 'San Luis Gonzaga'
(San Luis Gonzaga National University)
Address: Apartado 100 Bolivar 232, Ica

Universidad Nacional de Trujillo
(Trujillo National University)
Address: Diego de Almagro 396, Apartado 315, Trujillo

Universidad Nacional Técnica de Cajamarca
(Technical University of Cajamarca)
Address: Apartado 289, Cajamarca

Universidad Nacional de Piura
(Piura National University)
Address: Calle Apurimac 461, Apartado 295, Piura

Universidad Particular 'San Martín de Porres'
(San Martín de Porres Private University)
Address: JR Camana 168, Lima

Universidad Particular Peruana 'Cayetano Heredia'
(Cayetano Heredia Private University)
Address: Apartado 5045, Lima

Pontificia Universidad Católica del Perú
(Pontifical Catholic University of Peru)
Address: Fundo Pando, Apartado Postal 1701 y 5729, Lima

Universidad Nacional Técnica del Callao
(Technical University of Callao)
Address: Apartado 138, Callao

Puerto Rico

Arecibo Observatory
Address: POB 995, Arecibo

Caribbean Regional Library
Address: POB 21927, University Station, San Juan

Institute of Tropical Forestry
Address: USDA Forest Service, POB AQ Río Piedras 00928

Puerto Rican Nuclear Center
Address: Bio-Medical Building, Caparra Heights Station, San Juan

University of Puerto Rico General Library
Address: Mayaguez Campus, Puerto Rico

Universities

Bayamón Central University
Address: POB 1725, Bayamón, PR 00619

Inter-American University of Puerto Rico
Address: GPO Box 3255, San Juan, PR 00936

Universidad Católica de Puerto Rico
(Catholic University of Puerto Rico)
Address: Ponce PR 00731

Universidad de Puerto Rico
(University of Puerto Rico)
Address: Rio Piedras, PR 00931

University of the Sacred Heart
Address: POB 12383, Loiza Station, Santurce PR 00914

Surinam

Centre for Agricultural Research in Surinam
Address: POB 1914, Paramaribo

Dienst Lands Bosbeheer – Afdeling Natuurbeheer
(Surinam Forest Service – Nature Conservation Department)
Address: Cornelis Jongbawstraat 10, PO Box 436, Paramaribo

Geologisch Mijnbouwkundige Dienst
(Geological Mining Service)
Address: 2–6 Kleine Waterstraat, Paramaribo

Landbouwproefstation
(Agricultural Experiment Station)
Address: POB 160, Paramaribo

Stichting Surinaams Museum
(Surinam Museum)
Address: POB 2306, Paramaribo

University

Universiteit van Suriname
(University of Surinam)
Address: Dr Sophie Redmondstraat 118, Paramaribo

Uruguay

Asociación de Ingenieros del Uruguay
(Association of Uruguayan Engineers)
Address: Avenida Brigadier General Lavalleja 1464, Piso 14, Montevideo

Asociación de Química y Farmacia del Uruguay
(Chemical and Pharmaceutical Association of Uruguay)
Address: Avenida Brigadier General Lavalleja 1464, Piso 14, Montevideo

Asociación Odontológia Uruguaya
(Odontology Association of Uruguay)
Address: Avenida Brigadier General Lavalleja 1464, Piso 13, Montevideo

Asociación Rural del Uruguay
(Rural Association of Uruguay)
Address: Uruguay 864, Montevideo

Ateneo de Clínica Quirúrgica
(Athenaeum of Clinical Surgery)
Address: Avenida Brigadier General Lavalleja, Piso 13, Montevideo

Biblioteca Nacional del Uruguay
(National Library of Uruguay)
Address: Avenida 18 de Julio 1790, Montevideo

Centro de Documentación Científica, Técnica y Económica
(Centre of Scientific, Technical and Economic Documentation)
Address: Biblioteca Nacional del Uruguay, Avenida 18 de Julio 1790, Montevideo

Centro de Estadísticas Nacionales y Comercio Internacional del Uruguay – CENCI Uruguay
(Centre of National Statistics and International Commerce of Uruguay)
Address: Misiones 1361, Casilla de Correos 1510, Montevideo

Comisión Nacional de Energía Atómica
(National Commission for Atomic Energy)
Address: Rincón 723, Piso 3, Montevideo

Consejo Nacional de Higiene
(National Hygiene Council)
Address: Avenida 18 de Julio 1892, Montevideo

Consejo Nacional de Investigaciones Científicas y Técnológicas
(National Council of Scientific and Technological Research)
Address: Sarandí 450, Piso 4, Montevideo

Dirección Nacional de Meteorología del Uruguay
(National Meteorological Directorate)
Address: Javier Barrios Amorin 1488, Casilla 64, Montevideo

Instituto de Endocrinología 'Profesor Dr Juan C. Mussio Fournier'
(Professor Dr Juan C. Mussio Fournier Institute of Endocrinology)
Address: Hospital Pasteur, Calle Larravide 74, Montevideo

Instituto de Investigaciones Biológicas 'Clemente Estable'
(Clemente Estable Institute of Biological Research)
Address: Avenida Italia 3318, Montevideo

Instituto de Oncología
(Oncology Institute)
Address: Avenida 8 de Octubre 3265, Montevideo

Instituto Geográfico Militar
(Military Geographical Institute)
Address: Avenida 8 de Octubre 2904, Montevideo

Instituto Geológico del Uruguay
(Geological Institute of Uruguay)
Address: Calle Hervidero 2853, Montevideo

Instituto Interamericano del Niño
(Inter-American Child Institute)
Address: Avenida 8 de Octubre 2904, Montevideo

Instituto Nacional de Pesca
(National Fisheries Institute)
Address: Constituyente 1497, Montevideo

Instituto Uruguayo de Normas Técnicas
(Uruguayan Institute of Technical Standards)
Address: Avenida Brigadier General Lavalleja 1464, Montevideo

Observatorio Astronómico
(Astronomical Observatory)
Address: Casilla de Correos 867, Montevideo

Oficina Regional de Ciencia y Tecnología de la UNESCO para America Latina y el Caribe
(UNESCO Regional Office of Science and Technology for Latin America and the Caribbean)
Address: Bulevar Artigas 1320, Casilla 859, Montevideo

Sociedad de Arquitectos del Uruguay
Address: Avenida Brigadier General Lavalleja 1464, Piso 14, Montevideo

Sociedad de Cirugía del Uruguay
(Uruguayan Surgical Society)
Address: Avenida Brigadier General Lavallejas 1464, Piso 13, Montevideo

Sociedad de Radiología del Uruguay
(Uruguayan Society of Radiology)
Address: Avenida Brigadier General Lavalleja 1464, Piso 13, Montevideo

Sociedad Uruguay de Patología Clínica
Address: Hospital de Clinicas Dr Manuel Quintela, Avenida Italia s/n, Montevideo

Sociedad Uruguaya de Pediatría
(Uruguayan Society of Paediatrics)
Address: Avenida Brigadier General Lavalleja 1464, Piso 13, Montevideo

Sociedad Zoológica del Uruguay
(Zoological Society of Uruguay)
Address: Casilla 399, Montevideo

Universities

Universidad de la República
(University of the Republic)
Address: Avenida 18 de Julio 1824, Montevideo

Universidad del Trabajo del Uruguay
(University of Labour of Uruguay)
Address: Calle San Salvador 1674, Montevideo

Venezuela

Academia de Ciencias Exactas, Físicas y Matemáticas
(Academy of Exact Sciences, Physics and Mathematics)
Address: Apartado 1421, Bolsa de San Francisco, Caracas

Academia Nacional de Medicina
(National Academy of Medicine)
Address: Bolsa de San Francisco, Caracas

Asociación Venezolana de Ingeniería Eléctrica y Mecánica
(Venezuelan Association of Electrical and Mechanical Engineering)
Address: Apartado 6255, Caracas 105

Asociación Venezolana para el Avance de la Ciencia (ASOVAC)
(Venezuelan Association for the Advancement of Science)
Address: Apartado del Este 61843, Caracas

Biblioteca Nacional
(National Library)
Address: Bolsa de San Francisco, Caracas

Biblioteca Técnica Científica Centralizada
(Central Scientific and Technical Library)
Address: Apartado 254, Caracas

Centro de Estudios Venezolanos Indígenas
(Venezuelan Centre of Indigenous Studies)
Address: Apartado 261, Caracas

Centro Nacional de Investigaciones Agropecuarias
(National Centre for Agricultural and Livestock Research)
Address: Apartado 4653, Maracay 200, Estado Aragua

Consejo Nacional de Investigaciones Agrícolas
(National Council for Agricultural Research)
Address: Torre Norte, Piso 14, Centro Simón Bolívar, Apartado 12844, Caracas

Consejo Nacional para el Desarrollo de la Industria Nuclear
(National Council for the Development of the Nuclear Industry)
Address: Apartado 68233, Caracas 106

Estacion Experimental de Café
(Experimental Coffee Station)
Address: Rubio, Bramon, Tachira

Estación Meteorológica
(Meteorological Station)
Address: Ciudad Bolívar

Fundacion La Salle de Ciencias Naturales
(La Salle Foundation of Natural Sciences)
Address: Avenida Boyaca, Apartado Postal 1930, Caracas

Instituto Agrario Nacional
(National Agrarian Institute)
Address: Quinta Barrancas, Avenida San Carlos, Vista Alegre, Caracas 102

Instituto de Medicina Experimental
(Institute of Experimental Medicine)
Address: POB 50-587, Sabana Grande, Ciudad Universitaria, Caracas 1051

Instituto Nacional de Nutrición
(National Institute of Nutrition)
Address: Apartado 2049, Caracas

Observatorio Naval 'Juan Manuel Cagigal'
(Juan Manuel Cagigal Naval Observatory)
Address: Colina del Calvario, Apartado 6745, Caracas

Office de la Recherche Scientifique et Technique Outre-Mer 'Mision Venezuela'
(Overseas Scientific and Technical Research Office, Venezuelan Section)
Address: Apartado 68183, Caracas 106

Sociedad Venezolana de Ciencias Naturales
(Venezuelan Society of Natural Sciences)
Address: Calle Arichuna y Cumaco, El Marquez, Apartado 1251, Caracas

Universities and Colleges

Instituto Universitario Politécnico
(Polytechnic College)
Address: Apartado Postal 539, Barquisimeto, Estado Lara

Universidad de Carabobo
(Carabobo University)
Address: Avenida Bolívar 125-39, Apartado Postal 129, Valencia

Universidad Católica Andrés Bello
(Andrés Bello Catholic University)
Address: Urb. Montalban, La Vega, Apartado 29068, Caracas 1021

Universidad Central de Venezuela
(Central University of Venezuela)
Address: Ciudad Universitaria, Los Chaguaramos, Caracas, Apartado Postal 104

Universidad Centro Occidental
(Central-Western University)
Address: Apartado 400, Barquisimeto, Lara

Universidad de los Andes
(University of the Andes)
Address: Avenida 3, Independencia Edificio Central, Zona Postal 802, Mérida

Universidad Metropolitana
(Metropolitan University)
Address: Apartado 76819, Caracas 107

Universidad Nacional Experimental Francisco de Miranda
(Francisco de Miranda Experimental University)
Address: Edificio Mirilina, Calle Zamora Esquina de Iturbe, Coro, Estado Falcon

Universidad Nacional Experimental de los Llanos Centrales 'Romulo Gallegos'
(Romulo Gallegos Central Plains Experimental University)
Address: Apartado Postal 102, San Juan de los Morros 2301-A, Estado Guarico

Universidad Nacional Experimental de Táchira
(Táchira Experimental University)
Address: Apartado 35, Universidad, Paramillo, San Cristóbal

Universidad de Oriente
(University of the East)
Address: Edificio Rectorado, Apartado Postal 094, Cumana, Sucre

Universidad Rafael Urdaneta
(Rafael Urdaneta University)
Address: Apartado 614, Maracaibo

Universidad de Santa María
(Santa María University)
Address: Frente Plaza Madariaga, El Paraiso, Caracas

Universidad Simón Bolívar
(Simón Bolívar University)
Address: 80659 Prados del Este, Caracas

Universidad de Zulia
(Zulia University)
Address: Apartado de Correos 526, Maracaibo 4011, Estado Zulia

Index of Establishments

Academia Brasileira de Ciências 35, 55, 269

Academia Chilena de Ciencias 74

Academia Chilena de Ciencias Naturales 74

Academia Chilena de la Historia 74

Academia Chilena de la Lengua 74

Academia Colombiana de Ciencias Exactas, Físicas y Naturales 89, 280

Academia Costarricense de la Lengua 98

Academia Costarricense de Periodoncia 98

Academia de Ciencias de Cuba 108

Academia de Ciencias Exactas, Físicas y Matemáticas 247, 311

Academia de Ciencias Médicas, Físicas y Naturales de Guatemala 146, 292

Academia de Estomatología del Perú 304

Academia de Geografía e Historia de Costa Rica 98

Academia de Guerra 70

Academia de la Investigación Científica 183, 295

Academia Dominicana de la Historia 117

Academia Dominicana de la Lengua 117

Academia Ecuatoriana de Medicina 127, 289

Academia Hondureña 157

Academia Nacional de Agronomía y Veterinaria 16

Academia Nacional de Ciencias de Bolivia 31, 267

Academia Nacional de Ciencias de Buenos Aires 16, 260

Academia Nacional de Ciencias – Mexico 183, 295

Academia Nacional de Ciencias – Panama 200, 302

Academia Nacional de Ciencias Exactas, Físicas y Naturales de Lima 220, 304

Academia Nacional de Ingeniería 235

Academia Nacional de Medicina – Argentina 16, 260

Academia Nacional de Medicina – Brazil 55, 269

Academia Nacional de Medicina – Colombia 89, 280

Academia Nacional de Medicina – Peru 220, 304

Academia Nacional de Medicina – Venezuela 247, 311

Academia Panameña de la Historia 200

Academia Peruana de Cirugía 304

Academia Puertoriqueña de la Lengua Española 224

Academy of Exact Sciences, Physics and Mathematics 247, 311

Academy of Scientific Research 183, 295

Administración Nacional de Combustibles, Alcoholes y Portland (ANCAP) 234

Administrative Department of National Statistics – Colombia 82, 281

Adolfo Lutz Institute 56, 270

African Economic Commissions 15

Agencia Nacional de Planificación 160

Agencia para el Desarrollo Internacional (AID) 129

Agencia para el Financiamiento de Estudios y Proyectos (FINEP) 34

Agency for International Development (AID) 129

Agency for the Financing of Studies and Projects 34

Agrarian Institute of Economic Studies 69, 278

Agrarian Institute of Nicaragua 193, 302

Agricultural and Food Technology Centre 56

Agricultural and Livestock and Industrial Sub-Systems 128

Agricultural and Livestock Development
Centre 113, 116
Agricultural and Livestock Research
Enterprise of the State of Pernambuco
53
Agricultural Experiment Station –
Surinam 227, 308
Agricultural Experimental Station of the
North 219, 305
AGRINTER System 236
AGRIS 236
Agro-Industrial Experimental Station 11,
261
Agustín Codazzi Geographical Institute
82, 281
AID, see Agency for International
Development
Alberto Boerger Centre of Agricultural
Research 234
Alcoa 226
Alvaro Castro J. Library 99
Amantes de la Luz Society Library 117,
288
Amazonas University 40, 272
Americas House 108, 286
ANACAFE 145
Anáhuac University 173, 299
Andean Council of Science and
Technology 129
Andean Group, or Cartagena Agreement
77, 221
Andrés Bello Catholic University 243,
247, 312
Anselmo Liorente y Lafuente
Ecclesiastical History Museum 99
Antilles-Guiana University Centre 137,
292
Archivo Nacional de Colombia 90, 280
Archivo Nacional de Costa Rica 99, 284
Archivos Históricos y Bibliotecas 184,
295
Arecibo Observatory 224, 308
Argentine Agrarian Institute of Rural
Culture 11, 262
Argentine Centre of Scientific and
Technological Information 14, 260
Argentine Institute of Standards 10, 262
Argentine University of John F.
Kennedy 20, 266
Argentine University of 'La Empresa' 20,
266
Artigas Institute of Foreign Affairs 232
Asociación Chilena de Sismología e
Ingeniería Antisísmica 74, 277
Asociación Costarricense de Cirugía 98, 284

Asociación Costarricense de Pediatría 98,
284
Asociación de Ingenieros del Uruguay 309
Asociación de Médicas Mexicanas AC
182, 295
Asociación de Productores de Aceite
Esenciales 144
Asociación de Química y Farmacia del
Uruguay 309
Asociación Médica de Santiago 117, 287
Asociación Médica Dominicana 117, 288
Asociación Mexicana de Geólogos
Petroleros 183
Asociación Nacional de Universidades e
Institutes de Enseñanza Superior
(ANUIES) 170-1, 296
Asociación Odontológica Uruguaya 235,
309
Asociación Pediátrica de Guatemala 146
Asociación Rural del Uruguay 309
Asociación Venezolana de Ingeniería
Eléctrica y Mecánica 247, 311
Asociación Venezolana para el Avance de
la Ciencia (ASOVAC) 247, 311
ASOVAC, see Asociación Venezolana para
el Avance de la Ciencia
Associação Brasileira de Escolas
Superiores Católicas 39
Associação Brasileira de Odontologia 57,
269
Associação Brasileira de Psiquiatria 57,
269
Associação Brasileira de Química 55,
269
Associação de Educação Católica de Brasil
39
Associação Internacional de Lunologia 54,
269
Associação Médica Brasileira 57, 269
Association of Catholic Education in
Brazil 39
Association of Catholic Higher Education
in Brazil 39
Association of Essential Oil Producers
144
Association of Uruguayan Engineers 309
Astronomical Observatory – Argentina 14,
263
Astronomical Observatory – Uruguay
235, 310
Astronomical Observatory of Quito 127,
290
Ateneo de Clínica Quirúrgica 309
Athenaeum of Clinical Surgery 309
Atlantic University 83, 282

Atomic Energy Commission of Costa Rica
98, 285
Atucha Nuclear Centre 12
Augusto Malaret Hispano-American
Lexicographic Institute 224
Autonomous Institute of the National
Library and Library Services 247
Autonomous Metropolitan University
176, 300
Autonomous National University of
Honduras 157, 158, 295
Autonomous University of Aguascalientes
172, 299
Autonomous University of Baja California
173, 299
Autonomous University of Chapingo 173,
299
Autonomous University of Chiapas 173,
299
Autonomous University of Chihuahua
173, 299
Autonomous University of Coahuila 173,
299
Autonomous University of Guadalajara
174, 299
Autonomous University of Guerrero 175,
300
Autonomous University of Hidalgo 175,
300
Autonomous University of Juarez 175,
300
Autonomous University of Latin America
86, 283
Autonomous University of Queretaro 177,
301
Autonomous University of San Luis de
Potosí 178, 301
Autonomous University of Santo
Domingo 112-13, 114-15, 116,
117, 288
Autonomous University of Sinaloa 178,
301
Autonomous University of the
North-West 176, 300
Autonomous University of the State of
Mexico 172, 299
Autonomous University of the State of
Puebla 177, 301

Bacteriological Institute of Chile 70, 278
Balseiro Institute in Bariloche 12
Banco Nacional para el Desarrollo
(BNDE) 34
Bank of America 112

Bank of Guatemala 144
Baron Bliss Institute 23, 267
Barrow Experimental Farm 11, 261
Bayamón Central University 223, 308
Bayano State Cement Company 199
Belize College of Arts, Science and
Technology 23, 267
Bella Vista Research Centre 12, 261
Benito Juárez University of Oaxaca 177,
301
Bernardo O'Higgins Military School 67,
278
Biblioteca 'Alvaro Castro J.' 99
Biblioteca 'Carlos Monge Alfaro' de la
Universidad de Costa Rica 99, 284
Biblioteca Central de la Pontificia
Universidad Católica del Perú 221,
304
Biblioteca Central de la Pontificia
Universidad Javeriana 280
Biblioteca Central de la Universidad de
Chile 72, 277
Biblioteca Central de la Universidad
Mayor de 'San Andrés' 31, 267
Biblioteca Central de la Universidad
Mayor de 'San Simón' 31, 268
Biblioteca Central de la Universidad
Nacional Autónoma de México 171,
184, 296
Biblioteca Central de la Universidad
Nacional de Colombia 90, 280
Biblioteca Central de la Universidad
Nacional Mayor de San Marcos de
Lima 221, 304
Biblioteca Central, Instituto de
Investigaciones Agropecuarias – FAO
72
Biblioteca Central 'Rubén Martínez
Villena' de la Universidad de la
Habana 108-9, 285
Biblioteca Central, Universidade de
Brasília 56, 269
Biblioteca de la Asemblea Legislativa 99
Biblioteca de la Sociedad Amantes de la
Luz 117, 288
Biblioteca de la Sociedad Científica del
Paraguay 206, 303
Biblioteca de la Universidad Autónoma
de Santo Domingo 117, 288
Biblioteca de la Universidad Católica de
Puerto Rico 225
Biblioteca de la Universidad Nacional de
San Agustín 221, 304
Biblioteca del Congreso Nacional –
Bolivia 31

Biblioteca del Congreso Nacional – Chile 72, 277
Biblioteca del Ministerio de Relaciones Exteriores 99
Biblioteca del Servicio Geológico Nacional 14, 260
Biblioteca do Ministério das Relações Exteriores 56
Biblioteca General de la Universidad de Puerto Rico 224, 308
Biblioteca Madre María Teresa Guevara 225
Biblioteca Municipal Mario de Andrade 56
Biblioteca Nacional – Argentina 14, 260
Biblioteca Nacional – Brazil 56, 269
Biblioteca Nacional – Chile 72, 277
Biblioteca Nacional – Costa Rica 99, 284
Biblioteca Nacional – Colombia 90, 281
Biblioteca Nacional – Dominican Republic 117
Biblioteca Nacional – Ecuador 127
Biblioteca Nacional – El Salvador 136, 291
Biblioteca Nacional – Guatemala 146
Biblioteca Nacional – Honduras 158, 294
Biblioteca Nacional – Nicaragua 193
Biblioteca Nacional – Panama 201
Biblioteca Nacional – Peru 221, 304
Biblioteca Nacional – Uruguay 236, 309
Biblioteca Nacional – Venezuela 247, 311
Biblioteca Nacional de Antropología e Historia 'Dr Eugenio Davalos' 184
Biblioteca Nacional de México 171, 184, 296
Biblioteca Nacional 'José Martí' 108, 285
Biblioteca Regional de Medicina (BIREME) 236
Biblioteca 'Rubén Martínez Villena' 109, 285
Biblioteca Técnica Científica Centralizada 311
Biblioteca y Archivo Nacional – Paraguay 206
Bibliothèque Nationale d'Haiti 154, 294
Biomedical Research Institute – Dominican Republic 116
Birchner-Benner Institute 219, 305
BNDE, see Banco Nacional Para el Desarrollo
Board of Economic and Social Planning – Panama 196
Board of Industry – Panama 199
Board of Marine Resources – Panama 200
Board of Natural Resources – Panama 199

Board of Nutrition – Panama 200
Board of Regional Coordination – Panama 201
Board of Science and Technology – Bolivia 26, 28
Board of Standards and Technology – Bolivia 28
Bolivian Catholic University 29, 269
Bolivian Commission for Nuclear Energy 30, 268
Bolivian Geological Service 31, 268
Bolivian Institute of Agriculture and Livestock Technology 30, 268
Bolivian National Academy of Sciences 31, 267
Bolivian Oil Institute 30, 268
Bolivian Pontifical University 86, 283
Botany Institute – Cuba 108, 286
Brazilian Academy of Sciences 35, 55, 269
Brazilian Association of Chemistry 55, 269
Brazilian Centre for Educational Research 39
Brazilian Centre for Physics Research 39
Brazilian Coffee Institute 53, 271
Brazilian Geographical and Statistics Institute 34
Brazilian Institute for Education, Science and Culture 39
Brazilian Institute for Gastroenterological Study and Research 56, 270
Brazilian Institute of Forestry Development 53, 270
Brazilian Medical Association 54, 269
Brazilian Odontology Association 53, 269
Brazilian Petroleum Institute 52, 271
Brazilian Psychiatry Association 53, 269
Brazilian Society for Scientific Progress 55, 272
Bruynzeel 227
Buenos Aires Technological Institute 263
Butantan Institute 56, 271

Caldas University 83, 282
Camaguey University 107, 287
CAME, see Consejo de Ayuda Mutua Económica
Cancer Research Centre 53, 272
Carabobo University 243, 247, 312
Caribbean Committee for Development and Cooperation 117
Caribbean Food and Nutrition Institute (CFNI) 161, 295
Caribbean Institute of Anthropology and Sociology 246

Caribbean Regional Library 224, 308
CARICOM 148
CARIS, see Current Agricultural Research
  Information System
Carlos J. Finlay Historical Museum of
  Medical Sciences 109
Carlos Monge Alfaro Library of the Costa
  Rica University 99, 284
Carlos Pelligrini Higher School of
  Commerce 16, 264
Carnegie School of Home Economics 148
Casa de las Américas 108, 286
Catholic University of Chile 66, 72, 280
Catholic University of Córdoba 21, 266
Catholic University of Cuenca 124, 290
Catholic University of Cuyo 21, 266
Catholic University of La Plata 21, 266
Catholic University of Minas Gerais 44,
  274
Catholic University of Our Lady of the
  Assumption 205, 304
Catholic University of Paraná 46, 275
Catholic University of Pelotas 46, 275
Catholic University of Pernambuco 47,
  275
Catholic University of Petrópolis 47, 275
Catholic University of Puerto Rico 223,
  308
Catholic University of Salta 22, 267
Catholic University of Salvador 49, 276
Catholic University of Santa Fé 22, 267
Catholic University of Santiago de
  Guayaquil 125, 290
Catholic University of Santiago del Estero
  22, 267
CATIE, see Centro Agronómico Tropical
  de Investigación y Enseñanza
Cauca University 83, 282
Cayetano Heredia Private University 218,
  307
CCT, see Consejo Científico
  Tecnológico
CEAGANA, see Division de Ganadería y
  Boyada
CECT, see Comité Estatal de Ciencia y
  Técnica
CEDARE, see Centro de Datos y
  Documentación
CEMEC, see Centro para el
  Mejoramiento de la Enseñanza de las
  Ciencias
CENA, see Centro de Energia Nuclear na
  Agricultura
CENDA, see Centro de Desarrollo
  Agropecuario

CENDES, see Centro de Desarrollo
  Industrial del Ecuador and Centro de
  Estudios del Desarrollo
CENDOP, see Centro Nacional de
  Documentación sobre Población
CENTA, see Centro Nacional de
  Tecnología Agropecuaria
Central American and Panamanian
  Institute of Nutrition 146, 292
Central American Common Market 192
Central American Historical Institute 192
Central American Institute of Commercial
  Administration 192, 302
Central American Institute of Industrial
  Research and Technology (ICAITI)
  96, 97, 144, 292
Central American Institute of Public
  Administration 98
Central American School of Geography
  96
Central American University 192, 302
Central Bank – Dominican Republic 116
Central Library – Belize 23
Central Library of San Andrés University
  31, 267
Central Library of San Simón University
  31, 268
Central Library of the Greater National
  University of San Marcos de Lima
  221, 304
Central Library of the Institute of Land
  and Livestock Research 72
Central Library of the Javeriana
  Pontifical University 90, 280
Central Library of the National
  Autonomous University of Mexico
  171, 184, 296
Central Library of the National
  University of Colombia 90, 280
Central Library of the Pontifical Catholic
  University of Peru 221, 304
Central Library of the University of
  Brasília 56, 269
Central Library of the University of Chile
  72, 277
Central National University of the
  Province of Buenos Aires 16, 264
Central Nuclear de Atucha 12
Central Scientific and Technical Library
  311
Central University of Ecuador 124, 128,
  290
Central University of the East 115, 288
Central University of Venezuela 243, 247,
  312

Central-Western University 243, 312
Centre for Agricultural Research in
    Surinam 227, 308
Centre for Documentary Research of the
    National Institute of Industrial
    Technology 14, 260
Centre for Geographical Research of
    Brazil 39
Centre for Geotechnical Research 135,
    291
Centre for Higher Studies in Exact
    Sciences 260
Centre for Mining and Metallurgical
    Research 65
Centre for Research and Teaching of
    Tropical Agronomy 95, 97, 284
Centre for Research in Agriculture and
    Livestock Farming in Tropical Zones
    52, 270
Centre for Research into Scientific and
    Technological Resources (CIRCYT)
    183
Centre for the Development of National
    Research Capacity 200, 302
Centre for the Improvement of Science
    Teaching 96
Centre National d'Etudes Spatiales
    (CNES) 137, 138
Centre of Agricultural and Livestock
    Information and Documentation 106
Centre of Data and Documentation 201
Centre of Development Studies –
    Venezuela 242
Centre of Economic Development Studies
    89
Centre of Educative Studies 171, 296
Centre of Electronics and Computing
    (CCE) 86
Centre of Historical-Military Studies of
    Peru 220
Centre of Hydraulic Studies and
    Research 86
Centre of Hygiene and Educational
    Station for Tropical Diseases 182
Centre of Marine Research 94, 97
Centre of Measurement and Quality
    Certification 63, 68
Centre of Medical Research 16, 264
Centre of National Statistics and
    International Trade of Uruguay 236,
    309
Centre of Nuclear Energy in Agriculture
    54, 270
Centre of Pedagogic Documentation 72,
    277

Centre of Petroleum Technology 30
Centre of Population and Social Studies
    96
Centre of Research for the Mining and
    Metallurgical Industry 106, 286
Centre of Scientific Documentation 8
Centre of Scientific, Technical and
    Economic Documentation 236, 309
Centre of Technological Research 234
Centre of Urban and Regional Studies
    16, 260
Centre Spatial Guyanais 137, 291
Centre Technique Forestier Tropical 138,
    292
Centre Universitaire Antilles-Guyane 137,
    292
Centro Agronómico Tropical de
    Investigación y Enseñanza (CATIE)
    95, 97, 284
Centro Argentino de Información
    Científica y Tecnológica – UNESCO
    14, 260
Centro Brasileiro de Pesquisas
    Educacionais 39
Centro Brasileiro de Pesquisas Físicas 39
Centro Cultural Costarricense-
    Norteamericano 98, 99
Centro de Altos Estudios de Ciencias
    Exactas 260
Centro de Asistencia Técnica 115
Centro de Ciencia Tropical 97, 285
Centro de Datos y Documentación
    (CEDARE) 201
Centro de Desarrollo Agropecuario
    (CENDA) 113, 116
Centro de Desarrollo Industrial del
    Ecuador (CENDES) 130
Centro de Documentação e Informação da
    Câmara dos Deputados 56
Centro de Documentación Científica,
    Técnica y Económica 236, 309
Centro de Documentación e Información
    del Instituto Colombiano para el
    Fomento de la Educación Superior 90
Centro de Documentación Pedagógica –
    UNESCO 72, 277
Centro de Documentación Scientífica 8
Centro de Energia Nuclear na Agricultura
    (CENA) 54, 270
Centro de Estadísticas Nacionales y
    Comercio Internacional del Uruguay
    (CENCI) 236, 309
Centro de Estudio, Medición y
    Certificación de Calidad (CESMEC)
    63, 68

Centro de Estudios de la Realidad Social
Dominicana (CERESD) 113
Centro de Estudios del Desarrollo
(CENDES) 242
Centro de Estudios Educativos 171, 296
Centro de Estudios Historico–Militares
del Perú 220
Centro de Estudios Médicos 'Ricardo
Moreno Canas' 98, 285
Centro de Estudios sobre Desarrollo
Economico 89
Centro de Estudios Sociales y de
Población 96
Centro de Estudios Urbanos y Regionales
(CEUR) 16, 260
Centro de Estudios Venezolanos
Indígenas 246, 311
Centro de Higiene y Estación de
Adiestramiento en Enfermedades
Tropicales 182
Centro de Información Científica y
Humanística de la Coordinación para
la Investigación Científica (CICH) –
UNAM 184, 296
Centro de Información de las Naciones
Unidas – Argentina 14, 260
Centro de Información de las Naciones
Unidas – El Salvador 136
Centro de Información y Documentación
Agropecuaria (CIDA) 106
Centro de Investigación de Biología
Marina 11, 260
Centro de Investigación de Recursos
Científicas y Tecnológicos (CIRCYT)
183
Centro de Investigación Documentaria del
Instituto Nacional de Tecnología
Industrial – UNIDO 260
Centro de Investigación y de Estudios
Avanzados del Instituto Politécnico
Nacional 171, 296
Centro de Investigaciones 205, 304
Centro de Investigaciones Agrícolas
'Alberto Boerger' 234
Centro de Investigaciones Agrícolas de la
Nación 234
Centro de Investigaciones Bella Vista 12,
261
Centro de Investigaciones de Biología
Marina (CIBIMA) 113, 116
Centro de Investigaciones de Recursos
Naturales 10, 261
Centro de Investigaciones Geotécnicas
135, 291
Centro de Investigaciones Marinas 94, 97

Centro de Investigaciones Médicas 16,
264
Centro de Investigaciones para la
Industria Minero Metalúrgica 106,
286
Centro de Investigaciones Pesqueras 106,
286
Centro de Investigaciones Tecnológicas
234
Centro de Investigaciones Veterinarias
'Miguel C. Rubino' 234
Centro de Pesquisa Agropecuária do
Tropico Umido – EMBRAPA 52,
270
Centro de Pesquisas de Geografia do
Brasil 39
Centro de Pesquisas e Desenvolvimento
(CEPED) 52, 270
Centro de Pesquisas e Desenvolvimento
'Leopoldo A. Miguez de Mello' –
PETROBRAS 52, 270
Centro de Producción Ganadera 116
Centro de Servicios de Informacion
Agropecuarias 201
Centro de Tecnología Agrícola e
Alimentar – EMBRAPA 56, 270
Centro de Tecnología Petrolera 30
Centro Educativo Rural 19
Centro Espacial San Miguel –
Observatorio Nacional de Física
Cósmica 13, 261
Centro Espacial Vicente Lopez 13, 261
Centro Experimental de Ingeniería 199
Centro Interamericano de
Fotointerpretación 82
Centro Internacional del Mejoramiento de
Maíz y Trigo 181, 296
Centro Latinoamericano de Demografía
96
Centro Latinoamericano de Física 38
Centro Nacional de Ciencias y
Tecnologías Marinas 182, 296
Centro Nacional de Documentación
Científica y Tecnológica 31, 268
Centro Nacional de Documentación e
Información de Medicina y Ciencias
de la Salud (CEMDIN) 236
Centro Nacional de Documentación e
Información Educativa 31, 268
Centro Nacional de Estudios Nucleares
70
Centro Nacional de Evaluacion de
Tecnologías 239
Centro Nacional de Información de
Ciencias Médicas 107, 286

Centro Nacional de Información Educativa – UNESCO 14, 261
Centro Nacional de Información y Documentación (CENID) – UNESCO 72, 277
Centro Nacional de Investigaciones Agropecuarias (CNIA) – Dominican Republic 113, 116
Centro Nacional de Investigaciones Agropecuarias – Venezuela 245, 311
Centro Nacional de Investigaciones Científicas 106
Centro Nacional de Pesquisa de Mandioca e Fruticultura – EMBRAPA 52, 270
Centro Nacional de Radiación Cósmica 8
Centro Nacional de Tecnología Agropecuaria (CENTA) 135, 136, 291
Centro Nacional de Tecnología y Productividad Industrial (CNTPI) 234
Centro para el Desarrollo de la Capacidad Nacional de Investigación 200, 302
Centro para el Mejoramiento de la Enseñanza de las Ciencias (CEMEC) 96
Centro Paraguayo de Estudios de Desarrollo Económico y Social 206, 303
Centro Paraguayo de Ingenieros 206, 303
CEPA, see Comisiones Económicas para Africa
CEPAL, see Comisiones Económicas para America Latina
CEPED, see Centro de Pesquisas e Desenvolvimento
CERESD, see Centro de Estudios de la Realidad Social Dominicana
Cerro Tololo Interamerican Observatory 73, 279
CESMEC, see Centro de Estudio, Medición y Certificación de Calidad
CEUR, see Centro de Estudios Urbanos y Regionales
CFNI, see Caribbean Food and Nutrition Institute
Chacra Experimental de Barrow 11, 261
Charles Darwin Foundation 126
Charles Darwin Research Station 126, 289
Chemical and Pharmaceutical Association of Uruguay 309
Chemistry Society of Mexico 183, 298
Chihuahua Regional Institute of Technology 180, 298
Chile Institute 74, 278

Chilean Academy of History 74
Chilean Academy of Languages 74
Chilean Academy of Natural Sciences 74
Chilean Academy of Sciences 74
Chilean Antarctic Institute 65, 278
Chilean Association of Seismology and Anti-Seismological Engineering 74, 277
Chilean Biology Society 74, 279
Chilean Chemistry Society 74, 279
Chilean Commission for Nuclear Energy 70, 72, 277
Chilean Scientific Society 74, 279
CIBIMA, see Centro de Investigaciones de Biología Marina
CICH, see Centro de Información Científica y Humanística de la Coordinación para la Investigacion Científica
CIECC, see Consejo Interamericano para la Educación, Ciencia y Cultura
Ciencias y Letras de Nayarit 176
CIID-IDRC 201
CIRCYT, see Centro de Investigación de Recursos Científicos y Tecnológicos
Cítricos de Chiriqui 200
Citrus Fruits of Chiriqui 200
Claudio Gay Chilean Scientific Society 74, 279
Clemente Estable Institute of Biological Research 232, 310
CNCT, see Consejo Nacional de Ciencia y Tecnología – Cuba
CNEA, see Comisión Nacional de Energía Atómica
CNEN, see Comissão Nacional de Energia Nuclear
CNES, see Centre National d'Etudes Spatiales
CNI or CONI, see Consejo Nacional de Investigación
CNIA, see Centro Nacional de Investigaciones Agropecuarias
CNPq, see Conselho Nacional de Pesquisas and Conselho Nacional de Desenvolvimento Científico e Tecnológico
Coastal Institute for Veterinary Research 126
COLCIENCIAS, see Fondo Colombiano para Investigaciones Científicas
Colegio de Ingenieros de Guatemala 146
Colegio de la Purisima Concepción 174
Colegio de Médicos y Cirujanos de Nicaragua 193

Colegio del Estado 174
Colegio Mayor de Nuestra Señora del Rosario 87, 282
Colegio Nacional de Agricultura 'Luis A. Martínez' 126, 290
Colegio Nacional de Buenos Aires 16, 264
Colima University 174, 299
College of Arts, Science and Technology 160-1, 295
College of Mexico City 172
College of Our Lady of the Rosary 87, 282
Colombian Academy of Exact, Physical and Natural Sciences 89, 280
Colombian Agricultural and Livestock Institute 88, 281
Colombian Cancer Society 89, 281
Colombian Institute of Technical Standards 88, 281
Colombian Pathology Society 89, 282
Colombian Psychiatry Society 89, 282
Colombian Society of Chemists and Chemical Engineers 89, 282
Colombian Society of Engineers 89, 281
COMECON 102
Comisión Boliviana de Energía Nuclear 30, 268
Comisión Chilena de Energía Nuclear - IAEA 70, 72, 277
Comisión de Ciencia y Tecnología 238
Comisión de Ciencias del Mar y Pesquería 94, 97
Comisión de Desarrollo Científico 121
Comisión de Desarrollo Tecnológico 121
Comisión de Energía Atómica de Costa Rica 98, 285
Comisión de Investigaciones 196
Comisión de los Centros Atómicos de Constituyentes Eseiza 12
Comisión Ecuatoriana de Energía Atómica 123, 127, 289
Comisión Nacional de Energía Atómica (CNEA) - Argentina 9, 12, 261
Comisión Nacional de Energía Atómica - Bolivia 30
Comisión Nacional de Energía Atómica - Uruguay 235, 309
Comisión Nacional de Investigaciones Científicas y Tecnológicas (CONICYT) - Chile 62, 71, 72, 277
Comisión Nacional de Investigaciones Espaciales (CNIE) 13, 261
Comisión Nacional de Protección al Medio Ambiente y Conservación de los Recursos Naturales 102-3

Comisión Nacional para el Uso pacifico de la Energía Atómica 102
Comisión Panamena de Normas Industriales y Técnicas 199
Comisiones Económicas para Africa (CEPA) 15
Comisiones Económicas para America Latina (CEPAL) 15
Comisiones Sectoriales de Ciencia y Tecnología 120
Comissão Nacional de Energia Atómica 38
Comissão Nacional de Energia Nuclear (CNEN) 54, 270
Comité de Desarrollo y Cooperación del Caribe 117
Comité de Investigaciones Tecnológicas (INTEC-UNIDO) 72, 277
Comité Estatal de Ciencia y Técnica (CECT) 102, 104, 105
Comité Estatal de Normalización 102
Comité Nacional de Cooperación Técnica y Asistencia Económica 128, 129
Commission for Atomic Centres of the Eseiza 12
Commission for Geothermic Energy 65
Commission for Science and Technology 238
Commission for Scientific Development 121
Commission of Sea and Fisheries 94, 97
Commission of Technological Development 121
Committee of Technological Research (INTEC-UNIDO) 72, 277
COMPENDEX 71
CONACYT, see Consejo Nacional de Ciencia y Tecnología - Ecuador, Mexico
CONADE, see Consejo Nacional de Desarrollo
CONCYTEC, see Consejo Nacional de Ciencia y Tecnología - Peru
Confederation of Science and Technology 111, 112
CONICET, see Consejo Nacional de Investigación Científica y Técnica Argentina
CONICIT, see Consejo Nacional de Investigaciones Científicas y Tecnólogicas - Costa Rica, Venezuela
CONICYT, see Comisión Nacional de Investigaciones Científicas y Tecnológicas - Chile, Colombia, Uruguay

Consejo Andino de Ciencia y Tecnologia 129

Consejo de Ayuda Mutua Económica (CAME) 109

Consejo de Desarrollo Científico y Humanístico 242

Consejo de Investigaciones Científicas 160

Consejo de Rectores de las Universidades Chilenas 62, 278

Consejo Estatal del Azúcar 113

Consejo General de Universidades del Gobierno de México 174

Consejo Interamericano para la Educación, Ciencia y Cultura (CIECC) 15

Consejo Nacional de Alta Educación de Bolivia 28

Consejo Nacional de Ciencia y Tecnología - Bolivia 28

Consejo Nacional de Ciencia y Tecnología (CNCT) - Cuba 102

Consejo Nacional de Ciencia y Tecnología (CONACYT) - Ecuador 120-1, 122, 128, 129, 130

Consejo Nacional de Ciencia y Tecnología (CONACYT) - Mexico 38, 99, 165, 168, 170, 183, 184, 296

Consejo Nacional de Ciencia y Tecnología (CONCYTEC) - Peru 209

Consejo Nacional de Desarrollo (CONADE) 120

Consejo Nacional de Higiene 235, 309

Consejo Nacional de Investigación (CNI or CONI) 209, 210, 214-15

Consejo Nacional de Investigación Científica y Técnica (CONICET) - Argentina 8, 73, 261

Consejo Nacional de Investigaciones Agrícolas 245, 311

Consejo Nacional de Investigaciones Científicas - Haiti 154

Consejo Nacional de Investigaciones Científicas y Tecnólogicas (CONICIT) - Costa Rica 94, 95, 99

Consejo Nacional de Investigaciones Científicas y Tecnológicas (CONICIT) - Venezuela 238, 242

Consejo Nacional de Investigaciones Científicas y Tecnológicas (CONICIT) - Venezuela 238, 242

Consejo Nacional de Investigaciones Científicas y Tecnológicas (CONICYT) - Uruguay 229, 232, 309

Consejo Nacional de la Universidad Peruana 215, 304

Consejo Nacional para el Desarrollo de la Industria Nuclear 246, 311

Consejo Nacional para la Investigación Científica y Tecnológica de Costa Rica 38

Consejo Superior de Investigaciones Científicas (CSIC) 74

Consejo Venezolano de Normas Industriales (COVENIN) 238-9

Conselho Científico Tecnológico (CCT) 35

Conselho Nacional de Desenvolvimento Científico e Tecnológico (CNPq) 34, 35, 37, 38, 99, 130

Conselho de Reitores das Universidades Brasileiras 39

Conselho Nacional de Pesquisas (CNPq) 34, 74

Consultative Organization for Industrial Development 115

Contraloría General de la República 196, 197

Copper Corporation 65

Córdoba University 83, 282

CORFO, see Corporación de Fomento

Corporación Azucarera la Victoria 199

Corporación de Fomento Nacional (CORFO) 62, 63, 65, 68, 69

Corporation of National Public Works 62, 63, 65, 68, 69

Cosio Foundation 12, 262

Costa Rica University 94, 96, 97, 285

Costa Rican Academy of Geography and History 98

Costa Rican Academy of Periodontics 98

Costa Rican Institute of Political and Social Sciences 98

Costa Rican Language Academy 98

Costa Rican-North American Cultural Centre 98, 99

Costa Rican Paediatrics Association 98, 284

Costa Rican Surgery Association 98, 284

Council for Mutual Economic Aid 109

Council of Rectors of Chilean Universities 62, 278

Council of Rectors of the Brazilian Universities 38

Council of Scientific and Humanistic Research 242

Council of the Armed Forces for Research and Experimentation 13

Coussirat Araújo Regional Institute of Meteorology 55, 272

COVENIN, see Consejo Venezolano de
Normas Industriales
CTPD 15
Cuban Academy of Science 108
Cuban Institute of Mining and
Metallurgical Research 105, 286
Cuenca University 124, 290
Cuenta Mar del Plata 15, 129
Cupertino Arteaga Cancer Institute 31
Current Agricultural Research
Information System (CARIS) 71

Departamento Administrativo Nacional de
Estadísticas 82, 281
Departamento de Ciencia y Tecnología
140
Departamento de Ciencia y Tecnología de
la Dirección de Planeamiento de la
Universidad de la Republica 229
Departamento de Ciencia y Tecnología
de la División de Cooperación
Internacional de la Secretaría de
Planeamiento, Coordinación y
Difusión 229
Departamento de Cooperacion Cultural,
Científica y Tecnológica (DCT) 37
Departamento de Estudios Históricos
Navales 13, 261
Departamento de Fomento 62
Departamento de Información y
Documentación de CONICIT 99
Departamento de Investigaciones
Agropecuarias e Industriales 144
Departamento de Patentes 199
Departamento Efectos del Medio
Contaminado, Dirección Nacional de
Estudios y Proyectos - UNEP 14,
261
Departamento Nacional de Meteorologia
54, 270
Departamento Nacional de Produção
Mineral 38
Department for Measuring the Effects of
Pollution, National Direction of
Studies and Projects 14, 261
Department of Agricultural, Livestock
and Industrial Research - Guatemala
144
Department of Agriculture, Animal
Husbandry and Fisheries - Surinam
227
Department of Cultural, Scientific and
Technological Cooperation - Brazil
37-8

Department of Development - Chile 62
Department of Historical Naval Studies -
Argentina 13, 261
Department of Patents - Panama 199
Department of Science and Technology -
Guatemala 140
Department of Science and Technology of
the Directorate of Planning of the
University of the Republic 229
Department of Science and Technology of
the International Cooperation
Division of the Secretariat of
Planning, Coordination and Diffusion
229
Desert Region Research Institute 178
Deutsche Forschungsgemeinschaft (DFG)
74
Development University for the State of
Santa Catarina 50, 276
Dienst Lands Bosbeheer - Afdeling
Natuurbeheer 227, 308
Dirección de Asistencia Técnica
Internacional 62
Dirección de Cartografia Nacional del
Ministerio de Obras Públicas 242
Dirección de Ciencia y Tecnología -
Bolivia 26, 28
Dirección de Geología del Ministerio de
Energía y Minas 242
Dirección de Industrias 199
Dirección de Información y
Documentación 62
Dirección de los Programas de
Investigación del Ministerio de
Agricultura y Pesca 232
Dirección de Minas del Ministerio de
Minas e Hidrocarburos 242
Dirección de Normas y Tecnología 28
Dirección de Nutrición 200
Dirección de Planificación 62
Dirección de Planificación Económica y
Social 196
Dirección de Planificación y
Coordinación Regional 201
Dirección de Presupuesto de la Nación
196, 197
Dirección de Recursos Marinos 200
Dirección de Recursos Naturales 199
Dirección Ejecutiva 120
Dirección General de Estadística -
Guatemala 143
Dirección General de Estadística -
Mexico 170
Dirección General de Estadísticas y
Censos - Costa Rica 95

Dirección General de Estadísticas y
Censos - Uruguay 232
Dirección General de Estadísticas y
Censos - Venezuela 242
Dirección General de Geología, Minas y
Petróleo 95, 285
Dirección General de Geología y Minas
123, 289
Dirección General de Hidrocarburos 123,
289
Dirección General de Institutos
Tecnológicos Regionales SEP 180
Dirección General de Meteorología del
Perú 220, 304
Dirección General de Recursos Naturales
Renovables - El Salvador 135
Dirección General de Recursos Naturales
Renovables - Guatemala 145
Dirección General de Sanidad y
Reproducción Animal 135
Dirección General de Servicios de Salud
145
Dirección General del Servicio
Meteorológico Nacional 183, 296
Dirección Nacional de Agroindustries 199
Dirección Nacional de Meteorología del
Uruguay 235, 309
Directorate-General of Geology, Mining
and Petroleum 95, 285
Directorate of Information and
Documentation - Chile 62
Directorate of International Technical
Assistance - Chile 62
Directorate of Planning - Chile 62
Directorate of Research Programmes of
the Ministry of Agriculture and
Fisheries - Uruguay 232
Division de Ganadería y Boyada
(CEAGANA) 113, 116
Division of Scientific and Technological
Policy - Panama 195-6
Dr Eugenio Davalos National Library of
Anthropology and History 184
Documentation and Information Centre of
the Chamber of Deputies 56
Documentation and Information Centre of
the Colombian Institute for the
Development of Higher Education 90
Dominican Cultural Institute 117, 288
Dominican History Academy 117
Dominican Institute for Industrial
Technology 112, 116
Dominican Language Academy 117
Dominican Medical Association 117, 288
Dominican Sugar Institute 116, 288

Durango Regional Institute of
Technology 180, 298

Ecole Polytechnique d'Haiti 153, 294
ECOM, see Empresa Nacional de
Computación e Información
Economic System of Latin America 129
Ecuador National Library 127, 128
Ecuadorean Academy of Medicine 127,
289
Ecuadorean Commission for Atomic
Energy 123, 127, 289
Ecuadorean Institute of Natural Sciences
127, 289
Edson Queiroz Educational Foundation
42
EEC, see European Economic
Commission
El Vergel School of Agriculture 67, 278
EMBRAPA 52, 53, 56, 270
Emprêsa de Pesquisa Agropecuária da
Bahia 53, 270
Emprêsa de Pesquisa Agropecuária de
Minas Gerais (EPAMIG) 53, 270
Empresa Estatal de Cemento Bayano 199
Empresa Nacional de Computación e
Información (ECOM) 71
Empresa Nacional de Maquinaria 199
Empresa Nacional de Semillas 199
Emprêsa Pernambucana de Pesquisa
Agropecuária 53
Endocrinology and Metabolism Centre
12, 264
Engineering Institute of São Paulo 55,
271
Engombe Experimental Farm 116
ENTEL, see National Firm of
Telecommunications
Enterprise for Agricultural and Livestock
Research for the State of Bahia 53,
270
Enterprise for Agricultural and Livestock
Research for the State of Minas
Gerais 53, 270
Escuela Agrícola 'El Vergel' 67, 278
Escuela Agrícola Panamericana 157, 294
Escuela Centroamericana de Geografía 96
Escuela de Administración y Finanzas y
Tecnologías 282
Escuela de Ingeniería Aeronáutica 19, 265
Escuela Graduado 'Joaquin V. González'
18
Escuela Interamericana de
Bibliotecnologia 83

Escuela Militar 'General Bernardo
   O'Higgins' 67, 278
Escuela Nacional Agrícola 136, 291
Escuela Nacional de Agricultura 173
Escuela Nacional de Agricultura y
   Ganadería 192, 302
Escuela Nacional de Artes Plasticas 199
Escuela Nacional de Educación Técnica
   No 1 261
Escuela Nacional de Educación Técnica
   No 4 261
Escuela Nacional de Música 157
Escuela Naútica de Panama 199
Escuela Politécnica Nacional 125, 127, 290
Escuela Practica de Agricultura y
   Ganadería 'María Cruz y Manuel L.
   Inchausti' 18
Escuela Superior de Administración
   Pública 85
Escuela Superior de Comercio de Carlos
   Pellegrin 16, 264
Escuela Superior Politécnica de
   Chimborazo 126, 290
Escuela Superior Politécnica del Litoral
   126, 290
Estación Altoandina de Biología y Reserva
   Zoo-Botánica de Checayani 219, 305
Estación de Investigación Charles Darwin
   126, 289
Estación de Investigaciones Marinas de
   Margarita 245
Estación Experimental Agrícola del Norte
   219, 305
Estación Experimental Agro-Industrial 11,
   261
Estación Experimental Agropecuaria de
   Salta – INTA 10, 262
Estación Experimental Agropecuaria de
   Tulumayo 219, 305
Estación Experimental Agropecuaria
   Mendoza – INTA 11, 262
Estacion Experimental de Café 245, 312
Estación Experimental de Pastos y
   Forrajes 'Indio Hatuey' 106
Estacion Experimental 'Las Vegas' de la
   Sociedad Nacional de Agricultura 69,
   278
Estación Experimental Regional
   Agropecuaria – INTA 11, 262
Estación Hidrobiológica 12, 262
Estación Meteorológica 246, 312
European Economic Commission (EEC)
   148, 160
European Observatory for the Southern
   Hemisphere 72, 279

European Organization for the
   Astronomical Research of the
   Southern Hemisphere 72
European Space Agency 137
Evandro Chagas Institute 57, 271
Executive Directorate – Ecuador 120
Experimental Agricultural and Livestock
   Research Institute of the Northeast
   53, 271
Experimental Chemistry and Biology
   Institute – Cuba 108, 287
Experimental Coffee Station 245, 312
Experimental Engineering Centre 199
Experimental Institute of Primary
   Education No. 1 215, 305
Externado University of Colombia 86

Facultad de Ciencias Aplicadas a la
   Industría 262
Faculty of Applied Science for Industry
   262
FAO, see Food and Agriculture
   Organization
Federación Médica Peruana 220, 305
Federal University of Acre 39, 272
Federal University of Alagoas 39, 272
Federal University of Bahia 40, 273
Federal University of Ceará 41, 273
Federal University of Espírito Santo 42,
   273
Federal University of Fluminense 42, 273
Federal University of Goiás 42, 273
Federal University of Maranhão 43, 274
Federal University of Minas Gerais 44,
   274
Federal University of Ouro Prêto 45, 274
Federal University of Pará 45, 274
Federal University of Paraíba 38, 45, 274
Federal University of Paraná 46, 275
Federal University of Pelotas 47, 275
Federal University of Pernambuco 47,
   275
Federal University of Piauí 47, 275
Federal University of Rio de Janeiro 38,
   49, 276
Federal University of Rio Grande do
   Norte 48, 275
Federal University of Rio Grande do Sul
   48, 276
Federal University of Santa Catarina 50,
   276
Federal University of Santa Maria 50, 276
Federal University of São Carlos 50, 276
Federal University of Sergipe 51, 277

Federal University of the Mato Grosso 44, 274
Federal University of Uberlândia 51, 277
Federal University of Viçosa 51, 277
Federico Santa María Technical University 67, 280
Federico Villarreal National University 216, 306
FEMCIECC, see Fondo Especial Multilateral de Consejo Interamericano para la Educacion, la Ciencia y la Cultivo
Fernando Aguado Rico Electronic Technology Institute 105, 287
Finca Experimental de Engombe 116
FINEP, see Agencia para el Financiamiento de Estudios y Proyectos
Fisheries Research Centre 106, 286
Fletcher College 23, 267
Fondo Colombiano para Investigaciones Científicas (COLCIENCIAS) 77, 81, 88, 90
Fondo Especial Multilateral de Consejo Interamericano para la Educacion, la Ciencia y la Cultivo (FEMCIECC) 129
Fondo Nacional de Ciencia y Tecnología 129
Food and Agriculture Organization (FAO) 15, 31, 72, 73, 146, 184, 223
Forestry Institute 69, 278
Foundation Centre for Research and Studies 39
Francisco de Miranda Experimental University 244, 313
Francisco de Paula Santander University 83, 282
Francisco José de Caldas University 84, 282
Francisco Marroquin University 144, 293
Fredericks School of Home Economics 148
Free University of Colombia 86, 284
Friedrich Ebert Foundation 127
Fruit and Citrus Research Institute 138, 292
Fundação Centro de Pesquisas e Estudos 39
Fundação Educacional Edson Queiroz 42
Fundação Instituto Tecnológico de Estado de Pernambuco 52
Fundação Norte Mineira de Ensino Superior 45, 274

Fundação Universidade do Rio Grande 48, 275
Fundación Cosio 12, 262
Fundación Gildemeister 70, 278
Fundación La Salle de Ciencias Naturales 245, 312
Fundación 'Lisandro Alvarado' 246
Fundación Universidad de Bogotá 'Jorge Tadeo Lozano' 86, 283

Gabriel René Moreno University 29, 269
Gama Filho University 42, 273
General Board of Educational, Scientific and Cultural Relations 170, 296
General Board of Health Services 145
General Board of Meteorology of Peru 220, 304
General Board of Regional Institutes of Technology 180
General Board of Statistics 170
General Board of the National Meteorological Service 183, 296
General Controllership of the Republic 196, 197
General Council of Universities of the Government of Mexico 174
General Directorate of Geology and Mines 123, 289
General Directorate of Hydrocarbons 123, 289
General Directorate of Statistics and Census 232
General Documentation Centre of the Navy 56, 272
General Office of Renewable Natural Resources 145
General Secretariat of the National Council of Economic Planning 140-1, 142
GEOBOL, see Geological Service
Geographical Information System (GIS) 25
Geography Institute – Cuba 105
Geological Institute of Uruguay 310
Geological Mining Service 227, 308
Geological Service (GEOBOL) 24
Geologisch Mijnbouwkundige Dienst 227, 308
Geology and Palaeontology Institute – Cuba 105
Geology Office of the Ministry of Energy and Mines 242
Geophysics and Astronomy Institute – Cuba 107, 286

Geophysics Institute of Peru 215, 305
Georgetown Hospital Compound 150
Georgetown Technical Institute 148
German Research Society 74
Gildemeister Foundation 70, 278
Gonçalo Moniz Central Laboratory 57, 272
Gorgas Memorial Laboratory of Tropical and Preventive Medicine 200, 303
Gran Colombia University 86, 283
Greater National University of San Marcos de Lima 217, 307
Grupo Nacional de Radiología 107, 286
Guadalajara University 174, 299
Guanajuato University 174, 300
Guatemala Academy of Medicine, Physics and Natural Sciences 146, 292
Guatemalan College of Engineers 146
Guatemalan National Library 146
Guatemalan Paediatrics Association 146
Guiana Space Centre 137, 291
Gulf Oil Corporation 226
Guyabau Technical Training Complex 150, 293
Guyana Industrial Training Centre 148
Guyana Medical Science Library 150, 151, 293
Guyana Museum 150, 293
Guyana School of Agriculture Corporation 150, 294
Guyana Society 150, 293
Guyana Sugar Corporation Limited 150, 293
Guyana Zoo 150, 293
Guyanan Institute of International Affairs 150, 293

Haematology Institute, National Health Institute 12, 262
Haiti State University 153, 294
Harvard University 192
Havana University 106, 108-9, 287
Head Office for Statistics and Census – Costa Rica 95
Hermilio Valdizán National University of Huánuco 216, 306
High Andes Biology Station and Zoo-botanical Reserve of Checayani 219, 305
Higher Council of Scientific Research – Spain 74
Higher Institute of Agriculture 115, 288
Higher Institute of Business Studies and Technology 218

Higher Institute of Sciences 263
Higher Polytechnic of Chimborazo 126, 290
Higher Polytechnic of the Coast 126, 290
Higher School of Public Administration 85
Higher Technical Institute of Haiti 153, 294
Historical Archives and Libraries 184, 295
Honduran Academy 157
Honduran Institute of Inter-American Culture 158
Honduras National Library 158, 294
Hydro-biological Station 12, 262
Hydrographical Institute of the Navy 70

IBDF, see Instituto Brasiléiro de Desenvolvimento Forestal
Ibero-American University 175, 300
IBGE, see Instituto Brasileño de Geografía y Estadística
IBTA, see Instituto Boliviano de Tecnología Agropecuaria
ICAITI, see Instituto Centroamericano de Investigación y Tecnología Industrial
ICAP, see Instituto Centroamericano de Administración Pública
ICONTEC, see Instituto Colombiano de Normas Técnicas
ICSOPRU 15
IDIAP, see Instituto de Investigaciones Agropecuarias
IDIEM, see Instituto de Investigaciones y Ensayo de Materiales
IFARHU, see Instituto para la Formación y Aprovechamiento de Recursos Humanos
IIAE, see Instituto de Investigación Aeronáutica y Espacial
IICA, see Instituto Interamericano de Ciencias Agrícolas
ILDIS, see Instituto Latinoamericano de Investigaciones Sociales
ILO 31
IMF, see International Monetary Fund
IMIT, see Instituto Mexicano de Investigaciones Tecnológicas AC
INACAP, see Instituto Nacional de Capacitación Professional
INCAE, see Instituto Centroamericano de Administración de Empresas
INDAF, see Instituto Nacional de Desarrollo y Aprovechamiento Forestales

Indio Hatuey Pasture and Forage
  Experimental Station 106
INDITECNOR, see Instituto Nacional de
  Investigaciones Tecnológicas y
  Normalización
INDOTEC, see Instituto Dominicano de
  Tecnología Industrial
Industrial Development Centre 130
Industrial Trade Union of Panama 200
INEC, see Instituto Nacional de
  Estadística y Censos
INFONAC, see Instituto de Fomento
  Nacional
Información, Innovación y Tecnología
  (INFOTEC) 183
Information and Documentation
  Department of CONICIT 99
Information Centre of the United Nations
  14, 260
Information Service and Technological
  Extension 200
INFOTEC, see Información, Innovación y
  Tecnología
INFOTERRA, see International Referal
  System Sources Environmental
  Information
Ing Eduardo Terra Arocena Geological
  Institute 232
INIA, see Instituto de Investigaciones
  Agropecuarias
INPE, see Instituto de Pesquisas
  Espaciais
Institut de Recherches Agronomiques
  Tropicales et des Cultures Vivrières
  138, 292
Institut de Recherches sur les Fruits et
  Agrumes 138, 292
Institut Pasteur 138, 292
Institut Supérieur Technique d'Haiti 153,
  294
Institute for Training and Improvement
  of Human Resources 196, 201
Institute of Aeronautic and Space
  Research 13, 262
Institute of Agricultural and Livestock
  Research - Chile 63, 68, 278
Institute of Agricultural and Livestock
  Research - Panama 199, 201
Institute of Agricultural Research and
  Development of the Caribbean 160
Institute of Agricultural Science and
  Technology - Guatemala 145
Institute of Agricultural Sciences - Cuba
  106
Institute of Agronomy - Brazil 53, 270

Institute of Allergy Research - Peru 219,
  305
Institute of Andean Biology 217, 219, 305
Institute of Animal Biology - Bolivia 30
Institute of Animal Health and
  Reproduction - El Salvador 135
Institute of Animal Science - Cuba 106
Institute of Applied and Structural
  Mechanics - Argentina 10, 263
Institute of Astronomy - Mexico 183, 296
Institute of Atomic Energy 54, 271
Institute of Biology and Experimental
  Medicine - Argentina 12, 262
Institute of Biomedical Studies -
  Dominican Republic 115
Institute of Communication Sciences -
  Peru 220, 305
Institute of Development Studies -
  Guyana 150, 294
Institute of Economic and Social
  Development - Argentina 16, 262
Institute of Engineers and Architects of
  Honduras 157, 294
Institute of Experimental Medicine -
  Venezuela 246, 312
Institute of Experimental Medicine of the
  National Health Service 70, 279
Institute of Fisheries Development 69,
  278
Institute of Fundamental Research on the
  Brain - Cuba 107, 286
Institute of Fundamental Technical
  Research - Cuba 105
Institute of Geological Research - Chile
  65, 279
Institute of Geology - Brazil 55, 271
Institute of Geology, Mining and
  Metallurgy - Peru 215, 305
Institute of Health Research 94, 97
Institute of Industrial Promotion -
  Colombia 88
Institute of Jamaica 161, 295
Institute of Librarianship - Argentina 16,
  264
Institute of Mathematics - Mexico 171
Institute of Medical Research - Argentina
  12, 262
Institute of Microbiology and Arable and
  Livestock Farming - Argentina 10,
  263
Institute of Military Cartography of the
  Armed Forces - Dominican Republic
  117, 288
Institute of Natural Resources - Chile 71
Institute of Nuclear Affairs 89, 281

Institute of Occupational Hygiene and Environmental Contamination 72, 278

Institute of Public Health and Tropical Diseases - Mexico 182, 296

Institute of Radio-astronomy 8

Institute of Regional and Urban Planning - Argentina 16, 263

Institute of Renewable Natural Resources - El Salvador 135

Institute of Research and Examination of Materials 68, 279

Institute of Research into Sugar Cane By-products - Cuba 106

Institute of Scientific and Technical Documentation and Information - Cuba 108, 286

Institute of Sea Fishing - Venezuela 245

Institute of Soil and Agrotechnology - Argentina 11, 263

Institute of Space Research - Brazil 54, 271

Institute of Studies for National Security 70

Institute of Technological Research - Brazil 52, 271

Institute of Technological Research - Chile 68, 71

Institute of Technological Research - Colombia 88, 281

Institute of Technological Research and Technical Standards - Peru 219

Institute of the Armed Forces for Scientific and Technical Research 13

Institute of Tropical Forestry 223, 308

Institute of Zoonosis and Livestock Research 219, 305

Instituto 'Adolfo Lutz' 56, 270

Instituto Agrario Argentino de Cultura Rural 11, 262

Instituto Agrario de Estudios Económicos (INTAGRO) 69, 278

Instituto Agrario de Nicaragua 193, 302

Instituto Agrario Nacional 245, 312

Instituto Agronómico 53, 270

Instituto Antártico Chileno 65, 278

Instituto Argentino de Racionalización de Materiales (IARM) 10, 262

Instituto Artigas del Servicio Exterior 232

Instituto Autónomo de la Biblioteca Nacional y Servicios de Bibliotecas 247

Instituto Azucarero Dominicano 116, 288

Instituto Bacteriológico de Chile 70, 278

Instituto Balseiro en Bariloche 12

Instituto Bibliotecnológico 16, 264

Instituto Birchner-Benner 219, 305

Instituto Boliviano de Tecnología Agropecuaria (IBTA) 30, 268

Instituto Boliviano del Petróleo (IBP) 30, 268

Instituto Brasileiro de Desenvolvimento Forestal (IBDF) 53, 270

Instituto Brasileiro de Educação, Ciência e Cultura 39

Instituto Brasileiro de Estudos e Pesquisas de Gastroenterologia 56, 270

Instituto Brasileiro de Petróleo 52, 271

Instituto Brasileiro do Café 53, 271

Instituto Brasileiro de Geografía y Estadística (IBGE) 34

Instituto Butantan 56, 271

Instituto Caribe de Antropología y Sociología 246

Instituto Cartográfico Militar de las Fuerzas Armadas 117, 288

Instituto Centroamericano de Administración de Empresas (INCAE) 192, 302

Instituto Centroamericano de Administración Pública (ICAP) 98

Instituto Centroamericano de Investigación y Tecnología Industrial (ICAITI) 96, 97, 144, 292

Instituto Chile (Composed of 6 national academies) 74, 278

Instituto Científico de Lebu 69, 278

Instituto Científico y Literario 175

Instituto Colombiano Agropecuario (ICA) 88, 281

Instituto Colombiano de Normas Técnicas (INCONTEC) 88, 281

Instituto Costarricense de Ciencias Políticas y Sociales 98

Instituto Cubano de Investigaciones Mineras y Metalúrgicas 105, 286

Instituto de Astronomía 183, 296

Instituto de Asuntos Nucleares 89, 281

Instituto de Biología Andina 217, 219, 305

Instituto de Biología Animal 30

Instituto de Biología y Medicina Experimental 12, 262

Instituto de Botánica 108, 286

Instituto de Cancerología 'Cupertino Arteaga' 31

Instituto de Ciencia Animal 106

Instituto de Ciencia y Tecnología Agricola 145

Instituto de Ciencias Agrícolas 106
Instituto de Ciencias de la Comunicación
  220, 305
Instituto de Ciencias Nucleares 127, 289
Instituto de Cultura Dominicano 117,
  288
Instituto de Desarrollo Económico y
  Social 16, 262
Instituto de Documentación e
  Información Científica y Técnica 108,
  286
Instituto de Endocrinología 'Profesor Dr
  Juan C. Mussio Fournier' 234, 310
Instituto de Energia Atómica 54, 271
Instituto de Engenharia de São Paulo 55,
  271
Instituto de Engenharia Nuclear 54, 271
Instituto de Enseñanza de Meccánica y
  Electrónica 'Profesor Dr José F. Arias'
  233
Instituto de Estudios Bio-médicos 115, 116
Instituto de Estudios de la Seguridad
  Nacional 70
Instituto de Fomento Industrial 88
Instituto de Fomento Nacional
  (INFONAC) 187
Instituto de Fomento Pesquero 69, 278
Instituto de Geofísica y Astronomía 107,
  286
Instituto de Geografía 105
Instituto de Geología y Paleontología 105
Instituto de Hematología, Instituto
  Nacional de la Salud 12, 262
Instituto de Higiene del Trabajo y
  Contaminación Atmosférica 72, 278
Instituto de Ingenieros y Arquitectos de
  Honduras 157, 294
Instituto de Investigación Aeronáutica y
  Espacial (IIAE) 13, 262
Instituto de Investigación de Recursos
  Naturales 65, 279
Instituto de Investigación de Zonas
  Desérticas 178
Instituto de Investigación Social 'Juan
  XXIII' 192
Instituto de Investigación Técnica
  Fundamental 105
Instituto de Investigaciones Agropecuarias
  (INIA) – Chile 63, 68, 278
Instituto de Investigaciones Agropecuarias
  (IDIAP) – Panama 199, 201
Instituto de Investigaciones Alérgicas
  219, 305
Instituto de Investigaciones Biológicas
  'Clemente Estable' 232, 310

Instituto de Investigaciones Científicas y
  Técnicas de la FFAA 13
Instituto de Investigaciones de la Caña de
  Azúcar 106, 286
Instituto de Investigaciones de los
  Derivados de la Caña de Azúcar 106
Instituto de Investigaciones en Salud 94,
  97
Instituto de Investigaciones Fundamentales
  del Cerebro 107, 286
Instituto de Investigaciones Geológicas
  (IIG) 65, 279
Instituto de Investigaciones Marinas de
  Punta Betín 88, 281
Instituto de Investigaciones Médicas 12,
  262
Instituto de Investigaciones Nucleares
  107, 286
Instituto de Investigaciones Tecnológicas
  (INTEC) – Chile 68, 71
Instituto de Investigaciones Tecnológicas –
  Colombia 88, 281
Instituto de Investigaciones Tecnológicas
  y Normas Técnicas 219
Instituto de Investigaciones Veterinarias
  del Litoral 126
Instituto de Investigaciones y Ensayo de
  Materiales (IDIEM) 68, 279
Instituto de Lexicografía
  Hispanoamericiano Augusto Malaret
  224
Instituto de Matemática, Cibernética y
  Computación 108
Instituto de Matemáticas 171
Instituto de Mecánica Aplicada y
  Estructuras 10, 263
Instituto de Medicina Experimental 246,
  312
Instituto de Medicina Experimental del
  SNS 70, 279
Instituto de Meteorología 108, 286
Instituto de Microbiología e Industrias
  Agropecuarias – INTA 10, 263
Instituto de Nutrición de Centroamerica y
  Panama 146, 292
Instituto de Nutrición y Alimentos del
  Caribe 161, 295
Instituto de Oncología 310
Instituto de Oncología y Radiobiología de
  la Habana 107, 286
Instituto de Pesquisas do Experimentação
  Agropecuário do Nordeste (IPEANE)
  53, 271
Instituto de Pesquisas Espaciais (INPE)
  54, 271

Instituto de Pesquisas Radioativas 54,
    271
Instituto de Pesquisas Tecnológicas 52,
    271
Instituto de Planeamiento Regional y
    Urbano (IPRU) 16, 263
Instituto de Química y Biología
    Experimental 108, 287
Instituto de Radioastronomía 8
Instituto de Recursos Naturales (IREN)
    71
Instituto de Salubridad y Enfermedades
    Tropicales 182, 296
Instituto de Suelos y Agrotécnica 11, 263
Instituto de Zoología 108, 287
Instituto de Zoonosis e Investigación
    Pecuaria 219, 305
Instituto del Mar del Perú 219, 305
Instituto Dominicano de Tecnología
    Industrial (INDOTEC) - 112, 116
Instituto Ecuatoriano de Ciencias
    Naturales 127, 289
Instituto Evandro Chagas 57, 271
Instituto Experimental de Educación
    Primaria No 1 215, 305
Instituto Florestal 69, 278
Instituto Florestal - Estado de São Paulo
    53, 271
Instituto Geofísico del Peru 215, 305
Instituto Geográfico 'Agustín Codazzi' 82,
    281
Instituto Geográfico Militar - Argentina
    13, 262
Instituto Geográfico Militar - Chile 71,
    278
Instituto Geográfico Militar - Paraguay
    206, 303
Instituto Geográfico Militar - Peru 220,
    305
Instituto Geográfico Militar - Uruguay
    235, 310
Instituto Geográfico Militar y de Catastro
    Nacional 28, 268
Instituto Geográfico Nacional - Costa
    Rica 95, 285
Instituto Geográfico Nacional - El
    Salvador 135, 291
Instituto Geográfico Nacional -
    Guatemala 143, 292
Instituto Geográfico Nacional (IGN) -
    Honduras 157, 294
Instituto Geológico - Brazil 55, 271
Instituto Geológico del Uruguay 310
Instituto Geologico 'Ing Eduardo Terra
    Arocena' 232

Instituto Geológico, Minero y
    Metalúrgico - Peru 215, 305
Instituto Hidrográfico de la Armada 70
Instituto Histórico Centroamericano 192
Instituto Hondureño de Cultura
    Interamericana 158
Instituto Interamericano de Agricultura
    Experimental 126, 289
Instituto Interamericano de Ciencias
    Agrícolas - El Salvador 135
Instituto Interamericano de Ciencias
    Agrícolas - Guatemala 145
Instituto Interamericano de Ciencias
    Agrícolas de la OEA (IICA) 97
Instituto Interamericano de la OEA 95
Instituto Interamericano del Niño 235,
    310
Instituto Latinoamericana de
    Investigaciones Científicas en
    Educación a Distancia 242
Instituto Latinoamericano de
    Investigaciones Sociales (ILDIS) 127
Instituto Medico Sucre 31
Instituto Meteorológico Nacional 98, 285
Instituto Mexicano de Investigaciones
    Tecnológicas AC (IMIT) 180-1, 296
Instituto Mexicano de Recursos
    Renovables 181, 297
Instituto Mexicano del Petróleo 181, 297
Instituto Miles de Terapéutica
    Experimental 182, 297
Instituto Nacional de Administración,
    Gestión y Altos Estudios
    Internacionales 153
Instituto Nacional de Astrofísica, Optica
    y Electrónica 183, 297
Instituto Nacional de Cancerología 281
Instituto Nacional de Capacitación
    Profesional (INACAP) 68
Instituto Nacional de Cardiología 182,
    297
Instituto Nacional de Carnes 233
Instituto Nacional de Desarrollo Pesquero
    11, 263
Instituto Nacional de Desarrollo y
    Aprovechamiento Forestales
    (INDAF) 106, 287
Instituto Nacional de Energía Nuclear -
    Guatemala 145, 292
Instituto Nacional de Energía Nuclear -
    Mexico 182
Instituto Nacional de Estadística - Bolivia
    28, 268
Instituto Nacional de Estadística y Censos
    - Argentina 14, 263

Instituto Nacional de Estadística y Censos
(INEC) - Ecuador 123, 128, 289
Instituto Nacional de Estudios del Teatro
16
Instituto Nacional de Estudos
Pedagógicos 39
Instituto Nacional de Física - Peru 210
Instituto Nacional de Geología y Minería
9
Instituto Nacional de Higiene - Cuba
107, 287
Instituto Nacional de Higiene - Mexico
182, 297
Instituto Nacional de Higiene 'Izquieta
Pérez' 126, 289
Instituto Nacional de Investigación
Agraria 219, 305
Instituto Nacional de Investigación
Matemática 210
Instituto Nacional de Investigaciones
Agrícolas 181, 297
Instituto Nacional de Investigaciones
Agropecuarias 126
Instituto Nacional de Investigaciones
Científicas 205, 303
Instituto Nacional de Investigaciones
Forestales 181, 297
Instituto Nacional de Investigaciones
Geológico Mineras 82, 281
Instituto Nacional de Investigaciones
Nucleares 182, 297
Instituto Nacional de Investigaciones
Pecuarias 181, 297
Instituto Nacional de Investigaciones
Tecnológicas y Normalización
(INDITECNOR) 68, 279
Instituto Nacional de la Reforma Agraria
(INRA) 106, 287
Instituto Nacional de Limnología 12, 263
Instituto Nacional de Medicina Legal 88,
281
Instituto Nacional de Meteorología e
Hidrografía 127, 289
Instituto Nacional de Microbiología 12,
263
Instituto Nacional de Neumología 182,
297
Instituto Nacional de Nutrición - Ecuador
126, 289
Instituto Nacional de Nutrición -
Venezuela 246, 312
Instituto Nacional de Parasitología 205,
206, 303
Instituto Nacional de Pesca - Ecuador
126, 289

Instituto Nacional de Pesca - Mexico 182,
297
Instituto Nacional de Pesca - Uruguay
234, 310
Instituto Nacional de Recursos Naturales
- Colombia 82
Instituto Nacional de Salud (INPES) -
Colombia 89, 281
Instituto Nacional de Salud - Peru 219,
306
Instituto Nacional de Sismología 8
Instituto Nacional de Sismología,
Vulcanología, Meteorología e
Hidrología 145, 293
Instituto Nacional de Sistemas
Automatizados y Técnicas de
Computación 102
Instituto Nacional de Tecnología 52, 272
Instituto Nacional de Tecnología
Agropecuaria (INTA) 9, 10, 11, 263
Instituto Nacional de Tecnología
Industrial (INTI) 9, 10, 263
Instituto Nacional de Vitivinicultura 11,
263
Instituto 'Nami Jafet' para o Progresso da
Ciência e Cultura 55, 271
Instituto Nicaraguense de Cine 192, 302
Instituto Nicaraguense de Tecnología
Agropecuaria (INTA) 192, 193
Instituto Oceanográfico de la Armada 124,
290
Instituto 'Oscar Freire' 57, 272
Instituto 'Oswaldo Cruz' 57, 272
Instituto para la Formación y
Aprovechamiento de Recursos
Humanos (IFARHU) 196, 201
Instituto Peruano de Energía Nuclear
220, 306
Instituto Peruano de la Ciencia 210
Instituto Politécnico 199
Instituto Politécnico Nacional -
Universidad Técnica 179, 298
Instituto Regional de Meteorología
'Coussirat Araújo' 55, 272
Instituto Salvadoreño de Fomento
Industrial y el Centro Nacional de
Productividad 135
Instituto Salvadoreño de Investigaciones
del Café 135, 291
Instituto Salvadoreño de Tecnología
Industrial 135
Instituto Superior de Administración y
Tecnología 218
Instituto Superior de Agricultura (ISA)
115, 288

Instituto Superior de Ciencias 263
Instituto Técnico de Capacitación y
Productividad (INTECAP) 144
Instituto Tecnológico de Buenos Aires
263
Instituto Tecnológico de Costa Rica
96-7
Instituto Tecnológico de Electrónica
'Fernando Aguado Rico' 105, 287
Instituto Tecnológico de Santo Domingo
115, 288
Instituto Tecnológico de Sonora 180,
290
Instituto Tecnológico 'Mártires de Girón'
105, 287
Instituto Tecnológico Regional de
Chihuahua 180, 298
Instituto Tecnológico Regional de
Durango 180, 298
Instituto Tecnológico Regional de
Mérida 180, 298
Instituto Tecnológico Regional de
Morelia 180, 298
Instituto Tecnológico Regional de Oaxaca
180, 298
Instituto Tecnológico Regional de
Querétaro 180, 298
Instituto Tecnológico Regional de Saltillo
180, 298
Instituto Tecnológico y de Estudios
Superiores de Monterrey 179, 298
Instituto Tecnológico y de Estudios
Superiores de Occidente 179
Instituto Torcuato di Tella 16, 263
Instituto Universitario de Tecnología del
Mar 245
Instituto Universitario Politécnico 245,
312
Instituto Uruguayo de Normas Técnicas
233, 310
Instituto Venezolano de Análisis
Económico y Social 243
Instituto Venezolano de Investigaciones
Científicas 242
Instituto Venezolano de Petroquímica 242
INTA, see Instituto Nacional de
Tecnología Agropecuaria and
Instituto Nicaraguense de Tecnología
Agropecuaria
INTEC, see Instituto de Investigaciones
Tecnológicas
INTECAP, see Instituto Técnico de
Capacitación y Productividad
INTEC-UNIDO, see Comite de
Investigaciones Tecnológicas

Inter-American Centre of Photo-
Interpretation 82
Inter-American Child Institute 235, 310
Inter-American Council for Education,
Science and Culture 15
Inter-American Development Bank 227
Inter-American Experimental Agriculture
Institute 126, 289
Inter-American Institute of Agricultural
Sciences 135, 145
Inter-American Institute of Agricultural
Sciences of the Organization of
American States 97
Inter-American Institute of the
Organization of American States 95
Inter-American School of Librarianship
83
Inter-American Technical Service for
Agricultural Cooperation 206, 303
Inter-American University of Puerto Rico
223, 308
Intercontinental University 175, 300
International Association of Lunology 54,
269
International Bauxite Association 148, 227
International Centre for the Improvement
of Maize and Wheat 181, 296
International Information System
Research Documentation (ISORID)
71
International Monetary Fund (IMF) 227
International Referal System Sources
Environmental Information
(INFOTERRA) 71
International Serials Data Systems (ISDS)
71
INTI, see Instituto Nacional de
Tecnología Industrial
IPEANE, see Instituto de Pesquisas do
Experimentasão Agropecuario do
Nordeste
IREN, see Instituto de Recursos Naturales
ISA, see Instituto Superior de Agricultura
ISDS, see International Serials Data
Systems
ISORID, see International Information
System Research Documentation
Itaúna University 43, 273
Izquieta Pérez National Institute of
Hygiene 126, 289

Jamaica Library Service 161
Jamaican Association of Sugar
Technologists 161, 295

Javieriana Pontificial University 86, 284
Joaquin V. González Graduate School 18
Jorge Tadeo Lozano University
  Foundation of Bogotá 86, 283
Jose Faustino Sánchez Carrión National
  University 216-17, 307
José Martí National Library 108, 285
Juan XXIII Institute of Social Research
  192
Juan Agustín Maza University 21, 266
Juan Manuel Cagigal Naval Observatory
  246, 312
Juan Misael Saracho University 29, 269
Juárez Autonomous University of
  Tabasco 178, 301
Juárez University of the State of Durango
  174, 299
Juiz de Fora Federal University 43, 273
Julio de Mesquita Filho State
  University 51, 276
Junta de Investigaciones y
  Experimentaciones de las FFAA 13
Junta Nacional de Planificación y
  Coordinación Económica 120
Junta Universitaria 124
Justus Liebig of Giessen 88

La Salle Catholic University 87, 284
La Salle de México University 175, 300
La Salle Foundation of Natural Sciences
  245, 312
Laboratorio Central Gonçalo Moniz 57,
  272
Laboratorio de Análisis Tecnológico del
  Uruguay (LATU) 233
LAFTA, see Latin American Free Trade
  Association
Landbouwproefstation 227, 308
Las Vegas Experimental Station of the
  National Society of Agriculture 69,
  278
Las Villas University 107, 287
Latin American Centre of Demography
  96
Latin American Centre of Physics 38
Latin American Economic Commissions
  15
Latin American Free Trade Association
  (LAFTA) 34, 59, 77, 205, 229, 238
Latin American Institute for Social
  Research 127
Latin American Institute of Scientific
  Research in Correspondence
  Education 242

Latin American Network of
  Technological Information 129
Lebu Scientific Institute 69, 278
Legislative Assembly Library 99
Leopoldo A. Miquez de Mello Research
  and Development Centre 52, 270
Library of the Autonomous University of
  Santo Domingo 117, 288
Library of the Catholic University of
  Puerto Rico 225
Library of the Ministry of Foreign
  Relations - Brazil 56
Library of the Ministry of Foreign
  Relations - Costa Rica 99
Library of the National Congress -
  Bolivia 31
Library of the National Congress - Chile
  72, 277
Library of the National University of San
  Agustín 221, 304
Library of the Scientific Society of
  Paraguay 206, 303
Liceo Naútico Pesquero 245
Liga Nacional contra Cancer 136
Liga Uruguaya contra la Tuberculosis
  235
Lima University 216, 306
Lisandro Alvarado Foundation 246
Livestock Division - Dominican Republic
  113, 116
Livestock Production Centre 116
Luis A. Martínez National College of
  Agriculture 126, 290
Luis Vargas Torres Technical University
  125, 291

Mackenzie University 43, 274
Madre María Teresa Guevara Library
  225
Madre y Maestra Catholic University
  115, 288
Magdalena Technological University 85,
  283
Manuel Foster Astrophysics Observatory
  72, 279
Mar del Plata Fund 15, 129
Margarita Marine Research Station 245
María Cruz and Manuel L. Inchausti
  School of Practical Agriculture and
  Animal Husbandry 18
Mariano Gálvez de Guatemala University
  144, 293
Marine Biology Research Centre 113, 116
Mario de Andrade Municipal Library 56

Mariscal Jose Ballivián University 29, 30, 269

Mártires de Girón Technology Institute 105, 287

Mathematics, Cybernetics and Computing Institute - Cuba 108, 286

Medellín University 87, 284

Medical Research Council 161

Medical Research Council Laboratories - Jamaica 161, 295

Mendoza Experimental Station of Arable and Livestock Farming 11, 262

Mérida Regional Institute of Technology 180, 298

Meteorological Office of Chile 73, 279

Meteorological Station - Venezuela 246, 312

Meteorology Institute - Cuba 108, 286

Metropolitan University, 243, 313

Mexican Association of Petroleum Geologists 183, 295

Mexican Association of Women Doctors 182, 295

Mexican Institute for the Conservation of Natural Resources 181, 297

Mexican Institute of Technological Research 180-1, 296

Mexican Petroleum Institute 181, 297

Mexican Society of Mathematics 183, 297

Mexican Society of Paediatrics 182, 297

Mexican Society of Parasitology 182, 297

Mexican Society of Public Health 182, 298

Miguel C. Rubino Centre of Veterinary Research 234

Miles Institute of Experimental Therapy 182, 297

Military Geographical Institute - Argentina 13, 262

Military Geographical Institute - Chile 71, 278

Military Geographical Institute - Paraguay 206, 303

Military Geographical Institute - Peru 220, 305

Military Geographical Institute - Uruguay 235, 310

Military Institute of Geography and National Property Register - Bolivia 28, 268

Mining Office of the Ministry of Mines and Hydrocarbons 242

Ministerio de Asuntos Exteriores 37

Ministerio de Comercio e Industrias 199

Ministerio de Desarrollo Agropecuario 199, 201

Ministerio de Educación Superior 102

Ministerio de Energía y Minas - Peru 306

Ministerio de Minas y Energía 38

Ministerio de Relaciones Exteriores 38

Ministry for Planning and Coordination of Economic and Social Development - El Salvador 132

Ministry of Agricultural and Livestock Development - Panama 199, 201

Ministry of Agriculture - Brazil 53

Ministry of Agriculture - Chile 63, 64, 69

Ministry of Agriculture - El Salvador 135

Ministry of Agriculture - Guatemala 145

Ministry of Agriculture - Mexico 181

Ministry of Agriculture and Fisheries - Uruguay 234

Ministry of Agriculture and Food - Peru 219

Ministry of Commerce and Industry - Panama 199, 200

Ministry of Communications and Public Works - Guatemala 143

Ministry of Culture and Education - Argentina 6

Ministry of Defence - Argentina 6, 13

Ministry of Development - Venezuela 242

Ministry of Economy - Chile 64

Ministry of Education - Belize 23

Ministry of Education - Bolivia 28, 31

Ministry of Education - Panama 196

Ministry of Education and Culture - Uruguay 231, 232

Ministry of Energy and Hydrocarbons - Bolivia 30

Ministry of Energy and Mines - Peru 306

Ministry of Foreign Affairs - Brazil 37

Ministry of Foreign Affairs - Chile 65

Ministry of Foreign Relations - Brazil 38

Ministry of Foreign Relations - Uruguay 232

Ministry of Health - Costa Rica 94, 97

Ministry of Health - Guyana 150, 151

Ministry of Health - Panama 200

Ministry of Higher Education - Cuba 102

Ministry of Industry - Peru 219

Ministry of Industry and Commerce - Bolivia 28

Ministry of Industry, Commerce and Integration - Ecuador 121

Ministry of Labour and Social Welfare - Panama 196

Ministry of Mine Works - Chile 64

Ministry of Mines and Energy - Brazil 38

Ministry of National Defence – Chile 64

Ministry of Natural Resources and
Energy – Ecuador 123

Ministry of Planning and Coordination –
Bolivia 26

Ministry of Planning and Economic
Policy – Panama 196, 200

Ministry of Public Education – Chile 64

Ministry of Public Health – Brazil 56

Ministry of Public Health – Chile 64

Ministry of Public Health – Cuba 107

Ministry of Public Health and Population
– Haiti 154

Ministry of Public Health and Social
Assistance – Guatemala 145

Ministry of Public Works – El Salvador
135

Ministry of the Economy – El Salvador
135

Ministry of the Economy – Guatemala
143, 145

Mission ORSTOM au Perou –
Cooperation auprès du Ministerio de
Energía y Minas 221, 306

Mogi das Cruzes University 44, 274

Montemorelos University 176, 300

Monterrey University 176, 300

Morelia Regional Institute of
Technology 180, 298

Museo de Artes e Industrias Populares
146

Museo de Historia Eclesiástica 'Anselmo
Liorente y Lafuente' 99

Museo Histórico de las Ciencias Médicas
'Carlos J. Finlay' 109

Museo Nacional – Costa Rica 99

Museum of Art and Traditional and
Ethnic Industry 146

Nami Jafet Institute for the Advancement
of Science and Culture 55, 271

Nariño University 84, 283

National Academy of Agronomy and
Veterinary Sciences 16

National Academy of Engineering 235

National Academy of Exact, Physical and
Natural Sciences of Lima 220, 304

National Academy of Medicine –
Argentina 16, 260

National Academy of Medicine – Brazil
55, 269

National Academy of Medicine –
Colombia 89, 280

National Academy of Medicine – Peru
220, 304

National Academy of Medicine –
Venezuela 247, 311

National Academy of Sciences – Buenos
Aires 16, 260

National Academy of Sciences – Mexico
183, 295

National Academy of Sciences – Panama
200, 302

National Administration for Science and
Technology 6

National Administration of Combustibles,
Alcohol and Cement 234

National Agrarian Institute – Venezuela
245, 312

National Agrarian Society – Peru 220,
306

National Agrarian University 215,
306

National Agrarian University of the Selva
215, 306

National Alliance against Cancer 136

National Archives of Colombia 90, 280

National Archives of Costa Rica 99,
284

National Association of Universities and
Institutes of Higher Education 170-1,
296

National Astronomic Observatory – Chile
72, 279

National Astronomic Observatory –
Colombia 89, 281

National Astronomy and Ionosphere
Center, Cornell University 224

National Autonomous University of
Heredia 96, 285

National Autonomous University of
Mexico 171, 183, 184, 298

National Autonomous University of
Nicaragua 192, 193, 302

National Board for Economic Planning
and Coordination 120

National Board of Agro-industry 199

National Budget Office 196, 197

National Cardiology Institute 182, 297

National Centre for Agricultural and
Livestock Research – Dominican
Republic 113, 116

National Centre for Agricultural and
Livestock Research – Venezuela 245,
311

National Centre for Agricultural and
Livestock Technology – El Salvador
135, 136, 291

National Centre for Cassava and Fruit Research 52, 270

National Centre for Educational Documentation and Information 31, 268

National Centre for Scientific and Technological Documentation 31, 268

National Centre of Agricultural Research – Uruguay 234

National Centre of Cosmic Radiation 8

National Centre of Educational Information 14, 261

National Centre of Information and Documentation 72, 277

National Centre of Marine Science and Technology 182, 296

National Centre of Medical Science Information 107, 286

National Centre of Medicine and Health Sciences Documentation and Information 236

National Centre of Nuclear Studies 70

National Centre of Population Documentation 200-1

National Centre of Scientific Research 106

National Centre of Space Studies 137, 138

National Centre of Technological Evaluation 239

National Centre of Technology and Industrial Productivity 234

National College of Buenos Aires 16, 264

National Commission for Atomic Energy – Argentina 6, 9, 12, 261

National Commission for Atomic Energy – Bolivia 30

National Commission for Atomic Energy – Brazil 38

National Commission for Atomic Energy – Uruguay 235, 309

National Commission for Environmental Protection and Conservation of Natural Resources 102-3

National Commission for Nuclear Energy – Brazil 54, 270

National Commission for the Peaceful Use of Atomic Energy – Cuba 102

National Commission of Scientific and Technological Research 62, 71, 72, 277

National Commission of Space Research 13, 261

National Committee of Technical Cooperation and Economic Assistance 128, 129

National Computing and Information Enterprise 71

National Council for Agricultural Research 245, 311

National Council for Higher Education in Bolivia 28

National Council for Science and Technology – Ecuador 120-1, 122, 128, 129, 130

National Council for Scientific and Technical Research – Argentina 8, 261

National Council for Scientific and Technological Research – Costa Rica 38, 94, 95, 99

National Council for Scientific and Technological Research – Venezuela 238, 242

National Council for the Development of the Nuclear Industry 246, 311

National Council of Science and Technology – Bolivia 28

National Council of Science and Technology – Cuba 102

National Council of Science and Technology – Mexico 38, 99, 165, 168, 170, 183, 184, 296

National Council of Science and Technology – Peru 209

National Council of Scientific and Technological Development – Brazil 34, 35, 37, 38, 99, 130

National Council of Scientific and Technological Research – Uruguay 229, 232, 309

National Council of Scientific Research – Haiti 154

National Department of Meteorology 54, 270

National Department of Mineral Production 38

National Development Bank 34

National Development Council 120

National Development Institute 187

National Firm of Telecommunications 71

National Fisheries Institute – Ecuador 126, 289

National Fisheries Institute – Mexico 182, 297

National Fisheries Institute – Uruguay 234, 310

National Foundation for Agricultural Research 245

National Fund for Science and Technology 129

National Geographical Institute - Costa
Rica 95, 285
National Geographical Institute - El
Salvador 135, 291
National Geographical Institute -
Guatemala 143, 292
National Geographical Institute
- Honduras 157, 294
National Geological Service 192, 302
National Geological Service Library 14,
260
National Health Institute - Colombia 89,
281
National Health Institute - Peru 219, 306
National Hygiene Council 235, 309
National Information System - Colombia
89
National Information System for
Agricultural and Livestock Sciences
89
National Information System for
Economics and Business Studies 90
National Information System for
Education 90
National Information System for Energy
Resources 90
National Information System for Health
Sciences 89
National Information System for Marine
Sciences 90
National Information System for the
Environment and Natural Resources
90
National Information System for the
Industrial Sector 90
National Information System in Science
and Technology 128
National Institute for Arable and
Livestock Research 126, 289
National Institute for Pedagogic Studies 39
National Institute of Administration,
Management and Higher
International Studies 153
National Institute of Agrarian Reform -
Cuba 106, 287
National Institute of Agrarian Research -
Peru 219, 305
National Institute of Agricultural
Research 181, 297
National Institute of Arable and Livestock
Technology 9, 10, 11, 263
National Institute of Astrophysics, Optics
and Electronics 183, 297
National Institute of Automated Systems
and Computation 102

National Institute of Cancer 88, 281
National Institute of Forensic Medicine
88, 281
National Institute of Forest Development
and Exploitation 106, 287
National Institute of Forestry Research
181, 297
National Institute of Geological and
Mineral Research 82, 281
National Institute of Geology and Mining
9
National Institute of Hygiene - Cuba
107, 287
National Institute of Hygiene - Mexico
182, 297
National Institute of Industrial
Technology 9, 10, 263
National Institute of Limnology 12, 263
National Institute of Livestock Research
181, 297
National Institute of Mathematical
Research 210
National Institute of Meteorology and
Hydrography 127, 289
National Institute of Microbiology 12,
263
National Institute of Natural Resources -
Colombia 82
National Institute of Nuclear Energy -
Guatemala 145, 292
National Institute of Nuclear Energy -
Mexico 182
National Institute of Nuclear Research -
Mexico 182, 297
National Institute of Nutrition - Ecuador
126, 289
National Institute of Nutrition -
Venezuela 246, 312
National Institute of Parasitology 205,
206, 303
National Institute of Physics - Peru 210
National Institute of Pneumonology 182,
297
National Institute of Professional
Training 68
National Institute of Scientific Research
205, 303
National Institute of Seismology 8
National Institute of Seismology,
Vulcanology, Meteorology and
Hydrology 145, 292
National Institute of Statistics - Chile
72
National Institute of Statistics and Census
- Argentina 14, 263

National Institute of Statistics and Census
 - Ecuador 123, 128, 289
National Institute of Technological
 Research and Standards 68, 279
National Institute of Technology 52, 272
National Institute of the Development of
 Fisheries 11, 263
National Institute of Theatre Studies 16
National Institute of Viticulture and
 Viniculture 11, 263
National Library - Argentina 14, 260
National Library - Brazil 56, 269
National Library - Chile 72, 277
National Library - Colombia 90, 281
National Library - Costa Rica 99, 284
National Library - Dominican Republic
 117
National Library - El Salvador 136, 291
National Library - Guyana 151, 293
National Library - Haiti 154, 294
National Library - Jamaica 161
National Library - Mexico 171, 184, 296
National Library - Nicaragua 193
National Library - Panama 201
National Library - Peru 221, 304
National Library - Uruguay 236, 309
National Library - Venezuela 247, 311
National Library and Archives - Paraguay
 206
National Library Service - Belize 23, 267
National Machinery and Engineering
 Company 199
National Meat Institute 233
National Meteorological Directorate of
 Uruguay 235, 309
National Meteorological Observatory
 145
National Meteorological Office 98, 285
National Meteorological Service 14, 264
National Museum - Costa Rica 99
National Observatory of Brazil 55, 272
National Office of Cartography of the
 Ministry of Public Works 242
National Office of Engineering -
 Venezuela 239
National Office of Inventions, Technical
 Information and Trade Marks 102
National Office of Statistics - Peru 215,
 306
National Pedagogic University 84, 283
National Peruvian University Council
 215, 304
National Planning Agency 160
National Planning Office 94
National Polytechnic 125, 127, 290

National Polytechnic Institute - Technical
 University 179, 298
National Radiology Group 107, 286
National Register of Industrial Projects -
 Venezuela 238
National Research Council - Brazil 34,
 74
National Research Council - Peru 209,
 210, 214-15
National School of Agriculture - El
 Salvador 136, 291
National School of Agriculture - Mexico
 173
National School of Agriculture and
 Livestock - Nicaragua 192, 302
National School of Music 157
National School of Plastic Arts 199
National School of Technical Education
 No 1 261
National School of Technical Education
 No 4 261
National Science Foundation 73
National Science Research Council 151,
 293
National Seed Company 199
National Service for the Surveying and
 Conservation of Soil 53, 272
National Society of Mining 220, 306
National Statistical and Census Office -
 Venezuela 242
National Statistics Institute 28, 268
National Sub-system of Agricultural and
 Livestock Information of Uruguay
 235-6
National System for Scientific and
 Technological Development 26
National System of Metrology, Standards
 and Quality Control 135
National System of Scientific and
 Technological Development 35
National University of Asunción 205,
 206, 304
National University of Catamarca 16-17,
 264
National University of Central Peru 216,
 306
National University of Colombia 83, 88,
 282
National University of Comahué 17,
 264
National University of Córdoba 14, 17,
 264
National University of Cuyo 17, 264
National University of El Salvador 135,
 136, 291

National University of Engineering 216, 307
National University of Jujuy 17, 264
National University of La Plata 17, 265
National University of Loja 124, 290
National University of Mar del Plata 18, 265
National University of Río Cuarto 18, 265
National University of Rosario 19, 265
National University of Salta 19, 265
National University of San Agustín 217, 307
National University of Santiago del Estero 19, 265
National University of Technology 19, 265
National University of the Littoral 18, 265
National University of the North-East 18, 265
National University of the Peruvian Amazon 216, 306
National University of the South 19, 265
National University of Tucumán 19, 265
Natural Resources Research Centre 10, 261
Natural Resources Research Institute 65, 279
Nautical School of Panama 199
Naval Oceanographic Institute 124, 290
Nayarit Autonomous University 176, 300
Nayarit Institute of Sciences and Humanities 176
Netherlands Foundation for the Advancement of Tropical Research 227
New Amsterdam Technical Institute 148, 150, 294
Nicaraguan College of Doctors and Surgeons 193
Nicaraguan Institute of Agriculture and Livestock Technology 192, 193
Nicaraguan Institute of the Cinema 192, 302
Nicaraguan Society of Psychiatry and Psychology 193
North Mineira Foundation for Higher Education 45, 274
Notarial University of Argentina 21, 266
Nuclear Engineering Institute 54, 271
Nuclear Research Institute - Cuba 107, 286
Nuclear Sciences Institute - Ecuador 127, 289

Nueva León Autonomous University 177, 301

OAS, see Organization of American States
Oaxaca Regional Institute of Technology 180, 298
Observatorio Astrofísico 'Manuel Foster' 72, 279
Observatorio Astronómico - Argentina 14, 263
Observatorio Astronómico - Uruguay 235, 310
Observatorio Astronómico de Quito 127, 290
Observatorio Astronómico Nacional - Chile 72, 279
Observatorio Astronómico Nacional - Colombia 89, 281
Observatorio Europeo Austral 72, 279
Observatorio Interamericano de Cerro Tololo 73, 279
Observatorio Meteorológico Nacional 145
Observatorio Nacional do Brasil 55, 272
Observatorio Naval 'Juan Manuel Cagigal' 246, 312
Observatorio 'San Calixto' 31, 268
Odontology Association of Uruguay 235, 309
Office de la Recherche Scientifique et Technique Outre-Mer - Centre ORSTOM de Cayenne 138, 292
Office de la Recherche Scientifique et Technique Outre-Mer 'Mision Venezuela' 312
Office for Overseas Scientific and Technical Research, (ORSTOM) Cayenne Centre 138, 292
Office of Planning and Budgeting - Uruguay 232
Office of Technical Research 144
Oficina Coordinadora de Negociaciones del Estado Venezolano 239
Oficina de Investigación Técnica 144
Oficina de Planeamiento y Presupuesto 232
Oficina de Planificación Nacional 94
Oficina Meteorológica de Chile 73, 279
Oficina Nacional de Estadística 215, 306
Oficina Nacional de Invenciones, Información Técnica y Marcas 102
Oficina Regional de Ciencia y Tecnología de la UNESCO para America Latina y el Caribe 236, 310

Oficina Sanitaria Panamericana 146
Oncology and Radiobiology Institute of
    Havana 107, 286
Oncology Institute – Uruguay 310
Open University 96, 285
Organización de Estudios Tropicales 96,
    285
Organización Pan Americana de la Salud
    146
Organization of American States (OAS)
    15, 26, 34, 59, 73, 77, 90, 95, 97,
    102, 111, 117, 120, 128, 129, 132,
    140, 153, 157, 160, 163, 170, 192,
    195, 197, 201, 202, 205, 209, 227,
    229
Organization of the Central State
    Administration 102
Organization of Tropical Studies 96,
    285
ORSTOM, see Office de la Recherche
    Scientifique et Technique Outre-Mer
ORSTOM Mission of Peru, cooperating
    with the Ministry of Energy and
    Mines 221, 306
Oruro Technical University 29, 268
Oscar Freire Institute 57, 272
Oswaldo Cruz Institute 57, 272
Overseas Scientific and Technical
    Research Office, Venezuelan Section
    312

Paediatrics and Child Welfare Society
    of Paraguay 206, 303
PAHO, see Pan American Health
    Organization
Pamplona University 84, 283
Pan American Health Organization
    (PAHO) 146
Pan American Sanitary Office 146
Pan American School of Agriculture 157,
    294
Panamanean Academy of History 200
Panamanean Commission of Industrial
    and Technical Standards 199
Paraguayan Centre for Engineers 206,
    303
Paraguayan Centre of Economic and
    Social Development Studies 206, 303
Passo Fundo University 46, 275
Pasteur Institute 138, 292
Pedagogic and Technological University
    of Colombia 85, 283
Pedro Henríquez Ureña National
    University 115, 116, 288

Pedro Ruiz Gallo National University
    217, 307
Pereira Technological University 85, 283
Peruvian Academy of Stomatology 304
Peruvian Academy of Surgery 304
Peruvian Institute of Nuclear Energy 220,
    306
Peruvian Institute of Science 210
Peruvian Marine Institute 219, 305
Peruvian Medical Federation 220, 305
PETROBRAS 38, 52
PETROVEN 238
Piura National University 218, 307
Polytechnic College – Venezuela 245, 312
Polytechnic Institute 199
Polytechnic of Haiti 153, 294
Polytechnic of Nicaragua 192, 302
Pontificia Universidad Católica del
    Ecuador 124, 290
Pontificia Universidad Católica del Perú
    218, 307
Pontificia Universidad Javieriana 86, 284
Pontifícia Universidade Católica de
    Campinas 41, 273
Pontifícia Universidade Católica de São
    Paulo 51, 276
Pontifícia Universidade do Rio de Janeiro
    38, 49, 276
Pontifícia Universidade do Rio Grande
    do Norte 48, 275
Pontifical Catholic University of
    Campinas 41, 273
Pontifical Catholic University of Ecuador
    124, 290
Pontifical Catholic University of Peru
    218, 307
Pontifical University of Rio de Janeiro
    38, 49, 276
Pontifical University of Rio Grande do
    Norte 48, 275
Pontifical University of São Paulo 51,
    276
Private Technical University of Loja 125,
    291
PROCYT, see Programa Regional de
    Desarrollo Científico y Tecnológico
    de la OEA
Professor Dr José F. Arias Institute of
    Teaching in Mechanics and
    Electronics 233
Professor Dr Juan C. Mussio Fournier
    Institute of Endocrinology 234, 310
Programa Regional de Desarrollo
    Científico y Tecnológico de la OEA
    (PROCYT) 15, 197

Puebla Autonomous University 177, 301
Puerto Rican Academy of the Spanish
  Language 224
Puerto Rican Nuclear Center 224, 308
Punta Betín Institute of Marine Research
  88, 281

Querétaro Regional Institute of
  Technology 180, 298
Quindío University 84, 283

Radioactive Research Institute – Brazil
  54, 271
Rafael Landívar University 144, 293
Rafael Urdaneta University 244, 313
Real Colegio Seminario de San
  Buenaventura de Merida 243
Red de Información Tecnológica
  Latinoamericana (RITLA) 129
Regiomontana University 177, 301
Regional Experimental Station of Arable
  and Livestock Farming 11, 262
Regional Medical Library 236
Regional Programme of Scientific and
  Technological Development of the
  Organization of American States 15,
  197
Regional University of Blumenau 40, 273
Regional University of the North-East 44,
  274
Registro Nacional de Proyectos
  Industriales 238
Research and Advanced Studies Centre of
  the National Polytechnic Institute
  171, 296
Research and Development Centre 52,
  270
Research Centre – Paraguay 304
Research Centre of Marine Biology 11,
  260
Research Commission 196
Research Institute for Tropical Agronomy
  and Cultivation of Foodstuffs 138,
  292
Ricardo Moreno Canas Centre for
  Medical Studies 98, 285
RITLA, see Red de Información
  Tecnológica Latinoamericana
Romulo Gallegos Central Plains
  Experimental University 244, 313
Royal Agricultural and Commercial
  Society 73, 150
Royal University of San Felipe 66

Rubén Martínez Villena Central Library
  of the University of Havana 108-9,
  285
Rubén Martínez Villena Library 109,
  285
Rural Association of Uruguay 309
Rural Centre of Education 19
Rural Federal University of Pernambuco
  47, 275
Rural Federal University of Rio de
  Janeiro 49, 276
Rural Federal University of Rio Grande
  do Sul 47
Rural University of Brazil 49
Rural University of the State of Minas
  Gerais 51

Salta Experimental Station of Arable and
  Livestock Farming 10, 262
Saltillo Regional Institute of Technology
  180, 298
Salvadorean Institute of Coffee Research
  135, 291
Salvadorean Institute of Industrial
  Development and the National Centre
  for Productivity 135
Salvadorean Institute of Industrial
  Technology 135
San Andrés University 28, 29, 30, 268
San Antonio de Abad National University
  217, 307
San Buenaventura de Merida Royal
  College 243
San Buenaventura University 87, 284
San Calixto Observatory 31, 268
San Cristóbal de Huamanga National
  University 217, 307
San Francisco Javier Royal and Pontifical
  University of Bolivia 29, 30, 268
San Juán Bosco National University of
  Patagonia 18, 265
San Juán Bosco University of Patagonia
  22, 267
San Juan de Dios Hospital 97
San Luis Gonzaga National University
  218, 307
San Martín de Porres Private University
  218, 307
San Miguel Space Centre – National
  Observatory of Cosmic Physics 13,
  261
San Simón University 29, 31, 268
Santa María Argentinian Catholic
  University of Buenos Aires 20, 266

Santa María la Antigua University 199, 303
Santa María University 244, 313
Santander Industrial University 84, 283
Santiago de Calí University 87
Santiago Medical Association 117, 287
Santo Tomás de Aquino University of the North 22, 267
Santo Tomás University 87, 284
São Paulo State Forestry Institute 53, 271
São Paulo University 38, 50, 57, 276
School of Administration, Finance and Technology 282
School of Aeronautical Engineering 19, 265
School of Agriculture 161, 295
Scientific and Humanistic Information Centre for the Coordination of Scientific Research 184, 296
Scientific and Literary Institute – Mexico 175
Scientific and Technological Council 35
Scientific Research Council 160, 295
Scientific Society of Paraguay 206, 303
Scientific Society of São Paulo 55, 272
Sección de Ciencia y Tecnología 120
Sección de Política Científica y Tecnológica 195–6
SECLA, see Sistema Económico Latinamericano
Secretaría de Estado de Ciencia y Tecnología 2, 8
Secretaría de Planificación (SEPLAN) 34, 35, 38
Secretaría General del Consejo Nacional de Planificación Económica 140–1
Secretariado Técnico 111
Secretariado Uruguayo de la Lana (SUL) 234
Secretariat of Agriculture 6
Secretariat of Industrial Development 6, 9
Secretariat of Planning 34, 35, 38
Secretariat of Planning, Coordination and Diffusion – Uruguay 231
Section of Science and Technology 120
Sector Ciencias de la Oficina de Planeamiento de la Universidad de la República 232
Sector of Sciences of the Office of Planning of the University of the Republic 232
Sectorial Commissions in Science and Technology 120
Seminario de San Luis 124
Seminary of San Luis 124

SEPLAN, see Secretaría de Planificación
SERCOTEC, see Servicio de Cooperacion Técnica
Service Centre for Agricultural and Livestock Information 201
Servicio de Cooperacion Técnica (SERCOTEC) 68
Servicio de Endocrinología y Metabolismo 12, 264
Servicio de Información Técnica 130
Servicio de Información y Extensión Tecnológica 200
Servicio Geológico de Bolivia 31, 268
Servicio Geológico Nacional 192, 302
Servicio Meteorológico Nacional 14, 264
Servicio Técnico Interamericano de Cooperación Agrícola 206, 303
Serviço de Documentação Geral da Marinha 56, 272
Serviço de Pesquisa e Experimentação de Cancer 53, 272
Serviço Nacional de Levantamento e Conservação de Solos – EMBRAPA 53, 272
SIEX, see Superintendencia de Inversiones Extranjeras
Simón Bolívar University 244, 313
SINDECYT, see Sistema Nacional para el Desarrollo Científico y Tecnológico
Sindicato de Industriales de Panama 200
Sistema de Transferencia de Tecnología Rural (STTR) 129
Sistema Económico Latinamericano (SECLA) 129
Sistema Nacional de Desarrollo Científico e Tecnológico 35
Sistema Nacional de Información Científica y Tecnológica 128
Sistema Nacional de Información (SNI) 89
Sistema Nacional de Metrología, Normalización y Control de Calidad 135
Sistema Nacional para el Desarrollo Científico y Tecnológico (SINDECYT) 26
Smithsonian Tropical Research Institute 200, 303
SNIAF, see Subsistema Nacional de Información Agropecuaria del Uruguay
Social Security Department – Panama 200
Social Studies Centre 113
Sociedad Chilena de Química 74, 279

Sociedad Científica Chilena 'Claudio Gay' 74, 279
Sociedad Científica de Chile 74, 279
Sociedad Científica del Paraguay 206, 303
Sociedad Colombiana de Cancerología 89, 281
Sociedad Colombiana de Ingenieros 89, 281
Sociedad Colombiana de Patología 89, 282
Sociedad Colombiana de Psiquiatría 89, 282
Sociedad Colombiana de Químicos e Ingenieros Químicos 89, 282
Sociedad de Arquitectos del Uruguay 310
Sociedad de Biología de Chile 74, 279
Sociedad de Cirugía del Uruguay 235, 310
Sociedad de Ingenieros del Perú 220, 306
Sociedad de Pediatría y Puericultura del Paraguay 206, 303
Sociedad de Radiología del Uruguay 235, 310
Sociedad Matemática Mexicana 183, 297
Sociedad Mexicana de Parasitología AC 182, 297
Sociedad Mexicana de Pediatría 182, 297
Sociedad Mexicana de Salud Pública 182, 298
Sociedad Nacional Agraria 220, 306
Sociedad Nacional de Minería 220, 306
Sociedad Nicaraguense de Psiquiatría y Psicología 193
Sociedad Química de México 183, 298
Sociedad Uruguay de Patología Clínica 310
Sociedad Uruguaya de Pediatría 235, 310
Sociedad Venezolana de Ciencias Naturales 247, 312
Sociedad Zoológica del Uruguay, 311
Sociedade Brasileira para o Progresso da Ciência 55, 272
Sociedade Científica de São Paulo 55, 272
Society of Engineers of Peru 220, 306
Sonora Institute of Technology 180, 298
Sonora University 178, 301
Southern University of Chile 65, 280
Special Multilateral Fund of the Inter-American Council for Education, Science and Culture 129
State Committee of Science and Technology 102, 104, 105
State Committee of Standards 102
State of Morelos Autonomous University 176, 300

State Secretariat of Science and Technology 2, 8
State Sugar Council 113
State Technical University 67, 280
State University of Campinas 38, 41, 273
State University of Guayaquil 124, 290
State University of Londrina 43, 274
State University of Maringá 43, 274
State University of Ponta Grossa 48, 275
State University of Rio de Janeiro 49, 276
State University of the Southern Mato Grosso 44, 274
Statistical Office - Guatemala 143
Stichting Surinaams Museum 227, 308
Stichting voor Wetenschappelijk Onderzoek van de Tropen 227
Subsecretaria de Planeamiento Ambiental - UNEP 14, 264
Sub-secretariat of Environment Planning 14, 264
Subsistema Nacional de Información Agropecuaria del Uruguay (SNIAF) 235-6
Subsistemas Agropecuario e Industrial 128
Sucre Medical Institute 31
Sugar Cane Research Institute 106, 286
Sugar Industry Research Institute 161
Sugar Research Department of the Association of Sugar Producers 161
Superintendence of Foreign Investment 238
Superintendencia de Cooperacion Internacional (SCI) 37, 38
Superintendencia de Inversiones Extranjeras (SIEX) 348
Superintendency of International Cooperation 37, 38
Suralco 226
Surinam Forest Service - Nature Conservation Department 227, 308
Surinam Museum 227, 308
Surinam State Oil Corporation 226
System of Transference of Rural Technology 129
Systems Development Corporation 71

Táchira Experimental University 244, 313
Tamaulipas Autonomous University 178, 301
Technical Assistance Centre 115
Technical Cooperation Service 68
Technical Information System 130

Technical Institute of Training and
Productivity 144
Technical Secretariat – Dominican
Republic 111
Technical University of Babahoyo 125,
290
Technical University of Cajamarca 218,
307
Technical University of Callao 218, 307
Technical University of Machala 125, 291
Technical University of Manabi 125, 291
Technical University of Panama 199
Technological Analysis Laboratory of
Uruguay 233
Technological and Higher Studies
Institute of Monterrey 179, 298
Technological Information and
Innovation 183
Technological Institute and Higher
Studies of the West 179, 298
Technological Institute of Costa Rica
96–7
Technological Institute of Santo Domingo
115, 288
Technological Institute of the State of
Pernambuco 52
Technological University of Santiago 115,
288
TELENET 71
Tolima University 85, 283
Tomas Frías University 29, 269
Torcuato di Tella Institute 16, 263
Tropical Forestry Technical Centre 138,
292
Tropical Science Centre 97, 285
Trujillo National University 218, 307
Tulumayo Agricultural and Livestock
Research Station 219, 305

UNAM, see Universidad Nacional
Autonóma de México
UNDP, see United Nations
Development Programme
UNEP, see United Nations Environment
Programme
UNESCO, see United Nations
Educational, Scientific and Cultural
Organization
UNESCO Regional Office of Science and
Technology for Latin America and
the Caribbean 236, 310
UNICEF 184
UNICYT, see Unidad de Ciencia y
Tecnología

Unidad de Ciencia y Tecnología
(UNICYT) 111, 112
Unidad de Consultoria de la Dirección
Industrial 115
UNIDO, see United Nations Industrial
Development Organization
United Nations Development Programme
(UNDP) 15, 73, 91, 115, 129
United Nations Educational, Scientific
and Cultural Organization
(UNESCO) 15, 38, 39, 73, 126, 128,
170, 184, 236, 247, 260, 261, 277, 310
United Nations Environment Programme
(UNEP) 72, 184, 261, 264
United Nations Industrial Development
Organization (UNIDO) 15, 184, 236,
260, 277
United Nations Information Centre 136
United Nations Information Service 72
United Nations Organization 26, 34, 38,
59, 77, 102, 111, 120, 132, 140, 148,
153, 157, 160, 163, 184, 192, 195,
204, 205, 209, 229, 238
United States of America Commission for
Atomic Energy 224
Universidad Anáhuac 173, 299
Universidad Argentina de la Empresa 20,
266
Universidad Argentina 'John F. Kennedy'
20, 266
Universidad Austral de Chile 65, 280
Universidad Autónoma 'Benito Juárez' de
Oaxaca 117, 301
Universidad Autónoma Chapingo 173,
299
Universidad Autónoma de Aguascalientes
172, 299
Universidad Autónoma de Baja California
173, 299
Universidad Autónoma de Chiapas 173,
299
Universidad Autónoma de Chihuahua
173, 299
Universidad Autónoma de Ciudad Juárez
175, 300
Universidad Autónoma de Coahuila 173,
299
Universidad Autónoma de Guadalajara
174, 299
Universidad Autónoma de Guerrero 175,
300
Universidad Autónoma de Hidalgo 175,
300
Universidad Autónoma de Nayarit 176,
300

Universidad Autónoma de Nueva León 177, 301
Universidad Autónoma de Puebla 177, 301
Universidad Autónoma de Queretaro 177, 301
Universidad Autónoma de San Luis de Potosí 178, 301
Universidad Autónoma de Santo Domingo 112-13, 114-15, 116, 117, 288
Universidad Autónoma de Sinaloa 178, 301
Universidad Autónoma de Tamaulipas 178, 301
Universidad Autónoma de Zacatecas 179, 302
Universidad Autónoma del Estado de México 172, 299
Universidad Autónoma del Estado de Morelos 176, 300
Universidad Autónoma del Noroeste 176, 300
Universidad Autónoma Juárez de Tabasco 178
Universidad Autónoma Latinoamericana 86, 283
Universidad Autónoma Metropolitana 176, 300
Universidad Boliviana 'Juan Misael Saracho' 29, 269
Universidad Boliviana 'Mariscal Jose Ballivián' 29, 30, 269
Universidad Boliviana Mayor de 'San Andrés' 28, 29, 30, 31, 268
Universidad Boliviana Mayor 'Gabriel René Moreno' 29, 269
Universidad Boliviana Mayor Real y Pontificia de San Francisco Javier 29, 30, 268
Universidad Boliviana 'Tomas Frías' 29, 269
Universidad Católica Andrés Bello 243, 247, 312
Universidad Católica Argentina 'Santa María de los Buenos Aires' 20, 266
Universidad Católica Boliviana 29, 269
Universidad Católica de Chile 66, 72, 280
Universidad Católica de Córdoba 21, 266
Universidad Católica de Cuenca 124, 290
Universidad Católica de Cuyo 21, 266
Universidad Católica de La Plata 21, 266
Universidad Católica de Nuestra Señora de la Asunción 205, 304

Universidad Católica de Puerto Rico 223, 308
Universidad Católica de Salta 22, 267
Universidad Católica de Santa Fé 22, 267
Universidad Católica de Santiago de Guayaquil 125, 290
Universidad Católica de Santiago del Estero 22, 267
Universidad Católica de Valparaiso 66, 280
Universidad Católica Madre y Maestra 115, 288
Universidad Central de Venezuela 243, 247, 312
Universidad Central del Ecuador 124, 128, 290
Universidad Central del Este 115, 288
Universidad Centroamericana 192, 302
Universidad Centro-Occidental 243, 312
Universidad de Antioquia 82, 282
Universidad de Belgrano 20, 266
Universidad de Buenos Aires 264
Universidad de Caldas 83, 282
Universidad de Camaguey 107, 287
Universidad de Carabobo 243, 247, 312
Universidad de Cartagena 83, 282
Universidad de Chile 66, 68, 71, 72, 280
Universidad de Colima 174, 299
Universidad de Concepción 66, 280
Universidad de Córdoba 83, 282
Universidad de Costa Rica 94, 96, 97, 285
Universidad de Cuenca 124, 290
Universidad de Gran Colombia 86, 283
Universidad de Guadalajara 174, 299
Universidad de Guanajuato 174, 300
Universidad de la Habana 106, 108-9, 287
Universidad de la Pampa 17, 265
Universidad de la Patagonia 'San Juán Bosco' 22, 267
Universidad de la República 232, 233, 311
Universidad de las Américas 172, 299
Universidad de las Villas 107, 287
Universidad de Lima 216, 306
Universidad de los Andes – Colombia 86, 89, 284
Universidad de los Andes – Venezuela 243, 247, 313
Universidad de Medellín 87, 284
Universidad de Mendoza 21, 266
Universidad de Montemorelos 176, 300
Universidad de Monterrey 176, 300
Universidad de Morón 21-2, 266
Universidad de Nariño 84, 283

Universidad de Oriente – Cuba 107, 287
Universidad de Oriente – Venezuela 244, 247, 313
Universidad de Pamplona 84, 283
Universidad de Panamá 196, 197, 198, 199, 200, 303
Universidad de Puerto Rico 223, 308
Universidad de Quindío 84, 283
Universidad de San Buenaventura 87, 284
Universidad de San Carlos de Guatemala 143, 144, 293
Universidad de San Gregorio Magno 124
Universidad de Santa María 244, 313
Universidad de Santo Tomás 87, 284
Universidad de Sonora 178, 301
Universidad de Tolima 85, 283
Universidad de Yucatán 179, 302
Universidad de Zulia 245, 247, 313
Universidad del Aconcagua 20, 266
Universidad del Atlántico 83, 282
Universidad del Cauca 83, 282
Universidad del Museo Social Argentino 20–1, 266
Universidad del Norte – Chile 67, 280
Universidad del Norte – Colombia 87, 284
Universidad del Norte – Mexico 177, 301
Universidad del Norte Santo Tomás de Aquino, 22, 267
Universidad del Salvador 21, 266
Universidad del Sudeste 178, 301
Universidad del Trabajo del Uruguay 233, 311
Universidad del Valle 85, 283
Universidad del Valle de Guatemala 144
Universidad del Valle de México 175, 300
Universidad Distrital 'Francisco José de Caldas' 84, 282
Universidad Dominicana de Santo Tomás de Aquino 124
Universidad Estatal a Distancia 96, 285
Universidad Estatal de Guayaquil 124, 290
Universidad Externado de Colombia 86
Universidad Federal Rural de Rio Grande do Sul 47
Universidad Femenina de México 172, 299
Universidad Francisco de Paula Santander 83, 282
Universidad Francisco Marroquin 144
Universidad Iberoamericana 175, 300
Universidad Industrial de Santander 84, 283
Universidad Intercontinental 175, 300

Universidad 'Juan Agustín Maza' 21, 266
Universidad Juárez del Estado de Durango 174, 299
Universidad La Salle de México 175, 300
Universidad Laica 'Vicente Rocafuerte' de Guayaquil 125, 291
Universidad Libre de Colombia 86, 284
Universidad Mariano Gálvez de Guatemala 144, 293
Universidad Mayor de 'San Simón' 29, 31, 268
Universidad Metropolitana 243, 313
Universidad Michoacana de San Nicolás de Hidalgo 176, 300
Universidad Nacional Agraria 215, 306
Universidad Nacional Agraria de la Selva 215, 306
Universidad Nacional Autónoma de Heredia 96, 285
Universidad Nacional Autónoma de Honduras 157, 158, 295
Universidad Nacional Autónoma de México (UNAM) 171, 183, 184, 298
Universidad Nacional Autónoma de Nicaragua 192, 193, 302
Universidad Nacional de Asunción 205, 206, 304
Universidad Nacional de Catamarca 16–17, 264
Universidad Nacional de Colombia 83, 88, 282
Universidad Nacional de Córdoba 14, 17, 264
Universidad Nacional de Cuyo 17, 264
Universidad Nacional de El Salvador 135, 291
Universidad Nacional de Huánuco 'Hermilio Valdizán' 216, 306
Universidad Nacional de Ingeniería 216, 307
Universidad Nacional de Jujuy 17, 265
Universidad Nacional de la Amazonía Peruana 216, 306
Universidad Nacional de la Patagonia San Juán Bosco 18, 265
Universidad Nacional de La Plata 17, 265
Universidad Nacional de Loja 124, 290
Universidad Nacional de Mar del Plata 18, 265
Universidad Nacional de Piura 218, 307
Universidad Nacional de Río Cuarto 18, 265
Universidad Nacional de Rosario 19, 265
Universidad Nacional de Salta 19, 265

Universidad Nacional de San Agustín 217, 307

Universidad Nacional de San Antonio de Abad 217, 307

Universidad Nacional de San Cristóbal de Huamanga 217, 307

Universidad Nacional de Santiago del Estero 19, 265

Universidad Nacional de Trujillo 218, 307

Universidad Nacional de Tucumán 19, 265

Universidad Nacional del Centro de la Provincia de Buenos Aires 17, 264

Universidad Nacional del Centro del Perú 216, 306

Universidad Nacional del Comahué 17, 264

Universidad Nacional del Litoral 18, 265

Universidad Nacional del Nordeste 18, 265

Universidad Nacional del Sur 19, 265

Universidad Nacional Experimental de los Llanos Centrales 'Romulo Gallegos' 244, 313

Universidad Nacional Experimental de Táchira 244, 313

Universidad Nacional Experimental Francisco de Miranda 244, 313

Universidad Nacional 'Federico Villarreal' 216, 306

Universidad Nacional 'Jose Faustino Sánchez Carrión' 216-17, 307

Universidad Nacional Mayor de San Marcos de Lima 217, 307

Universidad Nacional 'Pedro Henríquez Ureña' (UNPHU) 115, 116, 288

Universidad Nacional 'Pedro Ruiz Gallo' 217, 307

Universidad Nacional 'San Luis Gonzaga' 218, 307

Universidad Nacional Técnica de Cajamarca 218, 307

Universidad Nacional Técnica del Callao 218, 307

Universidad Notarial Argentino 21, 266

Universidad Particular Peruana 'Cayetano Heredia' 218, 307

Universidad Particular 'San Martín de Porres' 218, 307

Universidad Pedagógica Nacional 84, 283

Universidad Pedagógica y Tecnológica de Colombia 85, 283

Universidad Politécnica de Nicaragua 192, 302

Universidad Pontificia Bolivariana 86, 283

Universidad Popular Autónoma del Estado de Puebla 177, 301

Universidad Rafael Landívar 144, 293

Universidad Rafael Urdaneta 244, 313

Universidad Real de San Felipe 66

Universidad Regiomontana 177, 301

Universidad Rural del Brasil 49

Universidad Santa María la Antigua 199, 303

Universidad Santiago de Calí 87

Universidad Simón Bolívar 244, 313

Universidad Social Católica de La Salle 87, 284

Universidad Técnica de Babahoyo 125, 290

Universidad Técnica de Machala 125, 291

Universidad Técnica de Manabi 125, 291

Universidad Técnica de Oruro 29, 268

Universidad Técnica del Estado 67, 280

Universidad Técnica 'Federico Santa María' 67, 280

Universidad Técnica 'Luis Vargas Torres' 125, 291

Universidad Técnica Particular de Loja 125, 291

Universidad Tecnológica de Magdalena 85, 283

Universidad Tecnológica de Panamá 199

Universidad Tecnológica de Pereira 85, 283

Universidad Tecnológica de Santiago 115, 288

Universidad Tecnológica Nacional 19, 265

Universidad Veracruzana 178-9, 301

Universidade Católica de Minas Gerais 44, 274

Universidade Católica de Pelotas 46, 275

Universidade Católica de Pernambuco 47, 275

Universidade Católica de Petrópolis 47, 275

Universidade Católica do Paraná 46, 275

Universidade Católica do Salvador 49, 276

Universidade de Brasília 40, 273

Universidade de Caxias do Sul 41, 273

Universidade de Fortaleza 42, 273

Universidade de Itaúna 43, 273

Universidade de Mogi das Cruzes 44, 274

Universidade de Passo Fundo 46, 275

Universidade de São Paulo 38, 50, 57, 276

Universidade do Amazonas 40, 272
Universidade do Estado do Rio de Janeiro
  49, 276
Universidade do Vale do Rio dos Sinos
  51, 277
Universidade Estadual de Campinas 38,
  41, 273
Universidade Estadual de Londrina 43,
  274
Universidade Estadual de Maringá 43,
  274
Universidade Estadual de Mato Grosso
  do Sul 44, 274
Universidade Estadual de Ponta Grossa
  48, 275
Universidade Estadual Paulista 'Julio de
  Mesquita Filho' 51, 276
Universidade Federal da Bahia 40, 273
Universidade Federal da Paraíba 38, 45,
  274
Universidade Federal de Alagoas 39, 272
Universidade Federal de Goiás 42, 273
Universidade Federal de Juiz de Fora 43,
  273
Universidade Federal de Mato Grosso 44,
  274
Universidade Federal de Minas Gerais 44,
  274
Universidade Federal de Ouro Prêto 45,
  274
Universidade Federal de Pelotas 47, 275
Universidade Federal de Pernambuco 47,
  275
Universidade Federal de Santa Catarina
  50, 276
Universidade Federal de Santa Maria 50,
  276
Universidade Federal de São Carlos 50,
  276
Universidade Federal de Sergipe 51,
  277
Universidade Federal de Uberlândia 51,
  277
Universidade Federal de Viçosa 51, 277
Universidade Federal do Acre 39, 272
Universidade Federal do Ceará 41, 273
Universidade Federal do Espírito Santo
  42, 273
Universidade Federal do Maranhão 43,
  274
Universidade Federal do Pará 45, 274
Universidade Federal do Paraná 46, 275
Universidade Federal do Piauí 47, 275
Universidade Federal do Rio de Janeiro
  38, 49, 276

Universidade Federal do Rio Grande do
  Norte 48, 275
Universidade Federal Fluminense 42, 273
Universidade Federal Rural de
  Pernambuco 47, 275
Universidade Federal Rural do Rio de
  Janeiro 49, 276
Universidade Federale do Rio Grande do
  Sul 48, 276
Universidade Gama Filho 42, 273
Universidade Mackenzie 43, 274
Universidade para o Desenvolvimiento do
  Estado de Santa Catarina 50, 276
Universidade Regional de Blumenau 40,
  273
Universidade Regional do Nordeste 44,
  274
Universidade Rural do Estado de Minas
  Gerais 51
Université d'Etat d'Haiti 153, 294
Universiteit van Suriname 227, 309
University Foundation of Rio Grande
  48, 275
University Institute of Marine
  Technology 245
University of Aconcagua 20, 266
University of Antioquia 82, 282
University of Belgrano 20, 266
University of Belize 23
University of Brasília 40, 273
University of Buenos Aires 16, 264
University of Cartagena 83, 282
University of Caxias do Sul 41, 273
University of Chile 66, 68, 71, 72, 280
University of Concepción 66, 280
University of Fortaleza 42, 273
University of Guyana 148, 150, 294
University of La Pampa 17, 265
University of Labour of Uruguay 233,
  311
University of London 161
University of Mendoza 21, 266
University of Michoacana de San Nicolás
  de Hidalgo 176, 300
University of Morón 21-2, 266
University of Panama 196, 197, 198, 199,
  200, 303
University of Puerto Rico 223, 224, 308
University of Puerto Rico General
  Library 224, 308
University of Salvador 21, 266
University of San Carlos de Guatemala
  143, 144, 293
University of Surinam 227, 309
University of the Americas 172, 299

University of the Andes – Colombia 86, 89, 284
University of the Andes – Venezuela 243, 247, 313
University of the East – Cuba 107, 287
University of the East – Venezuela 244, 247, 313
University of the North – Chile 67, 280
University of the North – Colombia 87, 284
University of the North – Mexico 177, 301
University of the Republic 232, 233, 311
University of the Rio dos Santos Valley 51, 277
University of the Sacred Heart 223, 308
University of the Social Museum of Argentina 20-1, 266
University of the South-East 178, 301
University of the West Indies 160, 161, 295
University of the West Indies Extra-Mural Department 23, 267
University of Valle de Guatemala 144, 293
University of Valle de Mexico 175, 300
UNPHU, see Universidad Nacional 'Pedro Henríquez Ureña'
Uruguayan Institute of Technical Standards 233, 310
Uruguayan League against Tuberculosis 235
Uruguayan Society of Paediatrics 235, 310
Uruguayan Society of Radiology 235, 310
Uruguayan Surgical Society 235, 310
Uruguayan Wool Secretariat 234
USAID 223

Valle University 85, 283
Valparaiso Catholic University 66, 280
Venezuelan Association for the Advancement of Science 247, 311
Venezuelan Association of Electrical and Mechanical Engineering 247, 311
Venezuelan Centre of Indigenous Studies 246, 311
Venezuelan Council of Industrial Standards 238-9
Venezuelan Institute of Scientific Research 242
Venezuelan Institute of Social and Economic Analysis 243
Venezuelan Petrochemical Institute 242
Venezuelan Society of Natural Sciences 247, 312
Veracruz University 178-9, 301
Vicente López Space Centre 13, 261
Vicente Rocafuerte Lay University of Guayaquil 125, 291
Victoria Sugar Corporation 199

War Academy 70
Wesley College 23, 267
WHO, see World Health Organization
Women's University of Mexico 172, 299
World Health Organization (WHO) 15, 31, 146

Yucatán University 179, 302

Zacatecas Autonomous University 179, 302
Zooligical Society of Uruguay 311
Zoology Institute – Cuba 108, 287
Zulia University 245, 247, 313

Subject Index

This index covers the subject matter and geographical location of scientific and technological research and development. The terminology used is based on ROOT, *the British Standards Institution thesaurus (1981).*

University faculties and attached institutes are not indexed individually by subject as most establishments in the region do cover the usual range of disciplines. However, where a special or unique long-term project is mentioned, this has been included in the index.

agriculture
  Argentina 10–11
  Bolivia 30
  Brazil 47, 52–3
  Chile 62–3, 68–9, 73
  Colombia 80, 88
  Costa Rica 95, 97
  Cuba 106
  Dominican Republic 112, 116
  Ecuador 126
  El Salvador 135–6
  Guatemala 142, 144–5
  Guyana 150
  Jamaica 160–1
  Mexico 162–3, 165, 181–2
  Nicaragua 193
  Panama 194–5, 199–200, 201
  Paraguay 206
  Peru 208, 210, 219
  Surinam 227
  technology 47, 53, 62–3, 73, 88, 106, 116, 133, 145, 150, 160, 199, 201
  Uruguay 234, 236
  Venezuela 241, 245
  see also agronomy; animal husbandry
agronomy
  Argentina 2, 10–11, 16
  Belize 23
  Bolivia 25
  Brazil 33, 35, 52–3

  Chile 58, 69
  Colombia 76
  Costa Rica 93, 95, 97
  Cuba 100, 106
  Dominican Republic 110–11, 113, 116
  Ecuador 119, 126
  El Salvador 131, 133
  French Guiana 137–8
  Guatemala 139–40, 144–5
  Guyana 147
  Haiti 152
  Honduras 155–6
  Jamaica 159
  Mexico 163
  Nicaragua 187
  Panama 195
  Paraguay 204
  Peru 208, 220
  Puerto Rico 222
  Surinam 226
  Uruguay 233, 234
  Venezuela 237
air pollution
  Chile 62
air transport engineering
  Chile 73
almonds
  Guyana 148
animal breeding
  Argentina 11

Brazil 53
animal husbandry
　Argentina, 2, 10-11
　Bolivia 25, 30
　Brazil 33, 35, 52-3
　Chile 59, 69, 73
　Colombia 76, 80, 88
　Costa Rica 93, 95, 97
　Cuba 101
　Dominican Republic 111, 112, 113, 116
　Ecuador 119, 126
　El Salvador 132, 133, 135
　French Guiana 137
　Guatemala 140, 144-5
　Guyana 147
　Haiti 153
　Honduras 156
　Jamaica 159
　Mexico 181
　Nicaragua 187
　Panama 195
　Paraguay 204
　Peru 208, 219
　Surinam 226
　Uruguay 228, 231, 233, 234
　Venezuela 238
Antarctic
　Chile 62, 65, 74
antimony
　Haiti 153
　Honduras 156
astronomy
　Argentina 14
　Bolivia 26
　Brazil 55
　Chile 72-3
　Colombia 89
　Ecuador 127
　Mexico 183
　Uruguay 235
　Venezuela 246
　see also radioastronomy
astrophysics
　Bolivia 31
　Brazil 55
　Chile 72
　Mexico 183

balsa wood
　Ecuador 119
bananas
　Costa Rica 93
　Dominican Republic 111

Ecuador 119
Guatemala 140
Guyana 148
Haiti 152
Honduras 156
Jamaica 159
Panama 195
Puerto Rico 222
barley
　Colombia 76
　Mexico 163
bauxite
　Dominican Republic 111
　French Guiana 137
　Guyana 148
　Haiti 153
　Jamaica 159
　Panama 195
　Surinam 226
beans
　Mexico 163
　see also soya beans
beryllium
　Brazil 34
biology
　Argentina 12
　Bolivia 26
　Chile 73-4
　Colombia 80
　Cuba 108
　French Guiana 138
　high-altitude 26, 218, 219
　Peru 210, 218, 219
　Uruguay 232
botany
　Argentina 10
　Colombia 91
　Cuba 108
　Ecuador 126
　French Guiana 137
　Mexico 181
　Peru 219
　Venezuela 247
building science/technology, see construction

cancer, see neoplasms
cartography
　Argentina 8
　Bolivia 25, 28, 31
　Chile 65, 68
　Colombia 82
　Costa Rica 95-6
　Dominican Republic 117

cartography – *cont.*
  geological 8, 25, 28, 31, 65, 82, 227,
    242
  Guatemala 143
  Honduras 157
  military 117, 220, 235
  Peru 220
  Surinam 227
  Uruguay 235
  Venezuela 242
cattle
  Argentina 2
  Bolivia 25
  Brazil 33
  Costa Rica 93
  Cuba 101
  Haiti 153
  Panama 195
  Paraguay 204
  Uruguay 228
cedar
  Cuba 101
  Dominican Republic 111
  El Salvador 132
  Nicaragua 187
  Paraguay 204
  Venezuela 238
cellulose
  Chile 59
  Colombia 91
cement and concrete technology
  Haiti 153
  Jamaica 160
  Panama 195, 199
  Uruguay 234
ceramic and glass technology
  Dominican Republic 111
cereals
  Argentina 11, 15
  Brazil 53
  Chile 58
  Cuba 100, 106
  Ecuador 119
  Mexico 181
  Surinam 226
  Venezuela 237
chemical engineering
  Argentina 19
  Colombia 89
  Mexico 181
  Uruguay 233
chemicals industry
  Argentina 2
  Chile 63
  Colombia 77, 88

  Uruguay 229, 234
  see also petrochemicals
chemistry
  Brazil 55
  Chile 68, 73–4
  Colombia 80, 89
  Cuba 108
  Mexico 183
  Uruguay 233
chromium
  Brazil 33
  Colombia 76
  Cuba 101
citrus fruits
  Belize 23
  Brazil 33, 52
  Cuba 106
  French Guiana 138
  Guyana 148
  Jamaica 159
  Panama 200
  Puerto Rico 222
climatology, *see* meteorology and
    climatology
coal technology
  Colombia 90
  Mexico 163
cocoa
  Brazil 33
  Costa Rica 93, 96
  Cuba 106
  Ecuador 119
  Haiti 152
  Panama 195
coffee
  Brazil 33, 53
  Colombia 75, 76
  Costa Rica 93, 96
  Cuba 100, 106
  Dominican Republic 111
  Ecuador 119
  El Salvador 131, 135
  Guatemala 139–40, 145
  Haiti 152
  Honduras 156
  Jamaica 159
  Nicaragua 187
  Panama 195
  Peru 208, 220
  Puerto Rico 222
  Venezuela 237, 245
communicable diseases
  Argentina 6
  Brazil 57
  Mexico 182

Peru 219
communication systems, *see*
   telecommunication systems
computer technology
   Argentina 15
   Chile 73
   Cuba 102, 104, 108
   Mexico 183-4
construction
   Chile 63, 74
   Mexico 165
   Venezuela 241
construction materials
   Brazil 52
   Colombia 90
   Ecuador 123
   Guyana 148
copper
   Chile 59, 62-3, 73
   Cuba 101
   Ecuador 119
   El Salvador 132
   Nicaragua 187
   Panama 195, 201
   Surinam 226
cosmic radiation
   Argentina 8
cotton
   Argentina 15
   Bolivia 25
   Colombia 26
   Cuba 100
   El Salvador 131
   Guatemala 140
   Haiti 152
   Honduras 156
   Nicaragua 187
   Paraguay 204
   Peru 208, 220
   Puerto Rico 222
   Venezuela 237
crop production/cultivation, *see* agronomy

dentistry
   Brazil 57
   Costa Rica 98
   Cuba 107
   Uruguay 235
deserts
   Mexico 178
diamond
   Brazil 33
   Guyana 148
   Venezuela 238

drugs
   Chile 73
   Guyana 148
   Panama 201-2

earth sciences (generally)
   Colombia 80
   El Salvador 135
   Peru 210
   Uruguay 236
   Venezuela 242, 246
   see also geography; geology;
      meteorology and climatology;
      oceanography; seismology
ecology
   Chile 73
   Colombia 80, 90
   Costa Rica 95, 97
   Cuba 102-3
   Dominican Republic 116
   Ecuador 123
   Guatemala 142, 143
   Puerto Rico 223
   Surinam 227
   Venezuela 247
ecology (marine)
   Argentina 12
   Brazil 35
   Chile 62
   Colombia 88, 90-1
   Cuba 106
   Mexico 182
   Peru 219
   Uruguay 236
economic development, *see* socioeconomic
      development
electric power systems
   Brazil 35
electrical engineering
   Venezuela 247
electromagnetic radiation
   Argentina 15
electronic engineering
   Argentina 2, 6, 13
   Chile 63
   Cuba 105
   Jamaica 160
   Mexico 183
   Venezuela 241
energy technology (generally)
   Argentina 6, 12-13
   Brazil 35
   Chile 74
   Colombia 80, 90

energy technology – *cont.*
  Costa Rica 95
  Ecuador 123
  Mexico 165
  Peru 221
  Venezuela 241
engineering (generally)
  Brazil 55
  Colombia 89
  Cuba 105
  Guatemala 142–3, 144, 146
  Paraguay 206
  Peru 210, 220
  Uruguay 235, 236
environmental engineering
  Brazil 52
environmental health
  Brazil 56–7
  Costa Rica 95
  Cuba 107
  Ecuador 123
  Haiti 154
  Mexico 182
  Uruguay 233

fertilizers
  Cuba 106
  El Salvador 132
  Guyana 150
fisheries, *see* freshwater biology and
    fisheries; marine biology and
    fisheries
fluorite
  Mexico 163
food technology
  Argentina 2, 6
  Brazil 52, 53
  Chile 73
  Colombia 77, 80, 88, 90–1
  Costa Rica 95, 96
  Dominican Republic 116
  Ecuador 123
  El Salvador 132
  Guyana 151
  Honduras 156
  Jamaica 160–1
  Panama 195, 202
  Paraguay 204, 205
  Peru 208, 210, 219
  Surinam 226, 227
  Uruguay 229
  Venezuela 241
forensic medicine
  Brazil 57

  Colombia 88
forestry
  Belize 23
  Bolivia 25
  Brazil 53
  Chile 62–3, 69, 73
  Colombia 76, 80
  Costa Rica 93
  Cuba 101, 106
  Dominican Republic 111
  Ecuador 119
  El Salvador 132
  French Guiana 137, 138
  Guatemala 140, 145
  Guyana 148
  Honduras 156
  Mexico 163, 165, 181
  Nicaragua 187
  Panama 195
  Paraguay 204
  Peru 208
  Puerto Rico 223
  Surinam 227
  Venezuela 238
freshwater biology and fisheries
  Argentina 12
  Brazil 53
  Chile 73
  Panama 200
fruits, *see* bananas; citrus fruits
funding of research
  Argentina 6
  Bolivia 26–8
  Brazil 35–7
  Chile 63–4
  Colombia 79–82
  Costa Rica 95
  Cuba 102, 104–5
  Dominican Republic 112
  Ecuador 121–2, 129
  El Salvador 134
  Guatemala 142
  Mexico 168–70
  Panama 196, 197–8
  Peru 209, 212–14
  Uruguay 231–2
  Venezuela 241

geography
  Argentina 13
  Brazil 39
  Chile 71
  Colombia 82
  Costa Rica 95–6

Cuba 105
El Salvador 135
Guatemala 143
Honduras 157
Paraguay 206
geology
    Argentina 9
    Bolivia 24–5
    Brazil 55
    Chile 62, 65, 73–4
    Colombia 82
    Costa Rica 95
    Cuba 105
    marine 242
    Mexico 183
    Nicaragua 192–3
    Panama 201
    Surinam 227
    Uruguay 232
    Venezuela 242
geothermal power
    Chile 62, 65, 73
glass technology, *see* ceramic and glass
    technology
gold
    Bolivia 25
    Colombia 76
    Ecuador 119
    French Guiana 137
    Guyana 148
    Haiti 153
    Nicaragua 187
    Panama 195
    Venezuela 238
graphite
    Brazil 34

health services
    Colombia 80, 89, 90
    Costa Rica 94, 97
    El Salvador 133
    Guatemala 143, 145
    Mexico 165
    Panama 200
    Peru 210, 219
    Venezuela 241
high-altitude studies, *see under* biology
housing
    Argentina 6
    Colombia 80, 90
    Costa Rica 95
    Ecuador 123
    Guatemala 143
    Guyana 151

Peru 215
Venezuela 241
hydraulics
    Chile 62
hydroelectric power
    Guyana 148
    Surinam 227
hydrology, *see* water resources

infectious diseases, *see* communicable
    diseases
information science, *see* library and
    information science
insects
    French Guiana 138
    Mexico 181
iron
    Chile 59, 74
    Colombia 76
    Cuba 101
    El Salvador 132
    Mexico 163
    Peru 209
    Surinam 226
irrigation works
    Chile 69
    Cuba 106

lead
    Ecuador 119
    Honduras 156
    Peru 209
library and information science
    Argentina 5, 8, 14
    Bolivia 31
    Brazil 35, 37, 55–6
    Chile 63, 71–2
    Colombia 80, 89–90
    Costa Rica 99
    Cuba 102, 108–9
    Dominican Republic 117
    Ecuador 127–8, 129–30
    El Salvador 134, 136
    Guatemala 142, 146
    Guyana 151
    Haiti 154
    Honduras 158
    Jamaica 161
    Mexico 183–4
    Nicaragua 193
    Panama 200–1
    Paraguay 206
    Peru 212, 221

library and information science – *cont.*
  Puerto Rico 224–5
  Uruguay 231, 235–6
  Venezuela 247

magnesium
  Brazil 34
mahogany
  Cuba 101
  Dominican Republic 111
  El Salvador 132
  Guatemala 140
  Nicaragua 187
  Venezuela 238
maize
  Colombia 76
  Guatemala 140
  Honduras 156
  Jamaica 159
  Mexico 163
  Panama 195
  Paraguay 204
  Venezuela 237
manpower (research)
  Argentina 6–7
  Bolivia 26–8
  Brazil 37
  Chile 63–4
  Colombia 81
  Costa Rica 93–4
  Cuba 105
  Dominican Republic 112
  Ecuador 121
  El Salvador 134
  emigration 63
  Guatemala 142
  Guyana 151
  Mexico 168–9
  Panama 196–7
  Peru 210, 214
  Uruguay 231–2
  Venezuela 241
manufacturing industries
  Argentina 2
  Bolivia 26
  Chile 59, 62, 68
  Colombia 75, 76–7
  Dominican Republic 111
  Ecuador 122
  El Salvador 132
  Guatemala 144
  Guyana 148
  Haiti 153
  Honduras 156

  Jamaica 160
  Panama 195
  Paraguay 205
  Peru 210, 214, 219
  Uruguay 229, 234
  Venezuela 241
marble
  Jamaica 159
marine biology and fisheries
  Argentina 11–12, 15
  Brazil 33, 35
  Chile 59, 62–3, 69, 73
  Colombia 80, 90–1
  Costa Rica 93, 94, 95, 97
  Cuba 101, 106
  Dominican Republic 113, 116
  Ecuador 119, 126
  French Guiana 137
  Guatemala 140
  Guyana 146
  Honduras 156
  Jamaica 163
  Mexico 165, 182
  Nicaragua 187
  Panama 195, 200
  Peru 208, 210
  Surinam 226
  Uruguay 229, 234
  Venezuela 245
marine environment, *see* ecology (marine)
materials science (generally)
  Argentina 10, 13
  Brazil 52
  Chile 68
  Uruguay 233
measurement
  Chile 63, 68
  El Salvador 135
  Uruguay 233
  see also standards organizations
meat
  Argentina 2, 15
  Chile 73
  Paraguay 204
  Uruguay 228, 233
mechanical engineering
  Argentina 2
  Chile 68
  Venezuela 247
medical sciences
  Argentina 12, 16
  Bolivia 30–1
  Brazil 55, 56–7
  Chile 69–70, 73
  Colombia 88–9

Costa Rica 97–8
Cuba 107
Dominican Republic 116–17
Ecuador 126–7
El Salvador 136
French Guiana 138
Guatemala 145, 146
Guyana 150
Jamaica 161
Mexico 182
Nicaragua 193
Panama 200
Paraguay 206
Peru 210, 218, 219, 220
Uruguay 234–5
Venezuela 246, 247
metallurgy/metalworking
  Brazil 52
  Chile 63, 65, 73
  Colombia 88, 90–1
  Cuba 105–6
  Venezuela 241
meteorology and climatology
  Argentina 10, 14
  Bolivia 31
  Brazil 52, 54–5
  Chile 72–3
  Costa Rica 98
  Cuba 108
  Ecuador 122, 127
  French Guiana 138
  Guatemala 145
  Mexico 183
  Peru 215, 220
  Uruguay 235
  Venezuela 246
mica
  Brazil 33
  Guyana 148
microbiology
  Argentina 10, 12
  Guatemala 145
  Paraguay 206
military technology
  Argentina 13
  Chile 70–1
  Dominican Republic 117
  Peru 220
  Uruguay 235
  Venezuela 246
mineral resources/mineralogy
  Argentina 2, 8, 9, 15
  Bolivia 25
  Brazil 33–4, 38
  Chile 59, 73

Colombia 76
Costa Rica 93
Cuba 101
Dominican Republic 111, 112
Ecuador 119
El Salvador 132
French Guiana 137
Guyana 148
Haiti 153
Honduras 156
Jamaica 159, 160
Mexico 163
Nicaragua 187
Panama 195
Paraguay 205
Peru 209, 215
Puerto Rico 222
Surinam 226
Uruguay 232, 234
Venezuela 238
  see also entries for specific minerals
mining (generally)
  Chile 62–3, 65
  Colombia 75
  Costa Rica 95
  Cuba 105–6
  Ecuador 123
  Peru 210, 220
  Venezuela 242

neoplasms
  Bolivia 30–1
  Brazil 57
  Chile 70
  Colombia 88
  Cuba 107
  El Salvador 136
nickel
  Cuba 101
  Surinam 226
nuclear power
  Argentina 12
  Bolivia 30
  Brazil 35, 54
  Chile 62, 70
  Colombia 89
  Costa Rica 98
  Cuba 102
  Ecuador 123, 127
  Guatemala 145
  Mexico 182
  Puerto Rico 224
  Uruguay 235
  Venezuela 246

nuclear technology (non-energy
    applications)
  agriculture 54, 98, 127, 220
  Bolivia 30
  Brazil 54
  chemistry 127
  Chile 70, 73
  Colombia 89
  Costa Rica 98
  Cuba 102, 107
  Ecuador 127
  medicine 30, 98, 107, 127, 220, 224
  Mexico 182
  Peru 220
  Puerto Rico 224
  Uruguay 235
  Venezuela 246
nursing
  Argentina 15
nutrition
  Bolivia 30
  Brazil 56
  Colombia 80, 90–1
  Ecuador 123, 126
  El Salvador 133
  Guatemala 146
  Jamaica 160–1
  Mexico 165
  Panama 200
  Venezuela 241, 246

oak
  Panama 195
oceanography
  Argentina 8
  Chile 70–1, 73
  Cuba 106
  Ecuador 124
  Peru 210, 224
  Venezuela 245, 246
oil technology, see petroleum technology
olives
  El Salvador 131
optics
  Chile 73
  Mexico 183

paper technology
  Dominican Republic 111
  Panama 202
  Uruguay 229

patents
  Brazil 35
  Panama 199
petrochemicals
  Brazil 52
  Colombia 77
  Venezuela 241, 242
petroleum technology
  Argentina 2
  Bolivia 26, 30
  Brazil 35, 52
  Chile 59, 63, 74
  Colombia 76
  Costa Rica 95
  Cuba 101
  Ecuador 119, 123
  Honduras 156
  Jamaica 159–60
  Mexico 163, 181
  Nicaragua 187
  Panama 195
  Paraguay 205
  Peru 209, 210
  Surinam 226
  Uruguay 229
  Venezuela 238, 241, 242
phosphates
  Jamaica 159
physics
  Argentina 15
  Brazil 39
  Chile 73
  Colombia 80
  Guatemala 146
  Peru 210, 220
pigs
  Peru 208
pine
  Nicaragua 187
  Paraguay 204
  Venezuela 238
poultry
  Honduras 156
  Panama 195
  Peru 208
  Surinam 226
production engineering
  Argentina 15
  Chile 68
  Mexico 180–1
  Panama 201
  Peru 221
psychiatry
  Brazil 57
public health, see environmental health

quartz
  Brazil 33
  El Salvador 132

radioactive dating
  Argentina 8
radioastronomy
  Argentina 8
  Cuba 108
  Peru 224
remote sensing
  Bolivia 24-5
rice
  Belize 23
  Bolivia 25
  Colombia 76
  Costa Rica 93
  Cuba 106
  Dominican Republic 111
  Guyana 148
  Surinam 226
  Venezuela 237
rosewood
  Nicaragua 187
rubber (natural)
  Brazil 33
  Venezuela 238

seismology
  Argentina 8
  Bolivia 31
  Chile 74
  Cuba 108
  Ecuador 127
  Guatemala 145
  Peru 215
  Venezuela 246
sheep
  Argentina 2
  Honduras 156
  Nicaragua 187
  Peru 208
  Uruguay 228
silver
  Ecuador 119
  Haiti 153
  Honduras 156
  Nicaragua 187
  Paraguay 205
  Peru 209
social sciences
  Argentina 16
  Bolivia 28

Colombia 80
Costa Rica 95, 96
Dominican Republic 113
Ecuador 127
Guatemala 142-3
Peru 215, 220
Venezuela 242, 246
see also following entry
socioeconomic development
  Bolivia 26
  Chile 69
  Colombia 80, 89
  Costa Rica 98
  Dominican Republic 112
  Guatemala 143
  Mexico 165
  Paraguay 206
  Peru 209, 210
soil science
  Argentina 10-11
  Brazil 52-3
  Costa Rica 97
  Cuba 106
  Dominican Republic 116
  Ecuador 122
  French Guiana 138
  Guatemala 145
  Mexico 181
solar power
  Argentina 13
  Cuba 105
soya beans
  Paraguay 204
space technology
  Argentina 13
  Brazil 35, 54
  French Guiana 137-8
standards organizations
  Argentina 10
  Bolivia 28
  Chile 68
  Colombia 88
  Cuba 102
  El Salvador 135
  Panama 199
  Peru 219
  Uruguay 233
  Venezuela 239
steelmaking
  Chile 59
  Colombia 76
sugar beet
  Chile 63
sugar cane
  Belize 23

sugar cane – *cont.*
  Bolivia 25
  Brazil 53
  Colombia 75
  Costa Rica 96
  Cuba 100, 106
  Dominican Republic 110, 113, 116
  Ecuador 119
  El Salvador 131
  French Guiana 137, 138
  Guyana 147, 150
  Haiti 152
  Jamaica 161
  Nicaragua 187
  Panama 199
  Peru 208
  Puerto Rico 222
  Surinam 226
sulphur
  Colombia 76
  El Salvador 132
  Honduras 156
  Mexico 163
surgery
  Chile 70

tantalite
  French Guiana 137
technological cooperation 15, 37–8, 73–4,
    90–1, 99, 109, 117, 128–30, 146,
    184–5, 201–2, 221, 236
telecommunication systems
  Argentina 15
  Brazil 35, 54
  Chile 71, 73
  French Guiana 137–8
  Mexico 165
  Venezuela 241
textile technology
  Chile 62
  Colombia 96
  Dominican Republic 111
  Guyana 151
  Jamaica 160
  Panama 195
  Uruguay 229, 234
thorium
  Brazil 34
tin
  Bolivia 25
titanium
  Brazil 34
tobacco
  Colombia 75

  Cuba 100, 106
  Dominican Republic 111
  El Salvador 132
  French Guiana 138
  Guyana 148
  Honduras 156
  Jamaica 160
  Panama 195
  Paraguay 205
  Puerto Rico 222
  Uruguay 229
transportation
  Brazil 35
  Mexico 165
tungsten
  Bolivia 25

uranium
  Bolivia 25
  Chile 73

veterinary science
  Argentina 16
  Costa Rica 95
  Ecuador 126
  Mexico 181
  Uruguay 234, 236
viticulture/wines
  Argentina 2, 11
  Peru 220

water resources
  Bolivia 25
  Chile 73
  Ecuador 122, 127
  French Guiana 138
  Guatemala 145
  Mexico 183
  Venezuela 246
wines, *see* viticulture
wood technology
  Chile 59
  Colombia 76, 90
  Guyana 148
  Honduras 156
  Puerto Rico 223

zinc
  Ecuador 119
  El Salvador 132

Peru 209
zoology
  Chile 74
  Cuba 108

Ecuador 126
Guyana 150
Peru 219
Venezuela 247

Index by Ann Edwards